Advances in Two-Phase Flow and Heat Transfer
Volume II

NATO ASI Series

Advanced Science Institutes Series

A Series presenting the results of activities sponsored by the NATO Science Committee, which aims at the dissemination of advanced scientific and technological knowledge, with a view to strengthening links between scientific communities.

The Series is published by an international board of publishers in conjunction with the NATO Scientific Affairs Division

A	**Life Sciences**	Plenum Publishing Corporation
B	**Physics**	London and New York
C	**Mathematical and Physical Sciences**	D. Reidel Publishing Company Dordrecht and Boston
D	**Behavioural and Social Sciences**	Martinus Nijhoff Publishers Boston/The Hague/Dordrecht/Lancaster
E	**Applied Sciences**	
F	**Computer and Systems Sciences**	Springer-Verlag Heidelberg/Berlin/New York
G	**Ecological Sciences**	

Series E: Applied Sciences – No. 64

Advances in Two-Phase Flow and Heat Transfer

Fundamentals and Applications
Volume II

edited by

S. Kakaç

University of Miami
Coral Gables, Florida, USA

and

M. Ishii

Argonne National Laboratory
Argonne, Illinois, USA

1983 **Martinus Nijhoff Publishers**
A member of the Kluwer Academic Publishers Group
Boston / The Hague / Dordrecht / Lancaster
Published in cooperation with NATO Scientific Affairs Division

Proceedings of the NATO Advanced Research Workshop on the Advances in Two-Phase Flow and Heat Transfer
Spitzingsee, August 31 - September 3, 1982

Library of Congress Cataloging in Publication Data

NATO Advanced Research Workshop on the Advances in Two-Phase Flow and Heat Transfer (1982 : Spitzingsee, Germany)
Advances in two-phase flow and heat transfer.

(NATO advanced science institutes series. Series E, Applied sciences ; no. 63-64)
"Proceedings of the NATO Advanced Research Workshop on the Advances in Two-Phase Flow and Heat Transfer, Spitzingsee, BRD, August 31-September 3, 1982"--T.p. verso.
"Published in cooperation with NATO Scientific Affairs Division."
Includes index.
1. Two-phase flow--Congresses. 2. Heat--Transmission--Congresses. I. Kakaç, S. (Sadık) II. Ishii, M. III. North Atlantic Treaty Organization. Scientific Affairs Division. IV. Title. V. Series.
TA357.N383 1982 620.1'064 83-3994

ISBN 90-247-2826-6 (this volume)
ISBN 90-247-2689-1 (series)
ISBN 90-247-2827-4 (set)

Distributors for the United States and Canada: Kluwer Boston, Inc., 190 Old Derby Street, Hingham, MA 02043, USA

Distributors for all other countries: Kluwer Academic Publishers Group, Distribution Center, P.O. Box 322, 3300 AH Dordrecht, The Netherlands

Printed in The Netherlands

PREFACE

Over the past two decades, two-phase flow and heat transfer problems associated with two-phase phenomena have been a challenge to many investigators. Two-phase flow applications are found in a wide range of engineering systems, such as nuclear and conventional power plants, evaporators of refrigeration systems and a wide variety of evaporative and condensive heat exchangers in the chemical industry.

This publication is based on the invited lectures presented at the NATO Advanced Research Workshop on the Advances in Two-Phase Flow and Heat Transfer. The Workshop was attended by more than 50 leading scientists and practicing engineers who work actively on two-phase flow and heat transfer research and applications in different sectors (academia, government, industry) of member countries of NATO. Some scientific leaders and experts on the subject matter from the non-NATO countries were also invited. They convened to discuss the state-of-the-art in two-phase flow and heat transfer and formulated recommendations for future research directions. To achieve these goals, invited key papers and a limited number of contributions were presented and discussed. The specific aspects of the subject were treated in depth in the panel sessions, and the unresolved problems identified.

Suitable as a practical reference, these volumes incorporate a systematic approach to two-phase flow analysis.

Topics covered at the Workshop included hydrodynamic and heat transfer modeling in two-phase flows, momentum-, heat- and mass-transfer at phase interfaces, condensation and boiling, two-phase flow instabilities, developments in instrumentation for two-phase measurements, two-phase flow and heat transfer applications in nuclear reactor safety and in heat exchange equipment, and unresolved problems in two-phase flow and heat transfer.

The volumes present both basic and applied information on two-phase flow and heat transfer for engineers, scientists and students who work on two-phase flow research and applications.

The volumes identify the unresolved problem areas and provide suggestions for priority research topics in the field of two-phase flow and heat transfer. Perhaps these volumes will stimulate research and development and continuing education courses in this area. The editors appreciate the cooperation of Martinus Nijhoff Publishers in making the proceedings widely available.

The sponsorship of the NATO Scientific Affairs Division is acknowledged and greatly appreciated.

The planning for this Workshop was carried out primarily at the Lehrstuhl A für Thermodynamik, Technische Universität Munchen and at the University of Miami, Coral Gables, Florida. Special thanks are extended to Professors F. Mayinger and O. Yeşin for their close collaboration on the Workshop and their work on the Organizing Committee. Dr. S. Kakaç of the University of Miami wishes to acknowledge in particular the contributions of Prof. Mayinger and the support of the Department of Mechanical Engineering and the Clean Energy Research Institute at the University of Miami. Special thanks are extended to Drs. H. Yüncü, E. Paykoç, H. S. Ergür, A. Mertol and secretary Frau D. Schmitt for their efforts in making the Workshop a smooth-running success.

Our sincere appreciation is due to the session and panel chairpersons and co-chairpersons for their efforts in programming, organizing and execution of the sessions and panels of the Workshop.

And finally, our profound gratitude is extended to all the invited lecturers and authors, who provided the substance of the Workshop, and to the participants whose never-flagging interest, inquisitiveness and commentary were so vital to the goals of this Workshop.

S. Kakaç
M. Ishii

CONTENTS

VOLUME 1

VOLUME 2

A NOVEL APPROACH FOR THE DETERMINATION OF CRITICAL TWO-PHASE FLOW

Efstathios E. Michaelides

Mechanical and Aerospace Engineering
University of Delaware
Newark, Delaware 19711, U.S.A.

ABSTRACT

The knowledge of critical two-phase flow is very important in many engineering design problems. Several models have been developed in the past, in order to predict the critical flux of an evaporating fluid. These models are based on separated flow conditions in the fluid and use various criteria for the onset of critical flow. The present paper presents a method for the prediction of critical two-phase flow based on the positive entropy production principle, which emanates from irreversible thermodynamics. Use is being made of the Fanno lines and the onset of critical flow is characterized by the maximum entropy point, at constant mass flux. Vertical as well as horizontal pipes are examined. Also the cases of slip or nonslip conditions for the two fluids are incorporated in the model and the results are compared to experimental data. It appears that the proposed criterion with the other assumptions for the flow, yield very good agreement between the calculated and the experimental data.

1. INTRODUCTION

In many nuclear and chemical applications it is necessary to know accurately the mass flux of a two-phase fluid. The maximum flux would be achieved when the flow becomes critical; under critical conditions the mass flux does not increase if the external pressure is further lowered; then the flow is said to become "choked".

Several analytical models have been developed for the

prediction of the critical conditions. They use apparently different criteria for the determination of the critical mass flux: Fauske's [1] theory uses the criterion of maximum pressure gradient and postulated that the derivative of this gradient with respect to slip ratio is zero. Moody's [2] model is based on the maximization of the mass flux with respect to the variables pressure and slip ratio. Levy [3], used the principle of maximization of the "momentum specific volume" and the relationship between void and dryness fraction derived in [4], in order to predict the critical conditions. In parallel, there are several experimental results on equilibrium critical flow rates [5], [6], [7], which compare well with these three analytical models.

Every one of the above models makes use of the separated-fluid theory and employs as criterion an extremum property of a particular flow quantity (slip ratio, pressure gradient, mass flux) in order to obtain the critical state of the fluid. The work outlined here aims at the presentation of a new model for critical flow based on the principle of irreversible thermodynamics, that the entropy production is non-negative in all flows. Here it is proved analytically that this principle yields an expression for the critical mass flux. A method is also developed for the determination of the critical conditions. For that matter, the "Fanno lines" are used, a technique which is proven and useful in single-phase gas dynamics.

2. ASSUMPTIONS AND GOVERNING EQUATIONS

For the analysis of the phenomenon of critical flow in pipes (both horizontal and vertical) the following general assumptions will be made:

a) The two-phases are in thermodynamic equilibrium
b) The flow is one dimensional in a closed duct
c) The flow is in steady state
d) The slip ratio k, (the ratio of the average vapor to liquid velocities) is a function of the thermodynamic state of the fluid.

The general assumptions will be supplemented later with others, which apply to parts of the model alone.

The governing equations for the flow will be subsequently written according to the above assumptions. Space and time average quantities will be used for all the equations. The superscript ' will be used for the liquid phase and " for the vapor phase throughout the conservation equations, that follow:

A. Continuity equation:

$$\dot{m} = A' \rho' V' + A'' \rho'' V'' = \text{const.} \tag{1}$$

or in terms of the mass flux ($G = \dot{m}/A$) and the void fraction $\phi = A''/A$,

$$G = (1-\phi) \rho' V' + \phi \rho'' V''. \tag{2}$$

Using the slip ratio k the above eqn. becomes:

$$G = [(1-\phi)\rho' + k\phi\rho'']V' \tag{3}$$

The quantity in the brackets has the units of density and represents the density of an homogeneous fluid, which flows with the liquid average velocity and supplies mass flux G. It may be called a "discharge density" for the liquid phase, $\bar{\rho}$.

B. Momentum equation:

The momentum equation for the average flow variables may be written as follows:

$$-\frac{dP}{dz} = G^2 \frac{d}{dz} \left[\frac{(1-x)^2}{\rho'(1-\phi)} + \frac{x^2}{\rho''\phi}\right] + [(1-\phi)\rho' + \phi\rho''] \, g \sin\alpha + \frac{\tau_w}{\mu} \tag{4}$$

where α is the inclination of the pipe. Thus, for a vertical pipe $\alpha = 90°$ and for a horizontal pipe $\alpha = 0°$. The shear stress τ_W is usually given by a closure equation. Here, the momentum equation will be used only in section VI to determine the elevation of the fluid in vertical pipes.

C. Energy equation:

For a duct exchanging heat at a rate $\dot{Q}$, the first law of thermodynamics will yield the following relationship:

$$\dot{m} \frac{d}{dz} \left[(1-x)h' + xh'' + \frac{1-x}{2} V'^2 + \frac{x}{2} V''^2 + gz\sin\alpha\right] = \frac{\dot{Q}}{\mu} \tag{5}$$

For an adiabatic pipe the above equation may be integrated to yield:

$$h° = (1-x)h' + xh'' + \frac{1-x}{2} V'^2 + \frac{x}{2} V''^2 + gz\sin\alpha \tag{6}$$

or in terms of the static enthalpy and the mass flux:

$$h^\circ = h + \left[\frac{1-x}{2} + \frac{x}{2} k^2\right] \frac{G^2}{\bar{\rho}^2} + gz\sin\alpha \tag{7}$$

D. Entropy production equation:

The entropy change along the duct is the sum of the entropy flux (associated with any heat input $\dot{Q}$) and the rate of entropy production $\dot{\theta}$, due to the irreversibilities of the flow:

$$\dot{m} \frac{ds}{dz} = \frac{\dot{Q}}{\mu T} + \frac{\dot{\theta}}{\mu} \tag{8}$$

The work dissipation rate $\dot{W}_d$ is $T\dot{\theta}$.
The entropy of the fluid s is a function of its state only and may be considered as a canonical variable of the system. It may always be written as:

$$s = (1-x)\, s' + xs''. \tag{9}$$

E. Equations of state:

The fundamental equation of the system will yield equations of state for all the properties mentioned above such as s', s'', h', h'', ρ' and ρ''. These properties may be obtained easily from the thermodynamic tables of the fluid. Therefore, the static enthalpy and entropy of the substance may be written as usual in terms of the temperature and quality in the following functional form:

$$h = h(T,x), \text{ and } s = s(T,x). \tag{10}$$

These functional forms are obtained for water substance from correlations based on steam tables [8]. The thermodynamic equilibrium between the two phases implies that the Clausius-Clapeyron equation applies to the whole system:

$$\frac{dP}{dT} = \frac{s''-s'}{v''-v'}. \tag{11}$$

F. Closure equations:

The void fraction ϕ may be written in terms of the quality x and the densities as follows:

$$\phi = \frac{1}{1+\frac{1-x}{x}\frac{\rho''}{\rho'}k} \tag{12}$$

Since k is assumed to be a thermodynamic variable of the system it appears that ϕ is a function of the thermodynamic state of the fluid and we may write for it $\phi = \phi(x,T)$.

An equation for the slip ratio itself is also needed as a closure equation with the restrictions outlined above. The following considerations were taken into account for the development of a functional form for the slip ratio:

a) At the limits of saturation ($x = 0$, or $x = 1$) the flow appears homogeneous and, therefore, $k = 1$. Hence, the function for k must exhibit a maximum between $x = 0$ and $x = 1$, and approach the value 1 at the two limits of saturation states.
b) Some theoretical studies, [1], showed that k cannot exceed the quantity $(\rho'/\rho'')^{1/2}$ and others [9] that it is given simply as $k = (\rho'/\rho'')^{1/3}$, or $k = k(T)$.
c) After the examination of available data for ϕ in evaporative flows it was concluded that k is a function of both x and T.

Therefore, the following functional form for k was decided to be adopted for the purposes of this study:

$$k = \beta\, x \left[\sqrt{\frac{\rho'}{\rho''}} - 1\right] e^{(1-\beta x)} + 1, \tag{13}$$

where β is a constant greater than 1. This choice of slip ratio yields a function of ϕ which is plotted in Fig. 1 for $\beta = 10$ and $\rho'/\rho'' = 110$, ($T = 200°C$), together with other expressions for ϕ such as no-slip, or $k = \sqrt{\rho'/\rho'}$. It can be seen that the choice of $k(T,x)$ does not alter substantially the function $\phi(T,x)$. In addition, the latter function behaves appropriately at the two limits $x = 0$ and $x = 1$.

Some authors [10] have used the assumption $k = 1$ (homogeneous fluid) throughout the flow field. Although under this assumption the conservation equations are simplified considerably the homogeneous fluid model has only limited applications in practice and may only serve as a first step towards a realistic model, which would describe more accurately the flow of the two phases.

It is believed that the departures from the homogeneous model are due to the following two causes: a) the existence of slip as

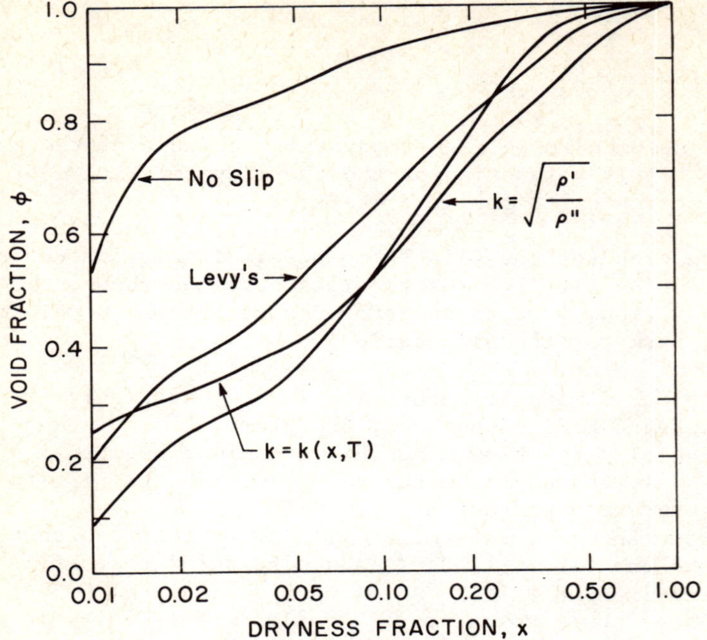

Fig. 1 The function $\phi(x)$ for T = 200°C.

the difference of the average phase velocities and b) thermodynamic nonequilibrium between the two phases. Both of them would have the same effect on the flow quantities, such as velocities, void fraction and critical mass flux. Here it is chosen for simplicity to consider only the slip ratio effect which is augmented in some cases in order to compensate for the nonequilibrium.

3. THE CRITICAL MASS FLUX

The continuity equation (3) may be written in terms of the discharge density $\bar{\rho}$ of the fluid as follows:

$$G = [(1-\phi)\rho' + \phi\ k\ \rho''] V' = \bar{\rho}\ V' \tag{14}$$

We may write also the change of static enthalpy in the direction of the flow from eqn. (7) as follows:

$$\frac{dh}{dz} = -\frac{G^2}{\bar{\rho}^2}\frac{d}{dz}\left(\frac{1-x}{2} + \frac{xk^2}{2}\right) - \left(\frac{1-x}{2} + \frac{xk^2}{2}\right)\frac{d}{dz}\left(\frac{G}{\bar{\rho}}\right)^2 - g, \tag{15}$$

which yields upon the substitution:

$$F = 1/2\ (1-x + k^2 x) \tag{16}$$

$$\frac{dh}{dz} = -\frac{G^2}{\bar{\rho}^2}\frac{dF}{dz} + \frac{2FG^2}{\bar{\rho}^3}\frac{d\bar{\rho}}{dz} - g \tag{15a}$$

The discharge density of the fluid $\bar{\rho}$ is a function of its thermodynamic properties since all ρ', ρ'', ϕ, and k are state functions. Therefore, it may be expressed as a function of two specific properties as follows:

$$\bar{\rho} = \bar{\rho}\ (h,s). \tag{17}$$

Consequently the derivative $d\bar{\rho}/dz$ may be written in terms of dh/dz and ds/dz:

$$\frac{d\bar{\rho}}{dz} = \left(\frac{\partial\bar{\rho}}{\partial h}\right)_s \frac{dh}{dz} + \left(\frac{\partial\rho}{\partial s}\right)_h \frac{ds}{dz}\ . \tag{18}$$

Substitution in eqn. (16) yields:

$$\frac{dh}{dz} = \frac{\dfrac{-G^2}{\bar{\rho}^2}\dfrac{dF}{dz} - g + \dfrac{2FG^2}{\bar{\rho}^3}\left(\dfrac{\partial\bar{\rho}}{\partial s}\right)_h \dfrac{ds}{dz}}{1 - \dfrac{2FG^2}{\bar{\rho}^3}\left(\dfrac{\partial\bar{\rho}}{\partial h}\right)_s} \tag{19}$$

The flow of two-phase mixtures in upward flowing or horizontal pipes is characterized by the decrease of the static enthalpy along the pipe. This is due to the fact that the static pressure decreases and this causes evaporation of the fluid and density decrease; a glance at the continuity equation will show that the velocities V' and V" must increase. Therefore, it follows from the energy equation (6), that the static enthalpy decreases along the flow, and, hence:

$$\frac{dh}{dz} \leq 0\ . \tag{20}$$

Also from the second law of thermodynamics, along the flow:

$$\frac{ds}{dz} \geq 0\ . \tag{21}$$

The choice of the discharge density function yields the usual conditions for the thermodynamic stability of the fluid:

$$(\partial\bar{\rho}/\partial h)_s \geq 0 \text{ and } (\partial\bar{\rho}/\partial s)_h \leq 0 \ .$$

A glance at eqn. (19) will prove that the numerator for dh/dz is always negative. Since the ratio must be negative the denominator has to be positive. Hence,

$$1 - \frac{2}{\bar{\rho}^3}\frac{FG^2}{} \left(\frac{\partial\bar{\rho}}{\partial h}\right)_s \geq 0 \ . \tag{22}$$

or

$$G \leq \sqrt{\frac{\bar{\rho}^3}{2F(\partial\bar{\rho}/\partial h)_s}} \ . \tag{23}$$

The quantity in the right hand side has the units of mass flux and represents the critical mass flux, G_{cr} . Hence

$$G \leq G_{cr} = \sqrt{\frac{\bar{\rho}^3}{2F(\partial\bar{\rho}/\partial h)_s}} \ . \tag{24}$$

Therefore, the mass flux in the pipe cannot exceed the critical flux as given by eqn. (24). This value for the critical mass flux G_{cr} coincides with the value obtained from the Fanno lines as will be shown in sections IV and V.

It must be emphasized that the above reasoning does not apply exactly to the critical flow in downward flowing pipes, where the pressure may be increased along the pipe, due to gravity. Here we have $g < 0$ and sometimes, the increased pressure may result to condensation and to an increase of the static enthalpy. The numerator of eqn. (19) is then of indeterminate sign and the same holds for the ratio dh/dz. In this case the same critical condition ($G = G_{cr}$) would yield a singularity for dh/dz, unless the numerator vanishes simultaneously.

4. THE FANNO LINES

When a two-phase mixture flows in an horizontal adiabatic duct, equations 3, 7, 10, 11, 12 and 13 apply simultaneously with the equations of state. The number of unknowns in the above equations is 7 (h, s, T, x, ϕ, V', k) for a given initial mass flux G. This implies that the system has one degree of freedom and, therefore, there is a relationship between any pair of the seven variables listed above. Analytically this would happen if all unknowns but two were eliminated in the system of equations. Then the remaining equation would have yielded the functional relationship between the pair of variables. This relationship can be represented as a line in the carterian plane of the two variables.

This method has been used in the past with the pair (h,s) to produce the so called "Fanno lines" in the theory of gas dynamics. The Fanno lines yield the critical flow in the single-phase fluids, which coincides with sonic conditions.

When vertical pipes are considered the number of unknowns is augmented by the presence of the height z of the pipe. Then the momentum equation may be used to eliminate z. This would require the use of another closure equation for the shear stress τ_w at the wall which is easily obtained from the theory of viscous fluids.

It will be attempted in this work to extend the technique of Fanno lines to two-phase flows in order to determine the critical conditions of the fluid.

5. THE CRITERION FOR CRITICAL FLOW

The second law of thermodynamics asserts that in any adiabatic system the entropy may only increase. For the flow through an adiabatic duct the principle of entropy increase implies that ds/dz is positive in the entire length of the pipe. The entropy production is the manifestation of the friction at the walls and the shear between the two phases. As explained in Section III the enthalpy gradient dh/dz is negative in horizontal and upward-flow vertical pipes. Hence, in an (h,s) diagram the slope of the line is always negative.

The laws of nature would not allow the flow to occur if $\dot{\theta} < 0$, in any portion of the pipe. Such flows are impossible and the mass flow rate would be adjusted so that $\dot{\theta} \geq 0$ throughout the pipe. The condition $\dot{\theta} = 0$ may only occur at the end of the pipe and under this condition, the mass flux may not be increased by further reduction of the outside pressure. This would imply the onset of critical conditions in the pipe.

Thus, in the pipe flow the condition $\dot{\theta} = 0$ is taken as the criterion for the onset of critical flow. In the h,s diagram this is manifested by the stationary value attained by entropy (s is a maximum and ds/dh = 0). The momentum equation (4) will yield the length of the pipe z_{cr} at which critical conditions are set. If the length of the pipe z* is lower than z_{cr} the mass flux can be sustained and increased further by a decrease of the outside pressure. If $z_{cr} = z^*$ the mass flux can be sustained but critical condition has been established at the end of the pipe and the mass flux cannot be further increased. If the length of the pipe is more than z_{cr} the flow conditions are impossible and the mass flux must attain lower values until $z^* \leq z_{cr}$. This situation is shown in Fig. 2 where the entropy of the fluid is depicted against the length of the flow. For the pipe of the length z^* shown all possible flows must obey the condition $G \leq G_2$. This phenomenon is observed in practice in the flashing geothermal wells where the mass flux cannot attain values higher than the critical mass flux. It is then said that the pipe is "choked".

6. THE SOLUTION OF THE EQUATIONS FOR HORIZONTAL AND VERTICAL PIPES

For the solution of the flow equations the reservoir conditions are prescribed (as T_o and x_o) and a constant mass flux G is assumed for the entire flow. The temperature T was taken for convenience as independent variable and equations 3, 7, 9, 10, 12, and 13 were solved simultaneously with an itterative procedure, in order to obtain the dependent variables, h, s, ϕ, x, V' and k. The Clausius Clapeyron equation (11) may be used for the determination of pressure and the momentum equation (4) is used for the calculation of the flow length z. Petrick's equation [11] is taken as the closure equation for the shear stress τ_w at the wall:

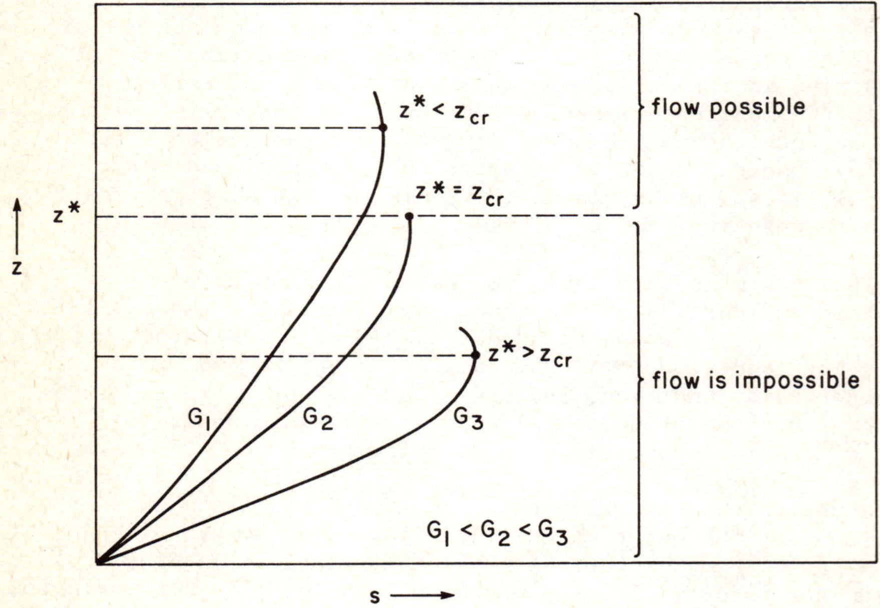

Fig. 2 Possible flow conditions in a pipe of length z.

$$\tau_{wTP} = \left(\frac{1-x}{1-\phi}\right)^2 \tau_{wSP}, \tag{25}$$

where τ_{wSP} is given in terms of the friction factor for liquid flow:

$$\tau_{wSP} = \frac{1}{2} f_{SP} \frac{1}{\rho'} \left(\frac{\dot{m}}{A}\right)^2 , \tag{26}$$

and the friction factor f_{SP} For the liquid is taken from the Blausius equation

$$f_{SP} = 0.079 \ Re'^{-0.25} , \tag{27}$$

for turbulent flow, or from the expression

$$f_{SP} = \frac{16}{Re'} , \tag{28}$$

for laminar flow.

The Fanno lines are shown in Fig. 3 for horizontal flow with initial condition $T_o = 300°C$ and $x_o = 0.0$. The critical flow point is depicted with a circle. Other pairs of dependent

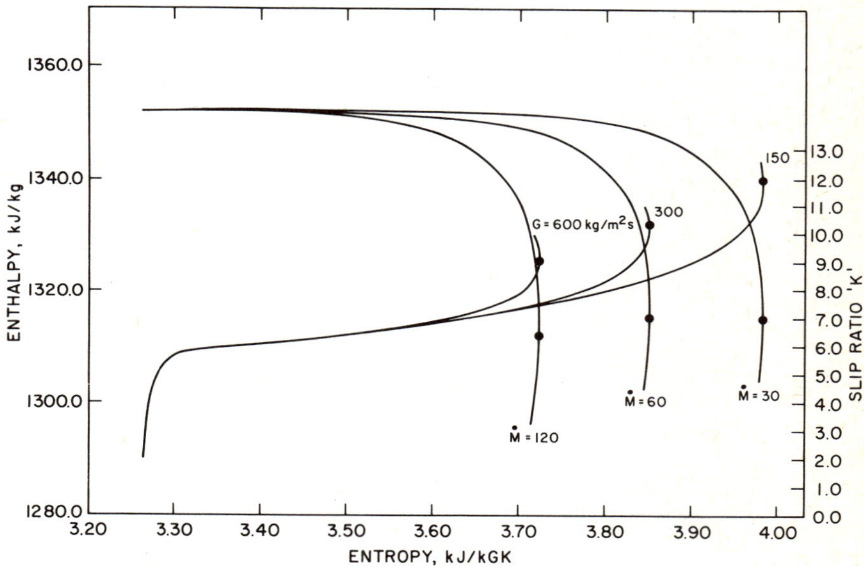

Fig. 3 Fanno lines with slip-ratio lines for various mass-flow rates in a pipe of area 0.2 m^2.

variables may be chosen instead of (h,s) such as (P,s), or (x,s). Figure 4 shows the curves obtained for these pairs with the initial condition: T_o = 200°C and x_o = 0.10. For a vertical pipe, the elevation z is plotted in Fig. 5 against static enthalpy for different mass flow rates; the critical length z_{cr} can be easily found for each one of the flow conditions. Figure 6 is also for a vertical pipe and depicts the enthalpy, pressure, quality and elevation of the fluid for reservoir conditions T_o = 200°C and x_o = 0. The latter figure would adequately describe the flow in a geothermal well. In this case the critical elevation for a given mass flux is of particular importance for the design of the geothermal power station. If the two-phase flow part of the well is deeper than z_{cr} the flow cannot be sustained at the specified flux. Then the mass flow rate will decrease to a lower value, compatible with the depth of the well.

7. COMPARISON WITH KNOWN DATA AND CORRELATIONS

A comparison was made between the data obtained by this model and other analytical or experimental results. Here, the theory developed by Moody [2] is taken as representative of the analytical models (in fact, there is very good agreement between all of them). Figures 7 and 8 show the comparisons made: Each one depicts the analytical curve of the critical mass flux predicted by Moody [2] for dryness fraction x = 0.2 and 0.6; the two figures also show

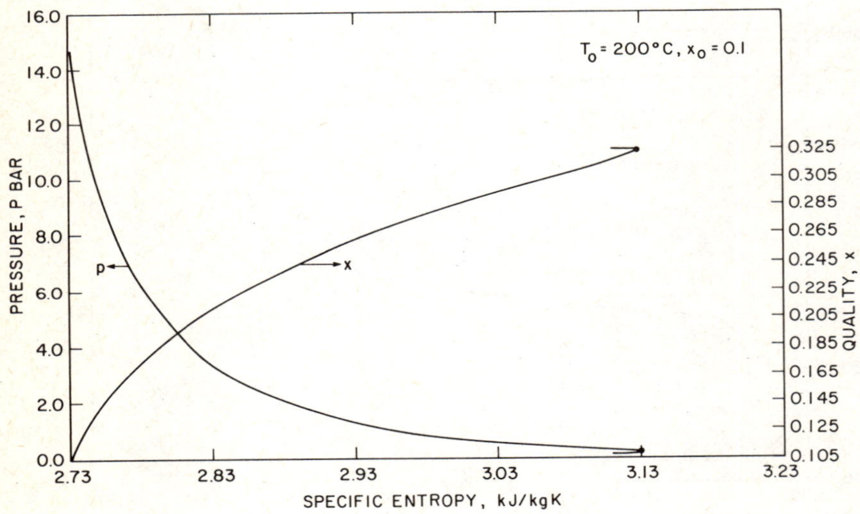

Fig. 4 (P,s) and (x,s) lines for G = 150 kg/m^{2}s, T_o = 200°C and x_o = 0.1.

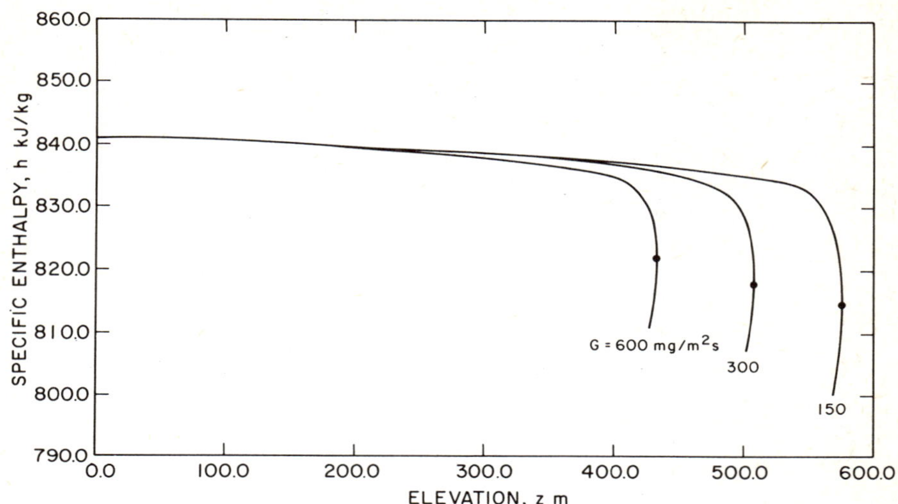

Fig. 5 Fanno lines for a vertical pipe. (h,z) diagram.

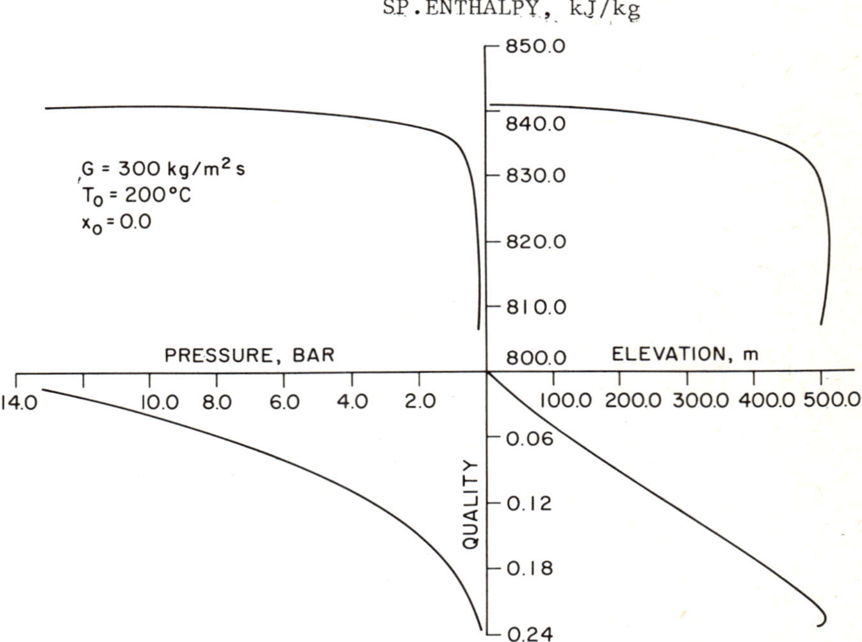

Fig. 6 Flow variables of interest in a vertical pipe.

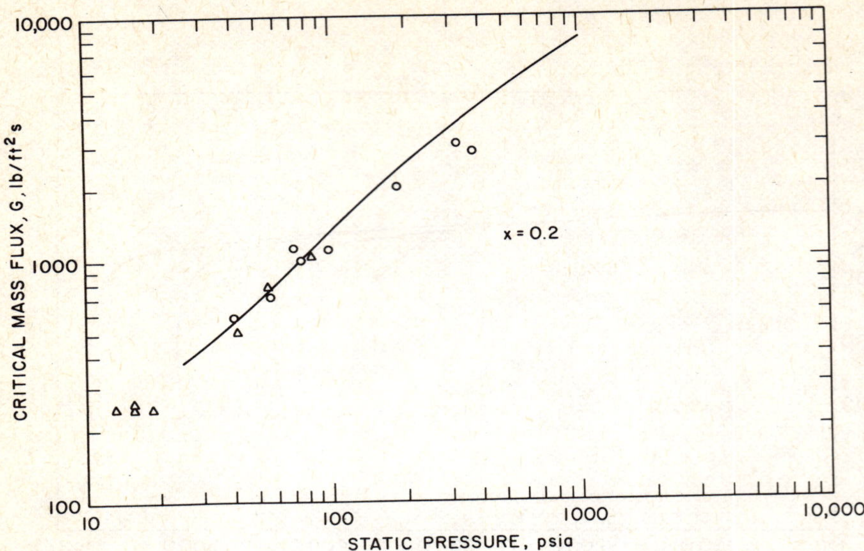

Fig. 7 Comparison of results for x = 0.2.

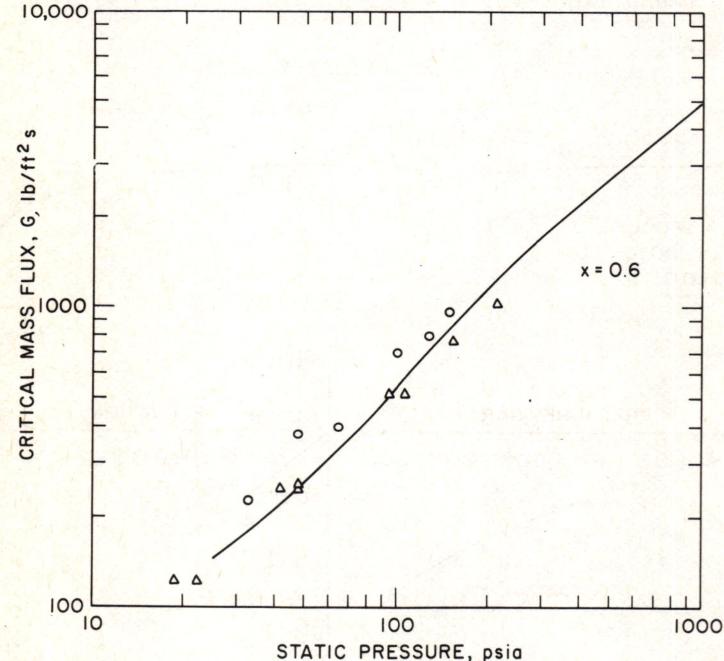

Fig. 8 Comparison of results for x = 0.6. The line represents data from [2], the o's from [1] and [5] and the Δ's the results of the Fanno lines.

experimental data obtained by Fauske [1] and Zaludek [5]. It may be seen that there is a very good agreement between the experimental results and the data obtained by the use of Fanno lines. There is also good agreement between the analytical results and this method.

The results for critical flow obtained by the use of Fanno lines were compared to other experimental and analytical results. Excellent agreement with them was observed too. It was, therefore, concluded that the criterion of entropy maximization employed here is valid for the determination of the critical flow rate in two-phase mixtures. This criterion emanates from the principles of irreversible thermodynamics and it is more general and unambigous than the other maximization criteria employed in the past. An advantage of the Fanno lines method is, that it yields the critical conditions for given reservoir (initial) conditions. A Fanno line would also describe exactly the flow at any point in the pipe, until the onset of critical conditions. This, in contrast to the other analytical models that would only yield the critical state and the critical mass flux, without any details of the flow.

8. CONCLUSIONS

A separated two-fluid model for two-phase flows is developed in order to determine the critical flow conditions.

The method of the Fanno lines is used for the determination of the fluid's state at the critical point. The principle of maximization of entropy at this point is used as the criterion for the onset of critical conditions.

The results obtained by this method show a remarkable agreement with other analytical and experimental results for critical flow. The agreement is an evidence for the validity of the model and justification for the criterion employed.

ACKNOWLEDGEMENTS

The author wishes to acknowledge the assistance of Mr. S. Parikh in some of the calculations and diagrams of this paper. The work was performed under partial support from the University of Delaware Research Foundation.

NOMENCLATURE

A	area, m^2
f	friction factor
F	defined in eqn. 16
g	gravitational acceleration, m/s^2
G	mass flux, kg/m^2s
h	enthalpy, kJ/kg
k	slip ratio
$\dot{m}$	mass flow rate, kg/s
P	pressure, MPa
$\dot{Q}$	heat flow rate, kJ/s
Re	Reynolds number
s	entropy, kJ/kgs
T	temperature, K
V	velocity, m/s
$\dot{W}_d$	rate of dissipated work, kJ/s
x	dryness fraction
z	axial coordinate, m

Subscripts

o	stagnation condition
SP	single-phase
TP	two-phase

Superscripts

'	liquid
"	vapor
$\bullet$	rate
$\circ$	total
$\bullet_{-}$	for superficial density

Greek

α	angle or pipe inclination
β	positive constant

θ entropy production, kJ/kg

μ hydraulic diameter, m

ρ density, kg/m^3

τ_w shear stress at the wall, MPa

ϕ void fraction

REFERENCES

1. Fauske, H. K., Contribution to the Theory of Two-Phase, One-Component Critical Flow, ANL-6633, Oct. 1962.

2. Moody, F. J., Maximum Flow Rate of a Single-Component Two-Phase Mixture, J. Heat Transfer, 87, (1), p. 134, 1965.

3. Levy, S., Prediction of Two-Phase Critical Flow Rate, Trans. ASME, J. of Heat Transfer, 87, p. 53, 1965.

4. Levy, S., Steam-Slip Theoretical Prediction from Momentum Model, Trans. ASME, J. of Heat Transfer, 82, p. 113, 1960.

5. Zaludek, F. R., The Low Pressure Critical Discharge of Steam Water Mixtures from Pipes, ASME Paper, HW-68934, 1961.

6. Fauske, H. K., The Discharge of Saturated Water through Tubes, Chem. Eng. Prog. Ser. 61, 59, p. 210, 1965.

7. Moy, J. E., Critical Discharges of Steam Water Mixtures, M.S. Thesis, University of Min., 1965.

8. Keenan, F. G., Keyes, F. G., Hill, P. G. and Moore, J. G., "Steam Tables" (International edition-metric units), Wiley, 1969.

9. Zivi, S. M., Estimation of the Steady-State Steam Void Fraction by Means of the Principle of the Minimum Entropy Production, Trans. ASME, J. of Heat Transfer, 86, p. 247, 1964.

10. Delor, M., Study of Choking in Two-Phase Flow, M.S. Thesis, Brown University, 1981.

11. Petrick, M., Two-Phase Air Water Flow Phenomena, ANL-5787, March 1958.

PLUG FLOW IN VERTICAL RISERS

Leif N. Persen

Institute of Applied Mechanics
Norwegian Institute of Technology
Trondheim, Norway

ABSTRACT

The background for the presentation is the need to understand more completely the flow of two non-reacting components (water/air, oil/gas) flowing in a pipeline which from an almost horizontal part continues through a bend to a vertical riser. This is the situation in many cases in connection with the oil production on the Norwegian continental shelf in the North Sea.

Based on observations from a film showing the different flow regimes in the vertical riser, the following conclusions for promising future experimentation are drawn:

1.) *In a situation where the flow towards the bend at the bottom of the riser is stratified flow, the length of the plugs as well as the length of the gas pockets between plugs in the vertical portion of the pipe are greatly influenced by the geometry of the bend as well as the gas volume in the horizontal part of the pipe. Experimentation to systematically investigate this influence is suggested.*

2.) *The plugs in the vertical riser are not of constant length after their formation. The plug accepts mass in the front from the liquid film on the wall and deposits a different quantity of mass to the liquid film in the rear. Simple solutions to the motion of plugs of variable length are presented. The mechanism of transition from plug flow to churn flow is given.*

3.) *It will be demonstrated that situations may occur in which transition from one flow regime (churn-flow) in one part of the pipe to another (plug flow) further downstream occurs. This indicates another short-coming of the flow-regime maps.*

4.) *A short description of the test facilities for two-phase flow at Trondheim, Norway will be given.*

1. INTRODUCTION

The phenomena called "slugging" in two-phase pipe flow arises either as a result of hydrodynamic instability or as socalled "terrain-induced" slug flow. It is the latter case which will be considered here, and the situation is supposed to be as shown schematically in Fig. 1. The flow is supposed to come from an almost horizontal part (A) of the pipe and then continue up a

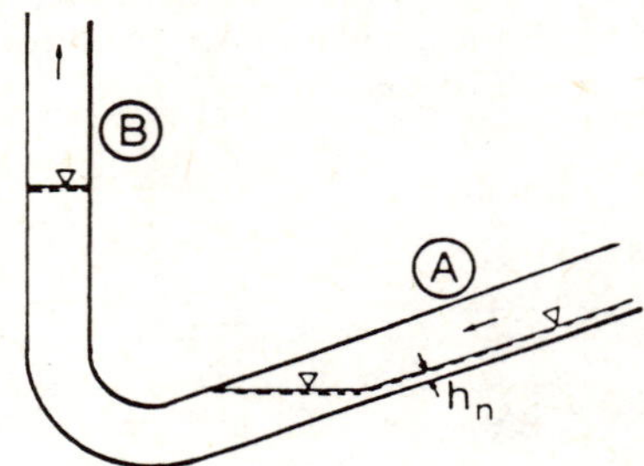

Fig. 1. Conditions at the bend inducing severe slugging.

vertical riser (B). In the A-section of the pipe the flow is supposed to be stratified; it is supposed to have its normal depth h_n as determined by the author [1] for given conditions.

When the bend is just "filled", the flow of gas will stop and the pressure in the gas will build up unless the gas has a high enough velocity so that it "blows the bend free". In this case one must expect annular flow to occur in the riser, a situation which is not examined here.

With a "filled" bend the liquid level in (A) and (B) will increase due to the steady flow of liquid. The difference in liquid levels will depend on the pressures in the two parts of the pipe, p_A and p_B respectively, and these in turn will depend on the conditions upstreams of (A) and downstreams of (B). The main emphasis will here be the study of the motion of a slug once it has been formed, and the way in which the conditions at (A) and (B) influence this motion.

2. THE ONSET OF SLUG FLOW

The scenario for the onset of slug flow can be described as follows:

1.) The vertical flow of liquid in the riser is less than what is transported towards the bend by the stratified flow. This leads eventually to the filling up of the bend (Stage 1).

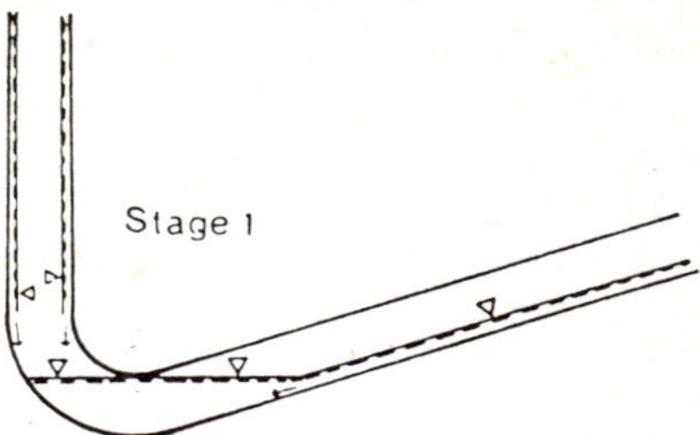

Fig. 2. The closing of the bend.

2.) The stratified flow in the almost horizontal pipe is affected by the "closing" of the tube. Since the flow entering the almost horizontal pipe is unchanged the liquid level in the bend will increase as will also the pressure in the trapped gas. The pressure in the vertical riser will however decrease as the slug (which one will have to assume is moving upwards after having been previously created) continues to move upwards. Both those effects tend to increase the difference in free surface (liquid) level in the riser and the almost horizontal pipe. (Stage 2).

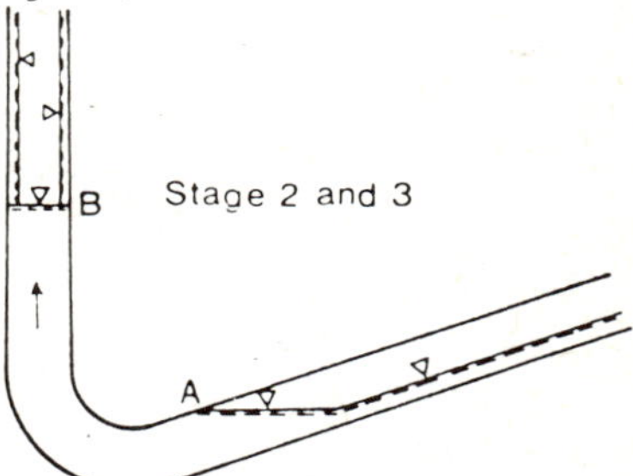

Fig. 3a. Accumulation of liquid in the bend.

3.) The liquid level A increases initially, then as the pressure difference increases the net force on the liquid plug becomes so great that it starts moving and the liquid level A will obtain the velocity of the plug which is identical to the velocity of the liquid level B. The motion of the plug can easily be described through the Bernoulli equation of ordinary single phase flow. (Stage 3).

4.) The last stage before reaching stage 1 again is now the development of stage 3. If the pressure difference is so great (and so little affected by the motion of A) that the liquid level is reaching the bend, the pressure in the gas phase in the riser and the almost horizontal pipe will equalize and finally stage 1 will again be reached. If that is not the case the plug will oscillate until finally the liquid surface A reaches the bend. Also now the stage 1 will eventually be established.

5.) The only motion not involved in this scenario of 4 sequences is the motion of the plug once it has been formed. First, it is supposed that the size of the plug does not change. This means that whatever amount of liquid enters the plug from the wall film the same amount must leave the plug at the lower end of the plug and this amount is then contained in the film beneath the plug.

From the description of the 4 (5) stages of the onset of slug flow it is obvious that the gas can no longer be conceived of as incompressible. When building up the pressure in the gas phase of the almost horizontal pipe, the volume of the gas V_G is being kept almost constant whereas the mass is steadily increasing. It will be assumed that the process takes place isothermally, i.e. whatever temperature change is taking place is so rapidly counteracted by a heat flux either to or from the pipe that the temperature may be considered to be constant.

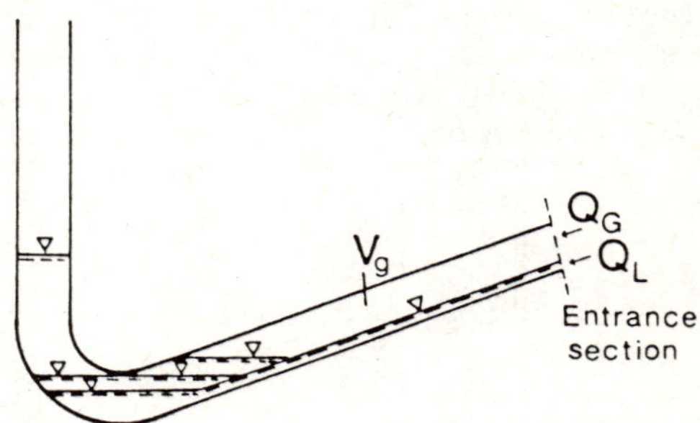

Fig. 3b. The 3 stages of liquid level in the bend.

Fig. 3 shows 3 stages of liquid level in the bend. The volume of the gas V_G (counted from the entrance section) is of course dependent on this liquid level, but because the distance from the bend to the surface is very small compared to the total length from the bend to the entrance, it is clear that V_G may be considered approximately constant.

3. THE BASIC EQUATION OF A PLUG'S MOTION

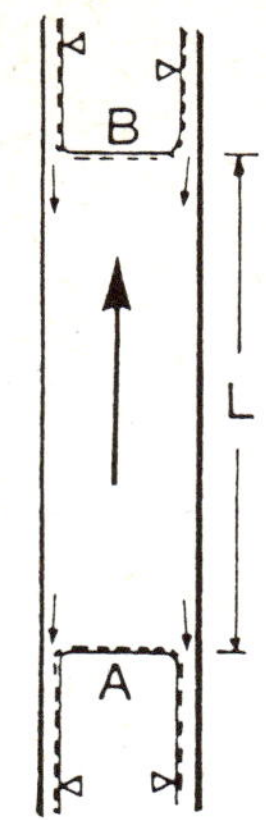

Fig. 4. A plug of length L in the riser.

The motion of the plug (of length L) in the vertical riser can be adequately described either as a rectilinear motion with variable mass or by means of the non-stationary Bernoulli's equation.

Bernoulli's equation from A to B: (Fig. 4)

$$\frac{v_A^2}{2g} + \frac{p_A}{\gamma} + z_A = \frac{1}{g}\int_A^B \frac{\partial v}{\partial t} ds + \frac{v_B^2}{2g} + \frac{p_B}{\gamma} + z_B + h_m \tag{3.1}$$

We shall have a difference in velocity at A and B due to the fact that the amount of mass "absorbed" by the plug at B is different from the amount "discharged" at A. This further implies that the length of the plug L is not a constant (Fig. 4).

In order to make all mathematical expressions "correct" it is now necessary to operate with separate velocities at A and B, and consequently an axial velocity distribution in the plug must be assumed. Since the cross-section of the tube is constant, the average velocity in each cross-section must be the same, thus the velocity in the plug is $v=v_A$ and the fact that v_B is different is then taken as a discontinuity at B due to the "incoming" fluid to the plug from the film above it. This means that the acceleration term in (3.1) may be expressed as

$$\frac{1}{g}\int_A^B \frac{\partial v}{\partial t}\, ds = \frac{L}{g}\frac{dv_A}{dt} \tag{3.2}$$

The head loss h_m consists of two terms:

1) loss due to turbulent "friction" $(h_m^{(1)})$

2) loss due to change in mass. $(h_m^{(2)})$

Now,

$$h_m^{(1)} = \lambda \frac{v^2}{2g} \cdot \frac{l}{d} \tag{3.3}$$

where

v = the mean velocity of the plug: $v = v_A$

$l = L$ = length of plug

d = diameter of the pipe

Fig. 5. A consecutive sequence of plugs.

The head loss due to the change in the mass of the plug can be expressed through the expression

$$h_m^{(2)} = c \frac{v_A^2}{2g} \tag{3.4}$$

It is now obvious that the motion of the plug is affected by the pressures p_A and p_B as well as by the flow in the film on the walls. This means, however, (with reference to Fig. 5) that the motion of plug (1) is determined both by the motion of plug (0) and of plug (2). The volume of gas trapped between the plugs is changing with the motion of its two neighboring plugs, and thus interacting with the plug motion through the pressure. This rather complex situation reveals that at least the initial motion of a plug will depend very much on time.

The subsequent sections will now deal with the motion of the plug based on these considerations but starting with simple cases and relaxing the constraining assumptions to take into account cases with increasing complexity.

4. THE MOTION OF A PLUG OF CONSTANT LENGTH

Assume the plugs' mass to be constant. Assume that the film's motion is independent of z. Then:

$$v = v_A = v_B \text{ , } L = constant \tag{4.1}$$

Equation (3.1) then gives:

$$\frac{L}{g}\frac{dv}{dt} + \frac{p_B - p_A}{\gamma} + L + \frac{\lambda}{2g} v^2 \cdot \frac{L}{d} = 0 \tag{4.2}$$

If such a situation is to become feasible, the pressure in the gas "pockets" between the plugs must have more or less equalled out, i.e. $p_B = p_A$ (This is however only true to the extent that the head loss of the gas flow in the pocket can be neglected). Under any circumstances the pressure difference may be considered to be constant. One now finds rather easily that

$$\frac{dv}{dt} + g + \frac{\lambda}{2d} v^2 = 0 \tag{4.3}$$

If one assumes v to be a function of z and observes that $\dot{z} = v$, one may re-formulate this equation:

$$\frac{dv}{dz} \cdot v + g + \frac{\lambda}{2d} v^2 = 0 \tag{4.4}$$

Then

$$\frac{d}{dz}(v^2) + 2g + \frac{\lambda}{d} v^2 = 0 \tag{4.5}$$

will give the solution:

$$v^2 = - 2g \frac{d}{\lambda} + Ce^{-\frac{\lambda}{d}z} \tag{4.6}$$

The initial condition is here taken to be:

$$v = v_o \quad for \quad z = z_o \tag{4.7}$$

which then gives the final solution in the form:

$$\frac{v^2}{v_o^{\,2}} = \frac{P}{\beta}\left[e^{-\beta(\frac{z}{z_o} - 1)} - 1\right] + e^{-\beta(\frac{z}{z_o} - 1)} \tag{4.8}$$

where

$$P = \frac{2gz_o}{v_o^2} \tag{4.9}$$

$$\beta = \frac{\lambda z_o}{d} \tag{4.10}$$

These are the two parameters of the solution. The first one is the ratio between the initial velocity v_o and the "Toricelli's velocity" at the location z_o where v_o occurs. The second indicates the importance of the friction under the assumptions made.

Examples have been computed and are shown in the diagram in Fig. 6. A general description of the behaviour is that the slug's velocity decreases initially almost proportional to the distance travelled and then almost abruptly decreases to zero.

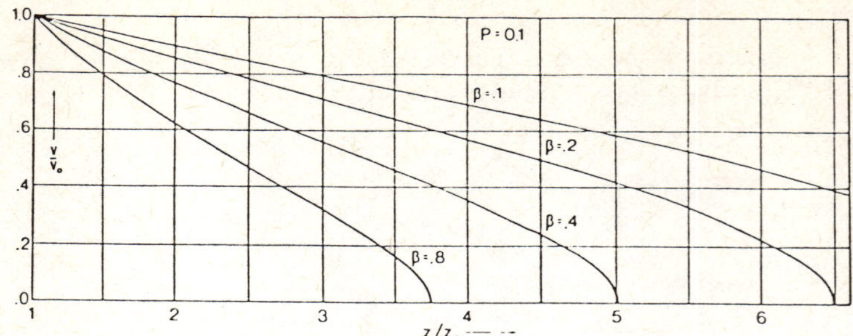

Fig. 6. The velocity v of a plug of constant length as function of the distance z travelled by the plug (Parameter $P = 0.1$, β is varied).

It ought to be stressed that the solution shown is found under the assumption that the slug's velocity v_o at a given position z_o is known. Any other initial condition will change the solution accordingly.

5. THE INITIAL MOTION OF A "CONSTANT LENGTH PLUG".

The motion of the plug after having reached a stage where the pressure difference between the front and the rear end of the plug is constant is solved in Sec. 4. Although the difference is supposed to be zero a constant pressure difference is basicly accounted for in (4.2) and will make no theoretical difference. At present the case when p_A is variable and where p_B may or may not be variable will be considered.

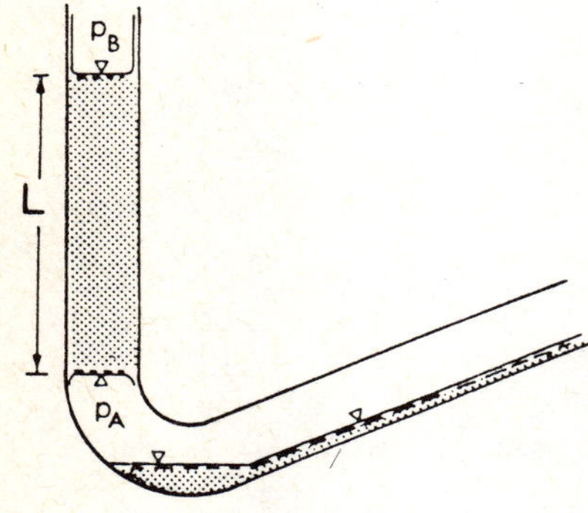

Fig. 7. The situation after formation of a plug.

Fig. 7 shows a sketch of the situatio immediately after the plug has been formed. The Bernoulli's equation (4.2) is still valid, i.e. the interaction of the plugs with the liquid film on the walls is also now neglected. The pressure p_A is at its peak immediately after the plug has been formed and since the volume of the gas behind the plug is now also comprising the volume in the almost horizontal part of the pipe, this volume increases so fast that the pressure p_A decreases. This happens in spite of the steady supply of gas at the entrance section

Assuming isothermal conditions and neglecting the gas entering the tube at the entrance one may to a first approximation assume $\bar{p}_A$ to be a linear function of the distance travelled by the plug:

$$p_A = p_o - p_1 \frac{z}{L} \tag{5.1}$$

where the constants p_o and p_1 will depend on the geometry of the system and will be discussed later. At this point it may be recognized that $p_A = p_B$ will be reached where the plug has travelled a distance z_1 given by

$$z_1 = L \frac{p_o - p_B}{p_1} \tag{5.2}$$

Beyond this point the pressure difference will no longer act to accelerate the plug, but will help viscosity to retard it. The differential equation (4.2) will still be valid, but now the pressure difference will not be zero and the equation will be:

$$\frac{L}{g}\frac{dv}{dt} + \frac{p_B - p_o}{\gamma} + \frac{p_1}{\gamma}\frac{z}{L} + L + \frac{\lambda}{2g}\frac{v^2 L}{d} = 0 \tag{5.3}$$

The solution to this equation is sought as before by realizing that the interest lies in finding the velocity of the plug as function of the distance z travelled by the plug, thus

$$\frac{dv}{dt} = \frac{dv}{dz} \cdot \dot{z} = v \frac{dv}{dz} \tag{5.4}$$

This substitution is made in (5.3) whereupon

$$\frac{L}{g} v \frac{dv}{dz} + (\frac{p_B - p_o}{\gamma L} + 1)L + \frac{p_1}{\gamma}\frac{z}{L} + \frac{\lambda}{2g} v^2 \frac{L}{d} = 0 \tag{5.5}$$

At this point the equation is non-dimensionalized:

The measure of length is taken to be the slug length L. Thus the non-dimensional length ζ will be:

$$\zeta = z/L \tag{5.6}$$

The measure of velocity is taken to be the velocity corresponding to the initial pressure difference which according to (5.1) will be $(p_o - p_B)$. The non-dimensional velocity ξ will then be:

$$\xi = v/v_o \qquad \text{where} \qquad v_o^2 = \frac{(p_o - p_B)2g}{\gamma} \tag{5.7}$$

Introduced into (5.5) these relations will render the following equation:

$$\frac{d}{d\zeta}(\xi^2) + \beta\xi^2 = \alpha - \delta\zeta \tag{5.8}$$

where the parameters are defined as follows:

$$\alpha = \left(\frac{p_o - p_B}{\gamma L} - 1\right)\frac{2gL}{v_o^2} \tag{5.9}$$

$$\beta = \lambda\frac{L}{d} \tag{5.10}$$

$$\delta = \frac{p_1}{\gamma}\frac{2g}{v_o^2} \tag{5.11}$$

These parameters have a special physical significance which may be explained as follows:

1.) α is a parameter which reflects the initial pressure difference. Introduction of (5.7) into (5.9) will transform the expression for α:

$$\alpha = 1 - \frac{\gamma L}{p_o - p_B} \tag{5.12}$$

This formulation reveals that α depends on the ratio between the pressure force acting on the slug and the weight of the slug. If this ratio is unity the value of α is zero. As the pressure force increases the value of α increases. The upper bound of α is 1. Thus: $0 \leq \alpha \leq 1$ is the range of variation of α.

2.) β is the parameter which reflects the influence of the frictional loss as the slug moves. It also depends on the ratio between the slug length and the diameter of the pipe. To give an idea of the order of magnitude of β one may recall that for rough pipes $\lambda = 0.03$ may be an average value leaving $\beta \cong 1$ for $L/d = 30$.

3.) δ is the parameter that reflects the influence of the "volume" of the almost horizontal part of the pipe. Returning to (5.7) one observes that δ may be reformulated as follows:

$$\delta = \frac{p_1}{p_o - p_B} \tag{5.13}$$

From (5.1) one deduces that p_1 is that fraction of the initial pressure difference between A and B with which this difference will be reduced per unit distance (measured in slug lengths) travelled by the slug. This fraction is dependent on the volume V_g (Fig. 3). It will be assumed here that this fraction is 0.1 - 0.2. Finally it is observed that δ may be related to z_1 as follows:

$$\zeta_1 = z_1/L = 1/\delta \tag{5.14}$$

The solution to (5.8) is easily found. It is supposed that the initial conditions are such that

$$\xi = 0 \qquad at \qquad \zeta = 0 \tag{5.15}$$

The solution satisfying this condition is:

$$\xi^2 = (\frac{\alpha}{\beta} + \frac{\delta}{\beta^2}) (1-e^{-\beta\zeta}) - \frac{\delta}{\beta}\zeta \tag{5.16}$$

This solution is illustrated in Fig. 8. It is seen how increasing values of α will give increasing values of the maximum velocity (ξ_{max}) and also increasing distances ζ_m travelled by the slug before its velocity changes sign. However, due to the existence of frictional loss the values of ξ_{max} and ζ_m will not tend to infinity with increasing pressure difference $p_o - p_B$ ($\alpha \to 1$). It is furthermore illustrated how increasing values of β reduces ξ_{max} and ζ_m. The situation is perhaps better illustrated by plotting the maximum value of the non-dimensional velocity ξ_{max} as a function of the parameters α, β and δ. Also the position ζ_m at which this value of the velocity occurs may be plotted as a function of the same parameters.

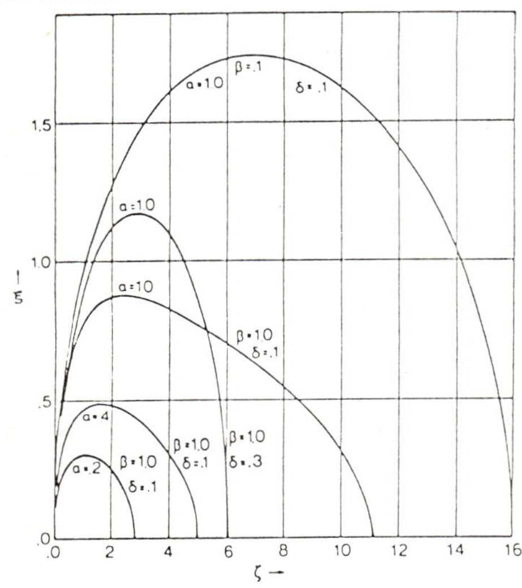

Fig. 8. ξ as a function of ζ showing a maximum ξ_{max} at $\zeta = \zeta_m$.

Fig. 9a and b show the situation. It ought to be stressed that the initial conditions used here differ from those applied in Sec. 4 where the motion of the plug after it has reached a given position with a given velocity is studied under the assumption of a constant pressure difference.

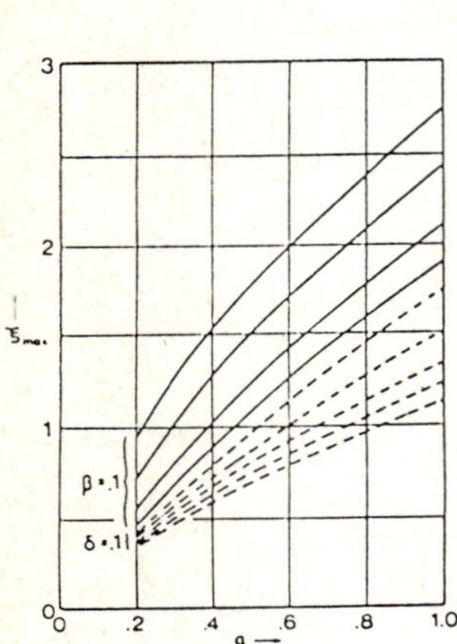

Fig. 9a) gives ξ_{max} as functions of α with β and δ as parameters.

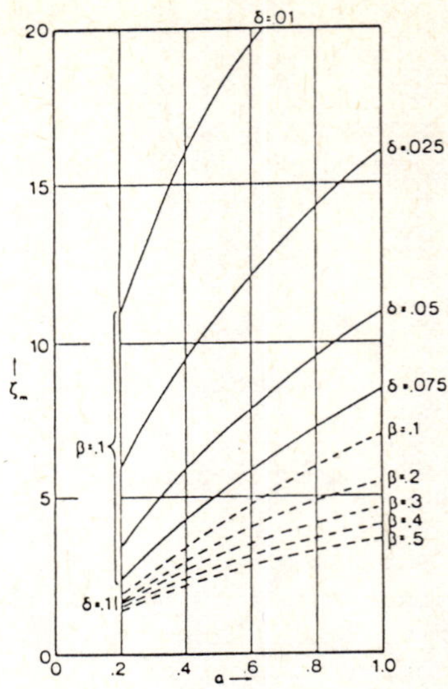

Fig. 9b) gives ζ_m as functions of α with β and δ as parameters.

6. MOTION OF A PLUG OF VARIABLE LENGTH

In Sec. 3 the general equation of motion for a plug of variable length is deduced. From (3.1) one obtains, by introducing (3.2):

$$\frac{L}{g}\frac{dv_A}{dt} + \frac{v_B^2}{2g} - \frac{v_A^2}{2g} + \frac{p_B - p_A}{\gamma} + L + \lambda\frac{v_A^2}{2g}\cdot\frac{L}{d} + h_m^{(2)} = 0 \qquad (6.1)$$

where also (3.3) has been observed. In this formulation the instantaneous length L of the plug appears and it is observed that the relation between the velocities v_A and v_B has been discussed in Sec. 3 with the following result:

$$v_B = v_A + \frac{dL}{dt} \qquad (6.2)$$

This means that the relation between v_A and v_B is only known when the way in which the mass of the plug changes has been specified. We shall here assume:

$$m = m_o(1 + \omega \frac{x_A}{L_o}) \tag{6.3}$$

which may be rewritten as

$$L = L_o(1 + \omega \frac{x_A}{L_o}) \tag{6.4}$$

where L_o is the initial length of the plug, and x_A is the instantaneous position of the point A. Thus:

$$\frac{dL}{dt} = \omega \frac{dx_A}{dt} = \omega v_A \tag{6.5}$$

and consequently:

$$v_B = v_A \; (1 + \omega) \tag{6.6}$$

This may be introduced into (6.1) whereupon:

$$\frac{L}{g}\frac{dv_A}{dt} + \frac{v_A^2}{2g}\,[(1+\omega)^2-1] + \frac{p_B-p_A}{\gamma} + L + \frac{v_A^2}{2g} \bullet \frac{\lambda L}{d} + h_m^{(2)} = 0 \tag{6.7}$$

The pressure difference will here be introduced under the same assumption as in Sec. 5, i.e.

$$p_A = p_o - p_1 \frac{x_A}{L_o} \tag{6.8}$$

It will furthermore be assumed that p_B = *constant*, thus:

$$\frac{p_B-p_A}{\gamma} = \frac{p_B-p_o}{\gamma} + \frac{p_1}{\gamma}\frac{x_A}{L_o} \tag{6.9}$$

The final equation of motion of the plug will then be:

$$\frac{L}{g}\frac{dv_A}{dt} + \frac{v_A^2}{2g}[(1+\omega)^2-1+c] + \frac{p_B-p_o}{\gamma} + L\;\lambda\frac{v_A^2}{2g}\frac{L}{d} + \frac{p_1}{\gamma}\frac{x_A}{L_o} = 0 \tag{6.10}$$

where it is to be observed that L is a function of the distance x_A travelled by the plug according to (6.4)

This equation is now non-dimensionalized in order to identify the parameters of the problem. In accordance with the procedure in Sec. 5 we shall introduce:

$$v_o^2 = \frac{p_o-p_B}{\gamma}\,2g \quad , \quad \text{and} \quad \xi = v_A^2/v_o^2 \tag{6.11}$$

We shall furthermore measure the distance travelled by the non-dimensional quantity:

$$\zeta = x_A/L_o \tag{6.12}$$

which gives

$$\frac{L}{L_o} = 1 + \omega\zeta \tag{6.13}$$

Equation (6.10) may now be reformulated as:

$$\left.\begin{aligned} & 2(1+\omega\zeta)\frac{dv_A}{dt} + \frac{v_A^2}{L_o}[(1+\omega)^2-1+c] - \frac{v_o^2}{L_o} + (1+\omega\zeta)2g + \\ & \frac{v_A^2}{L_o} \cdot \frac{\lambda L_o}{d}(1+\omega\zeta) + \frac{p_1 2g\zeta}{\gamma L_o} = 0 \end{aligned}\right\}$$

If as usual $v_A = v$ is conceived of as a function of x_A (or ζ), one obtains:

$$\frac{dv_A}{dt} = \frac{dv}{dx_A} \cdot \frac{dx_A}{dt} = \frac{dv}{d\zeta} \cdot v/L_o = \frac{1}{2L_o}\frac{d}{d\zeta}(v^2) \tag{6.14}$$

which introduced into the equation finally leaves it in the form:

$$\left.\begin{aligned} & (1+\omega\zeta)\frac{d\xi}{d\zeta} + \xi\left[(1+\omega)^2 -1+c+ \frac{\lambda L_o}{d}(1+\omega\zeta)\right] - \\ & 1 + (1+\omega\zeta)\frac{2gL_o}{v_o^2} + \frac{p_1 2g}{\gamma\, v_o^2} \cdot \zeta = 0 \end{aligned}\right\} \tag{6.15}$$

In order to more properly understand the physical significance of the parameters, the two parameters δ and β are introduced analogous to the case in Sec. 5:

$$\left.\begin{aligned} \delta &= \frac{p_1 2g}{\gamma v_o^2} \\ \beta &= \frac{\lambda L_o}{d} \end{aligned}\right\} \tag{6.16}$$

Furthermore the parameter α is defined as

$$\alpha = 1 - \frac{2gL_o}{v_o^2} = 1 - \frac{L_o\gamma}{p_o-p_B} \tag{6.17}$$

and the parameter c is replaced by

$$(1+\omega)^2 -1+c = \mu \tag{6.18}$$

whereupon

$$(1+\omega\zeta)\frac{d\xi}{d\zeta} + \xi(\mu + \beta(1+\omega\zeta)) = 1 - (1+\omega\zeta)(1-\alpha) - \delta\zeta \qquad (6.19)$$

This 1st order linear differential equation for ξ is easily solved:

$$\xi = C\,(1+\omega\zeta)^{-\mu/\omega}\,e^{-\beta\zeta} - \frac{1}{\beta}\left(1+\frac{\delta}{\omega}-\alpha\right) + \left[\left(1+\frac{\mu}{\beta}\right)\left(1+\frac{\delta}{\omega}\right) - \frac{\mu\alpha}{\beta}\right](1+\omega\zeta)^{-\mu/\omega}\,e^{-\beta\zeta}\int_0^{\zeta}(1+\omega z)^{-1+\mu/\omega}\,e^{+\beta z}\,dz \qquad (6.20)$$

The physical significance of the parameters:

(1) $\alpha = 1 - \dfrac{2gL_o}{v_o^2} = 1 - \dfrac{\gamma L_o}{p_o - p_B}$

α is always less than 1, $0<\alpha<1$

α gives the ratio between the initial length of the slug (L_o) and the pressure head between A and B (initially) $[(p_o - p_B)/\gamma]$

(2) $\beta = \lambda\dfrac{L_o}{d}$, *the initial length ($L_o$) of the slug divided by the diameter of the tube times the friction coefficient λ of the pipe.*

Ex.: $\lambda = 0.03$ (?), $L_o/d = 100$? $\beta = 3$

(3) $\delta = \dfrac{p_1 2g}{\gamma v_o^2} = \dfrac{p_1}{p_o - p_B}$ *The parameter represents the ratio between the pressure head p_1/γ and the initial pressure head between A and B.*

(4) $\omega = \dfrac{dL/dt}{v_A}$ = *the ratio between the rate at which the length L of the slug changes and the velocity v_A.*

Really, the parameter ω represents the change of slug length and may be both positive and negative

(5) $\mu = (1+\omega)^2 - 1 + c$ *The coefficient c represents the head loss due to change in mass of the slug. Since ω may be both positive and negative, the same holds also for μ:*

This exposition of the motion of a slug of variable mass implies certain limiting values. Thus equation (6.13) implies that the slug length L becomes zero after a certain distance ζ_{max} has been travelled by the slug.

$$\zeta_{max} = -\frac{1}{\omega}$$

Another characteristic distance is implied in equation (6.9). The pressure difference $(p_B - p_A)$ representing the driving force of the slug becomes zero when the slug has travelled a distance ζ_p:

$$\zeta_p = \frac{p_B - p_o}{p_1} = \frac{1}{\delta}$$

The solution found in (6.20) seems originally to contain 5 parameters. However, the solution can be reformulated in such a way that only 3 parameters appear explisitly:

The initial conditions are such that $\xi = 0$ for $\zeta = 0$ which leaves

$$C_1 = \frac{1}{\beta}\left(1 + \frac{\delta}{\omega} - \alpha\right) \tag{6.21}$$

The solution (6.20) may then be formulated as:

$$\left.\begin{aligned} &\frac{\xi\beta}{1+\delta/\omega-\alpha} = \left[(1+\omega\zeta)^{-\mu/\omega}\, e^{-\beta\zeta} - 1\right] + \\ &B(1+\omega\zeta)^{-\mu/\omega}\, e^{-\beta\zeta} \int_0^{\zeta} (1+\omega z)^{-1+\mu/\omega}\, e^{+\beta z}\, dz \end{aligned}\right\} \tag{6.22}$$

where

$$B = \frac{\left(\left(1 + \frac{\mu}{\beta}\right)\left(1 + \frac{\delta}{\omega}\right) - \frac{\mu\alpha}{\beta}\right)\beta}{\left(1 + \frac{\delta}{\omega} - \alpha\right)} \tag{6.23}$$

If now the variable of integration in the integral is changed so that

$$s = \omega z \tag{6.24}$$

the solution will contain only 3 parameters: μ/ω, β/ω, B/ω.

The final formulation of the solution:

$$\left.\begin{aligned} &\frac{\beta\xi}{1+\frac{\delta}{\omega}-\alpha} = \left[(1+\omega\zeta)^{-\mu/\omega}\, e^{-\frac{\beta}{\omega}\omega\zeta} - 1\right] + \\ &\frac{B}{\omega}(1+\omega\zeta)^{-\mu/\omega}\, e^{-\frac{\beta}{\omega}\omega\zeta} \int_0^{\omega\zeta} (1+s)^{-1+\mu/\omega}\, e^{+\frac{\beta}{\omega}s}\, ds \end{aligned}\right\}$$

where consequently a normalized velocity V_{norm} will be:

$$V_{norm} = \frac{\beta\xi}{1+\frac{\delta}{\omega}-\alpha} = F(\omega\zeta,\ \frac{\mu}{\omega},\ \frac{\beta}{\omega},\ \frac{B}{\omega}) \tag{6.25}$$

A numerical evaluation of this solution is shown in the following diagrams

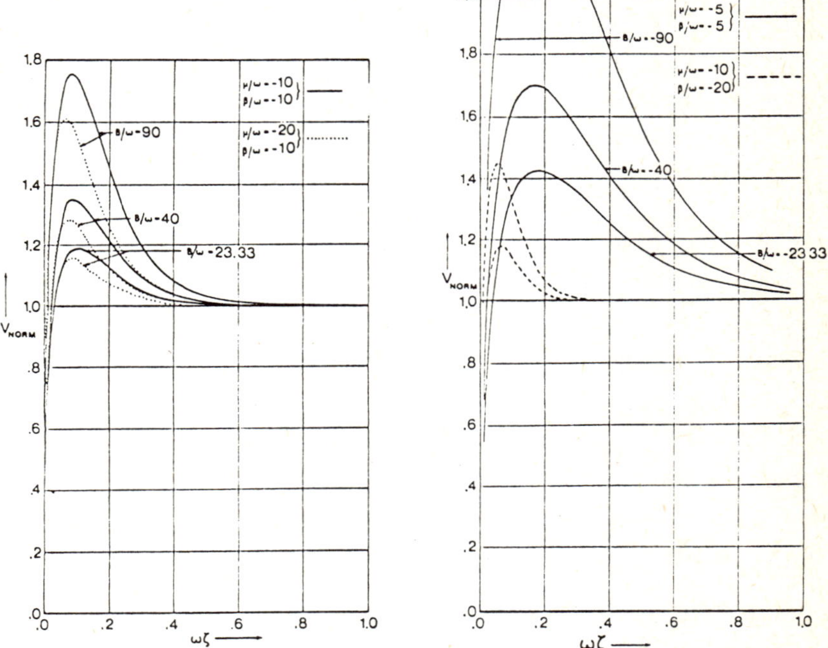

Fig. 10a and b. The normalized velocity V_{norm} as function of the normalized distance $\omega\zeta$.

They describe how the plug's velocity increases initially rather rapidly and then reduces to an assymptotic value 1. The cases Fig. 10a show how this value is reached at a situation where the plug's length is approximately 40% of its initial lenght. Fig. 10b however describes a situation where the plug's velocity has not reached its assymptotic value before it disappears, i.e. its length becomes zero.

The situation described by this solution is encountered in situations where the plug "disintegrates" and falls down along the walls as a film. (Churn flow is a more pronounced form of this mechanism). This increases the length of the following plug, and for this plug the situation may be as described by the same solution with a positive value of ω. It may however also be described by the case treated in the next section.

7. A DIFFERENT APPROACH TO THE MOTION OF A PLUG OF VARIABLE LENGTH

Again the basic equation for the motion of a plug of variable length is used in the form deduced in (3.1) and used in (6.1).

$$\frac{L}{g}\frac{dv_A}{dt} + \frac{v_B^2}{2g} - \frac{v_A^2}{2g} + \frac{p_B - p_A}{\gamma} + L + \lambda\frac{v_A^2}{2g}\frac{L}{d} + h_m^{(2)} = 0 \qquad (7.1)$$

The basic equation is reformulated in Sec. 6 by the introduction of several assumptions. Some of these will be kept here:

1) The pressure difference $p_B - p_A$ will be expressed as

$$\frac{p_B - p_A}{\gamma} = \frac{p_B - p_o}{\gamma} + \frac{p_1}{\gamma}\frac{x_A}{L_o} \qquad (7.2)$$

and the pressure p_B is assumed constant. The pressure p_A is then given as in (6.8)

2) The head loss $h_m^{(2)}$ is assumed given by (3.4)

3) The way in which the plug length L changes was given in (6.4). Here this length will be supposed to change with changing velocity:

$$L = L_o(1 + kv_A) \qquad (7.3)$$

This means that if the plug attains constant velocity, the length of the plug remains constant.

(A natural question to ask is whether either one or the other of these ways in which the plug length varies is correct or if perhaps none of them reflects reality. The answer is a yes/no. Experimental evidence indicates regions of constant length and constant velocity over rather long distances measured in number of plug lengths. However, it also indicates regions in which the plug length do change with distance at constant velocity. Thus the two cases treated here do have their practical relevance).

From (6.2) and (7.3) it is now possible to deduce the velocity v_B:

$$v_B = v_A + \frac{dL}{dt} = v_A + kL_o\frac{dv_A}{dt} \qquad (7.4)$$

These assumptions are now introduced in the basic equation (7.1) whereupon

$$\left.\begin{aligned}&\frac{L_o}{g}(1+kv_A)\frac{dv_A}{dt}+\frac{1}{2g}\left\{(v_A+kL_o\frac{dv_A}{dt})^2-v_A^2(1-c)\right\}+\\&\frac{p_B-p_o}{\gamma}+\frac{p_1}{\gamma}\frac{x_A}{L_o}+L_o(1+kv_A)\left[1+\frac{\lambda}{2gd}v_A^2\right]=0\end{aligned}\right\}\quad(7.5)$$

At this point a non-dimensional formulation of the problem must be sought. The velocity v_A is made dimensionless with the velocity v_o:

$$\xi=\frac{v_A}{v_o}=\textit{non-dimensional velocity}\quad(7.6)$$

The distance x_A is made dimensionless by means of the initial slug length L_o:

$$\zeta=\frac{x_A}{L_o}=\textit{non-dimensional distance}\quad(7.7)$$

The velocity v_o is supposed to be given through the initial pressure difference: (6.7)

$$-v_o^2=(p_B-p_o)2g/\gamma\quad(7.8)$$

Furthermore, the velocity v_A can be sought as a function of x_A instead of as a function of time:

$$\frac{dv_A}{dt}=\frac{dv_A}{dx_A}\frac{dx_A}{dt}=v_A\frac{dv_A}{dx_A}=\frac{v_o^2}{L_o}\xi\frac{d\xi}{d\zeta}\quad(7.9)$$

Finally, the previously used parameters are to be introduced to the extent which it is possible:

$$\left.\begin{aligned}&\beta=\frac{\lambda L_o}{d}\\&\delta=\frac{p_1}{\gamma}\frac{2g}{v_o^2}\\&\alpha=1-\frac{L_o\gamma}{p_o-p_B}=1-\frac{2gL_o}{v_o^2}\end{aligned}\right\}\quad(7.10)$$

In view of these formulations it is possible to reformulate (7.5) after some elementary algebra:

$$2(1+\omega\xi)\xi\frac{d\xi}{d\zeta}+(\xi+\omega\xi\frac{d\xi}{d\zeta})^2=(1-c)\xi^2+1-\delta\zeta+(1-\alpha)(1+\omega\xi)(1+\frac{\beta\xi^2}{1-\alpha})\quad(7.11)$$

This equation contains 5 parameters, namely c, α, β and δ to which one must add ω defined by:

$$\omega = kv_o \tag{7.12}$$

Due to the way in which the non-linearity of the equation occurs the solution cannot easily be expressed through elementary functions. A numerical Runge-Kutta method is introduced. For this purpose the equation (7.11) is simplified by a reformulation: The function $F(\xi,\zeta)$ is introduced

$$F(\xi,\zeta) = 1-c\xi^2 - \delta\zeta + (1-\alpha)(1+\omega\xi)(1+\frac{\beta\xi^2}{1-\alpha}) \tag{7.13}$$

Equation (7.11) then becomes:

$$\omega^2\xi^2 (\frac{d\xi}{d\zeta})^2 + 4\omega\xi^2 \frac{d\xi}{d\zeta} + 2\xi \frac{d\xi}{d\zeta} = F(\xi,\zeta)$$

which can be written:

$$(\omega\xi \frac{d\xi}{d\zeta})^2 + (4\xi + \frac{2}{\omega}) \omega\xi \frac{d\xi}{d\zeta} = F(\xi,\zeta)$$

Finally:

$$\frac{d\xi}{d\zeta} = - \frac{2\xi + \frac{1}{\omega}}{\omega\xi} \pm \frac{1}{\omega\xi} \sqrt{F(\xi,\zeta) + (2\xi + \frac{1}{\omega})^2} \tag{7.14}$$

In this form the equation is well suited for a Runge-Kutta procedure.

The starting conditions are given at $x_A = 0$ where:

$$\left\{\begin{matrix} v_A = 0 \\ L = L_o \end{matrix}\right\} \Rightarrow \xi = 0 \text{ for } \zeta = 0$$

Equation (7.13) then leaves: $F(0,0) = 1 + (1-\alpha) = (2-\alpha)$

This indicates that $d\xi/d\zeta\Big|_{\substack{\xi=0\\ \zeta=0}}$ must be infinite

The solution to the problem for a certain combination of the parameters is exhibited in Fig. 11. It is observed that these curves resemble to some extent both those for the plug of constant length in Fig. 6 and those in Fig. 8. The variation of the plug's length in the present cases is to some extent visible from Fig. 11 realizing that the change in the plug's length is proportional to the velocity [Eq.(.3)]. The velocity does not in these cases reach an assymptotic value, and the plugs' length is consequently constantly changing. As mentioned earlier the situation described here as well as in the preceding section is observed experimentally.

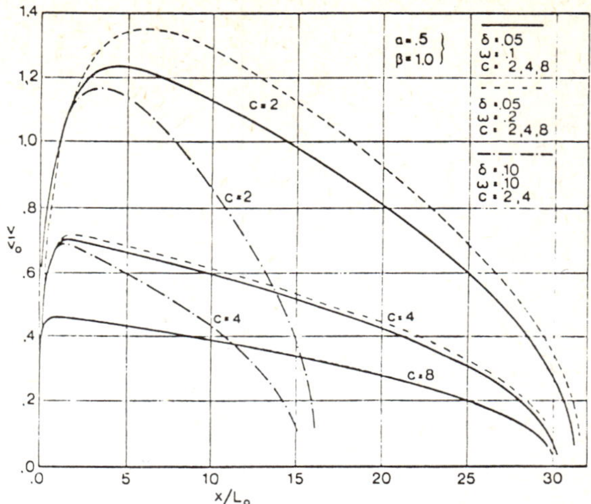

Fig. 11 The non-dimensional velocity v/v_0 is given as a function of the distance x/L_0

8. REMARKS

The deduction shown in this report are based on simple applications of the non-steady Bernoulli equation for pipe flow. To an astonishing degree they represent realities as they can be observed in the laboratory. The break-down of liquid plugs and the transition to churn flow seems to be well explained. The necessary analytic basis for linking together a series of plugs seems to be present. However, it is an observation from the film accompanying this report that the lengths of the plugs are not deterministicly defined, nor is their velocity or spacing. A statistical approach to the problem may be better suited for prediction purposes, and a future effort to pursue this line of reasoning is already in the making.

A final remark should be made. This study focuses on the motion of an individual plug. It is however shown how the motion of one plug is affected through the neighboring plugs. Thus the solutions given will represent appropriate sub-solutions which can be combined into specially designed programs for given configurations of a pipeline. It should be mentioned that the University of Trondheim through the local research organization SINTEF will be operating a large scale two-phase flow rig with a 400m long horizontal line and a 60m high riser as of November 1982.

REFERENCE

1. Persen, L. N., "Stratified Two-Phase Flow in Circular Pipes," Internal Report, Institute of Mechanics, January 1981.

FLOODING IN COUNTER-CURRENT TWO-PHASE FLOW

W. A. Ragland and E. N. Ganic*

Reactor Analysis and Safety Division
Argonne National Laboratory
Argonne, Illinois, U.S.A.

Department of Mechanical Engineering
University of Illinois at Chicago
Chicago, Illinois, U.S.A.

1. ABSTRACT

This summary presents the various definitions which have been proposed for the flooding phenomenon. A collection of experimental results for vertical flooding is presented to show that although the research has included many parameters, a large portion of the work has centered on air and water in tubes ranging from 38 mm to 50 mm in diameter. Some detail of individual experiments is presented to demonstrate various types of apparatus and instrumentation in use. The well known correlations of Wallis and Kutateladze as well as some early empirical correlations are given with a discussion of the limits of these correlations. Recent attempts by Crowley, Dilber, and Richter to provide an analytical base for flooding are discussed. In closing, the authors highlight an interesting work by Maron and Dukler with suggestions for extending this approach to provide a more complete analytical solution for flooding. For simplicity in this survey of flooding literature, the experimental aspects of flooding will be presented first as a summary of the range of experimental data available. The analytical aspects of the experimental works will be included with the purely analytical works.

*Presently with the Faculty of Mechanical Engineering, University of Sarajevo, Post fah. 107, Sarajevo, Yugoslavia.

2. INTRODUCTION

Flooding is a phenomenon which is best described as the transition from counter-current to co-current flow. Early notice was taken of this phenomenon in the chemical engineering industry. Verschoor [58] investigated flooding in packed columns and found that there was a limiting velocity for upward flowing vapor above which liquid could not be made to flow downward in the column. The phenomenon of flooding is important in the chemical industry because a number of chemical processes rely on the counter-current flow of a gas and a liquid to provide a highly active exchange interface for both heat and mass transfer [24]. Examples of flooding limited processes include wickless thermosyphons [28,29,35,49] and the emergency core cooling system (ECCS) of pressurized water nuclear reactors [3,4,14,15,16,41,42], and the behavior of molten fuel pin cladding of nuclear reactors during severe accident conditions [37].

In the situation of counter-current flow, there are two fluids in motion, a liquid moving under the influence gravity and a gas and moving due to a pressure gradient. Flooding may be vertical or inclined. This survey will deal only with vertical flooding. Figure 1 shows the typical topography for vertical flooding. To simply state the condition of vertical, counter-current flow: liquid flows down while the gas flows up in the same channel.

The phenomenon of flooding may then be described as the limit point at which any increase in either the liquid or gas flow rate results in less than all the liquid flowing downward. Several other definitions have also been used to describe flooding: (1) flooding occurs at the point that any of the liquid is in upward motion, even if the overall flow is downward; (2) flooding occurs when there is net flow reversal of the liquid; (3) flooding occurs at the onset of entrainment; and (4) flooding occurs when the liquid bridges the flow area. With these many definitions of flooding, it is not surprising that the experimental observations of flooding yield systematic differences in the measurement of the flooding point. A further complication of the definition of flooding arises from the converse situation of transition from upward co-current flow to counter-current flow. This is sometimes called deflooding. Ocassionally the conditions for de-flood have been analyzed as the condition for flooding.

The published literature on flooding includes both analytical and experimental work. For simplicity, the experimental aspects of flooding will be presented separately. The analytical aspects of the experimental works will be included with the purely analytical works.

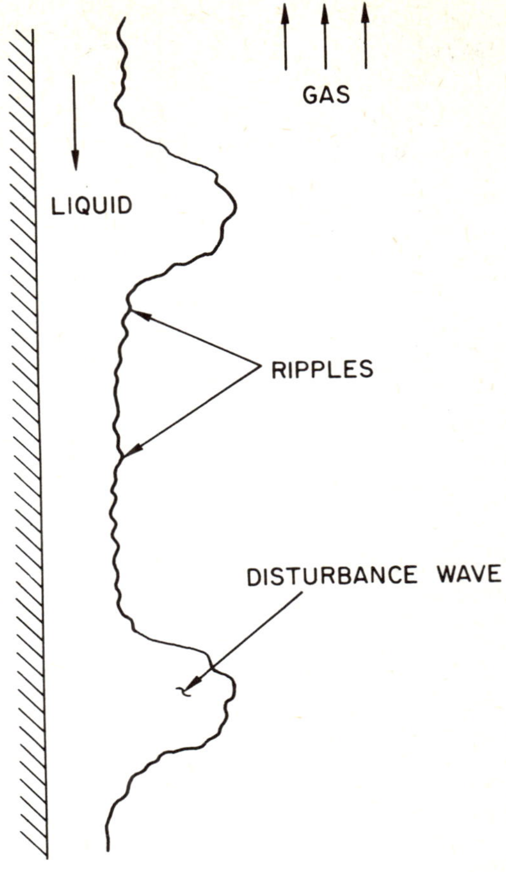

Figure 1. Typical Flooding Profile

3. EXPERIMENTS

Flooding has been studied experimentally in a variety of geometries, with several sets of fluid pairs, and over a range of flowrates for both liquid and gas. The most common geometry studied is that of the tube of circular cross section. Other geometries include rectangular channels, annulii, and multiple channels such as those formed by the fuel pins in in a nuclear reactor. The most common fluid pair is air and water with steam and water second. Other fluids include glycerol, glycerol and water solutions, isopropyl alcohol, methyl alcohol, n-neptane, ethylene glycol, diethylene glycol, aqueous sec.-octyl alcohol, silicon oil, and water with various surfactant agents. Nitrogen has been also used as the gas phase. The data collected has included: liquid holdup; pressure drop; average liquid film thickness; liquid surface charcteristics such as disturbance wave and ripple height, velocity, and speed; and the flow rate of both liquid and gas at the flooding point and entrainment.

The experimental data is organized first by the geometry of the test section. The second level of organization is by the fluid pair. Finally, the experiments are presented in chronological order.

3.1 Experiments in tubes

The air-water flooding in tubes has been studied over a wide range of diameters and lengths.

Wallis [59,60], Hewitt and Wallis [27] and Hewltt, Lacey, and Nichols [26] studied flooding in a 31.8 mm ID acrylic resin pipe. The water was introduced through a porous bronze sinter section and removed by a second sinter section [27]. Test sections lengths of 0.23 m, 0.46 m, 0.91 m, 1.82 m, and 3.65 m were used [26]. The water flow rate varied from 2.5 g/s to 75 g/s. The air flow rate varied from zero to 11.4 g/s. The experiment was performed at pressures of 137 kPa, 275 kPa, and atmospheric. The data recorded included air and water flow rates, pressure drop, system pressure, and film thickness using a conductance probe. High speed photography at 100 to 250 frames per second was used to study the liquid film [27]. Liquid film thickness was measured using a fluorescent dye irradiated by monochromic light [26]. The fluorescence was separated from reflected incident light by a spectrometer.

Clift, Pritchard, and Nedderman [12] studied flooding in a 31.8 mm ID, 3.7 m long tube. with a 76 mm long bronze sinter feed section in the middle of the tube. The velocity of the air varied from 8.25 m/s to 9.50 m/s. The water volumetric flow rate varied from 1 cm^3/sec to 52 cm^3/sec. In addition to flow rates, the pressure drop in the tube was also measured. Waves were visually observed as gas velocities approached the flooding point. As the flooding point was almost reached, the pressure drop increased sharply and the waves on the film became sufficiently large to bridge the tube. The bridges were broken by the air flow and the resulting drops were carried up by the air stream. Further increase in either the gas or liquid flow caused an increase in both the pressure drop and the height of the liquid slugs above the liquid entrance. The liquid below the entrance was less chaotic after flooding. Still further increase (after flooding) caused a slight decrease in the pressure drop.

Diehl and Koppany [17] studied flooding with water and air in tubes with diameters of 15.8 mm, 26.6 mm, and 52.5 mm. The flooding point was determined by noting the gas velocity at which there was a large increase in entrainment, and by checking the pressure drop increase as gas flow was increased. The results were reported only in terms of gas velocity and entrainment loads.

Pushkina and Sorokin [44] studied both flooding and liquid film breakdown. Flooding was defined as the onset of entrainment with some motion of a part of the liquid upward. This included cases in which the entrained droplets were deposited on the wall above the water injection point and flowed downward in a film with no exit of fluid from the top of the tube. Film breakdown was defined as the point at which liquid downflow ceases and all of the liquid is carried upward. Two different sets of apparatus were used in the study. One of these consisted of a 309 mm ID tube with a porous liquid feed insert. The porous section was 210 mm with a 3.75 meter section above and a 2.5 meter section below giving a total test section length of 6.46 meters. The second apparatus consisted of a set of varying diameter tubes which were top fed from a reservoir through a funnel. The air was introduced into a lower plenum. Tubes with an interior diameter of 6.2 mm, 8.8 mm, 9.0 mm, 12.2 mm, and 13.1 mm were used. The length of the tubes varied from 0.5 meter to 0.8 meter. The water flow rate was the minimum which would continously wet the tube around its entire circumference. For the porous feed apparatus, the air velocity was adjusted to produce both flooding and film flow reversal. For the top feed apparatus, the air velocity was adjusted to produce film flow reversal. The velocity of the air was in the range 14 m/s to 16 m/s. Measurements were made of the water and air flow rates.

Wallis and Makkenchery [66] investigated the transition region between the flooding point and upward co-current flow. In this region, there is no net film motion and hence the phenomenon is sometimes called the "hanging film" phenomenon. The experiments were carried out in tubes 6.35 mm and 140 mm in diameter. Liquid entered from the top of the tube either by overflowing a weir or by running down a tapered funnel. Measurements were made of the air and water flows.

Kusuda and Imura [35] performed experiments to determine the critical heat flux in an open two-phase thermosyphon. They confirmed that dryout occurs only with counter-current annular two-phase flow as a result of flooding rather than local film boiling. One tube was 1 meter in length and 28.4 mm in diameter. Other tubes were 0.2 and 0.3 meters in length and 10.7 mm in diameter.

Wallis and Kuo [67] performed flooding designed to study the interfacial effects of flooding. One experiment involved measuring the height to which water could be supported over the lip of a tube with air exiting the tube. Tubes of the following diameters were used: 6.4 mm, 19 mm, 31.2 mm, 38.2 mm, 63.5 mm, 113 mm, and 145 mm. Measurements were made at the water level and air flow. A second experiment involved the behavior of a toroidal liquid ring hanging around the inner wall of a glass tube with gas flow. The liquid ring was initally formed by allowing the liquid to build up over a teflon inset which formed a step in the tube.

Dukler and Smith [23] performed flooding studies in a 50.8 mm tube with very carefully controlled liquid and gas entry conditions. Their apparatus consisted of a 3.96 m long plexiglass tube with a 0.1 m porous, sintered metal inlet located at axial midplane. The gas entrance consisted of 304 mm I.D. section with a 50.8 mm nozzel for gas entry and an expanding taper on the main tube to pull the liquid film away from the stream of entering gas. Data was collected for liquid Reynolds numbers of 329.1, 822.8, 1645.5, and 3291.0 respectively. The superficial gas velocities varied from 6.5 m/s to 15.2 m/s and corresponded to gas Reynolds numbers from 22,250 to 59,971. The data recorded included liquid and gas flow rates. Pressure level and film thickness was measured at four axial locations. The film thickness was measured using conductivity probes. At the two lower liquid flow rates, waves on the falling film changed only slightly in appearance with increases in air flow. Just below the flooding point, fluctuations were noted in the film at both the air entrance and just below the water entrance. A small increase in the air flow caused these fluctuations to spread over the entire film creating a highly agitated surface. The film then climbed and entrainment took place simultaneously. For the higher water flow rates, droplets were observed striking the surface of the tube above the water entrance with gas flow just below the flooding point. The droplets formed a very thin film which ran down the tube. A small increase in the air flow, caused a disturbance in the liquid film 150 mm below the water entrance. At the flooding point, the upper section of the tube filled with an aerated liquid in churn flow. A net upflow of liquid resulted.

Imura, Kusuda, and Funatsu [28] studied flooding in 1 meter long acrilic resin tubes with varying diameters. The ranges of parameters studied for the air and water system are given in Table 1. The water was introduced over a bell mouth at the top of the tube. Air and water flow rates and pressure drop data were recorded. The flooding point was determined by visual observation and increase in the pressure drop.

TABLE 1. Range of Parameters studied by Imura et al [28]

Tube ID	Liquid Properties		Gas Properties	
	superficial flow	Re	Velocity	Re
(mm)	(mm/s)		(m/s)	(10^3)
11.2	37.1-532.	145-1820	1.62-7.17	1.62- 7.17
16.0	17.3-470.	77-2030	4.65-8.93	4.39- 9.14
21.0	49.1-278.	316-1520	4.45-7.94	5.68-10.2

Suzuki and Ueda [55] investigated flooding in transparent acrylic resin tubes of 10 mm, 18 mm, and 28.8 mm diameters. The lengths of the 10 mm and 18 mm ID tubes used were 0.5, 1.0, 1.5, and 2.0 meters. The lengths of the 28.8 mm tubes were 0.59, 1.54, and 1.83 meters. The water was introduced through a sintered section of all. The water flow rate was varied to provide liquid Reynolds numbers from 240 to 8180. Air flow was ranged to cover both the flooding point and the limit of upward co-current flow. Transistion points were recorded for both increasing and decreasing gas flow. Measurements were made of the liquid and gas flow rates. The waves on the liquid film were studied by using a contact probe to determine wave height and by using high speed photography to determine wave velocity.

Diehl and Koppany [17] performed single-component condensation studies in a 8.1 mm tube.

Wallis et al. [65] reported the results of experiments in a transparent 50.8 mm pipe. The pipe was 1.5 m in length with a large upper plenum (0.6 m high) and a large lower plenum (0.7 m high). The system was a top flood arrangement with water collection in the upper plenum. Both "square" and rounded transitions were used at both the upper and lower end of the pipe. Water at temperatures from 277 K to 372 K was supplied to the upper plenum through a feed ring near the bottom of the plenum. Saturated steam at a pressure of 41 kPa (gage) to 90 kPa (gage) was supplied to the lower plenum below a demister and turbulence suppressor. Steam was supplied at flow rates up to 23 g/s. The experiments were conducted in both steam first and water first sequences. Measurements were made of steam flow, water inlet flow, water flow to the lower plenum, inlet steam and water temperature, and upper and lower plenum temperatures and pressures.

English et al. [24] investigated two phase flow in a flooded, updraft, partial condenser. The apparatus consisted of a 1.8 m long, 19 mm diameter tube in a condenser. The fluid was n-propyl alcohol which entered a calming section at the lower end of the tube as a vapor. Various fractions of the vapor were condensed and exited the tube as condensate at the bottom and as entrained droplets at the upper end. The parameters measured included: gas mass flow rate; entrainment mass flow rate; liquid to gas fraction, and pressure drop. The gas flow rate varied from 3.39 Kg/s. m^2 to 13.56 Kg/s. m^2. Liquid to gas fractions from 0.2 to 0.8 were studied. An additional study addressed the effect of tapers (diagonal cuts across the entrance). Tapers of 0, 30, 45, and 75 degrees were studied.

Cliff, Pritchard, and Nedderman [12] also studied flooding using solutions of glycerol and water as shown in Table 2. With the more viscous solutions, waves were observed in the lower part of the tube with low or zero air flow. Increase in air flow caused the point of inception of the waves to move upward until the entire portion of the tube below the liquid entrance was wavy.

Diehl and Koppany [17] studied flooding using a hydrogen and diesel oil system in tubes with diameters of 16.8 mm, 18.7 mm, and 23.7 mm.

TABLE 2. Range of Glycerol/water mixtures [12]

Solution (%Glycerol	%Water)	Viscosity (Pa-s $.10^{-3}$)	Density (g/ml)	Surface Tension (N/m)
82	18	82.5	1.21	0.0650
77	23	46.0	1.20	0.0658
70	30	23.4	1.18	0.0664
59	41	10.4	1.15	0.0680
25	75	2.18	1.06	0.0702
--	100	1.32	1.00	0.0720

Kusuda and Imura [35] used a number of fluids in their thermosyphon. Ethyl-alcohol, nandmal-hexane and carbon-tetra-chlandide were used.

Imura, Kusuda, and Funatsu [28] investigated flooding with ethyl-alcohol, ethylene-glycol, and n-Heptane. The tubes were 11.2 mm, 16.0 mm, 21.0 mm and 24.2 mm in diameter. All of the tubes were 1 meter in length. The range of the experimental parameters are given in Table 3.

Wallis and Kuo [67] investigated the conditions of the "hanging film" phenomenon using silicon oil and air.

Suzuki and Ueda [55] studied flooding using aqueous glycerol solutions and aqueous sec.-octyl alcohol solutions in 18 mm and 28.8 mm tubes. Air was used as the gas in all cases. The glycerol solutions allowed the viscosity of the fluid to be varied while maintaining an almost constant surface tension. The liquid Reynolds numbers varied from 7 to 3620. Table 4 gives the viscosities and surface tension of the various solutions and the dimensions of the apparatus used. The sec.-alcohol solutions allowed the surface tension to be varied while maintaining a constant viscosity equal to that of water. Table 4 give the data concerning the range of surface tension, viscosity and liquid Reynolds number.

TABLE 3. Parameters studied by Imura et al [28]

Tube ID	Liquid Properties		Gas Properties	
	Superficial Flow	Re	Velocity	Re
(mm)	(mm/s)		(m/s)	(10^{-3})
Ethyl-alcohol				
11.2	13.9-489.	24.- 971	2.84-6.44	1820- 4810
11.6	6.0-552.	15 -1580	2.59-7.70	2410- 8400
21.0	11.0-389.	36 -1380	3.71-7.32	3710- 7320
Ethylene-glycol				
11.2	26.1-629.	5.7- 152	1.65-5.67	810- 3710
16.0	10.8-840.	2.7- 303	1.10-8.00	900- 7810
21.0	9.5-632.	2.7- 303	1.75-8.48	1830-11400
24.2	12.7-539.	3.5- 182	2.43-7.90	3370-11700
n-Heptane				
11.2	20.4-393.	101 -2080	0.86-4.60	560- 3240
16.0	14.3-488.	104 -3750	0.56-6.13	510- 6070
21.0	7.8-415.	62 -3610	0.80-11.9	800-11900
24.2	8.1-375.	87 -3610	1.42-7.72	2070-12100

TABLE 4. Parameters studied by Suzuki and Ueda [55]

Viscosity	Surface Tension	Liquid Reynolds Number	
(Pa.s .10^{-3})	(N/m)	18 mm tube	28.8 mm tube
Water			
0.9	0.068	390-5240	820-8180
Aqueous glycerol solutions (308K)			
2.1	0.056	140-2840	320-3620
5.1	0.053	30-1030	80-1200
10.0	0.053	25- 380	50- 510
24.0	0.053	7- 100	35- 140
Aqueous sec.-octyl alcohol solutions (298 K)			
0.9	0.055	390-3280	820-4010
0.9	0.047	390-3930	820-8180
0.9	0.037	390-2620	820-8180

Grolmes, Lambert, and Fauske [25] investigated flooding in a nitrogen and water system. The tube diameters ranged from 4 mm to 25 mm ID. Length to diameter ratios from 10 to 100 were studied. Measurements included gas and liquid flow rates and pressure drop. The onset of flooding was determined by both visual observation and pressure drop. For the larger tubes (6, 13, 25 mm), the liquid film waviness grew steadily in amplitude over the entire length of the tube as gas velocity increased. However, for the 4 mm tube, the transition to flooding was described as "a closing of the gas annulus by a single wave crest." Intermittent slug flow developed following the flooding transition.

3.2 Flooding in anulli

Stainthorp [52,53] investigated both co-current and counter-current air-water flow with emphasis on the effect of the air flow on the wave properties of the falling liquid film. The test section consisted of a tube 34.5 mm ID and 0.7 m long. A second 19.8 mm OD tube concentrically mounted formed an annular region and provided a housing for photo-transistors located on the tube axis. Since the inner tube was approximately half the diameter of the outer tube and water was introduced only on the surface of the outer tube, the goemetry could also be considered that of a tube with a "probe". It is considered annular here for comparison with other work done in annular geometries which also have similar diametrical ratios. Water was introduced through the outer wall by using a 0.1 m long section of porous ceramic tubing of the same bore as the tube. Liquid flow was adjusted to provide liquid Renolds from 10 to 180. Air velocities ranged from zero to 10 m/s. Measurements were made of the air and water flow rates. Film thickness and wave velocity was measured using a 1.5 g/l solution of Nigrosine and the change in light transmittance was measured by a pair of photo-transistors illuminated by a pair of 0.25 mm diameter beams 16.5 mm apart.

Naff and Whitbeck [40] invesitgated counter-current flow in an annular geometry using a plexiglass vessel with aluminum inserts to simulate the reactor core barrel and downcomer region of a pressurized water reactor (PWR). Figure 2 shows the geometry. Tests were conducted for radial gaps of 9.4 mm, 12.4 mm, and 16.0 mm with downcomer lengths of 0.15 m, 0.63 m, 0.86 m, and 1,72 m. The gas was introduced into the top of the "core barrel" and entered the lower plenum through bottom perforations. Water entered through a nozzel located near the top of the vessel wall and the air-water mixture exited through another nozzel located 180 degrees opposite. The incoming liquid velocity varied between 0.54 m/s and 2.2 m/s. The gas velocities were adjusted to obtain flooding by controlling the exit pressure so that a constant lower plenum pressure could be maintained. Tests were conducted at a system pressure of 69 kPa and 206 kPa (gage) to provide gas densities of 16 and 32

kg/m^3 Measurements were made of the air and water flow rates and the filling rate of the lower plenum. Visual observations were also made.

A similar set of experiments were performed at Battelle Columbus Laboratories by Segev and Collier [48]. In these experiments a 2/15 scale model of a PWR plenum was used. Table 5 shows some of the physical dimensions of both the 2/15 scale model and a 1/15 scale model as well a the dimensions of the PWR.

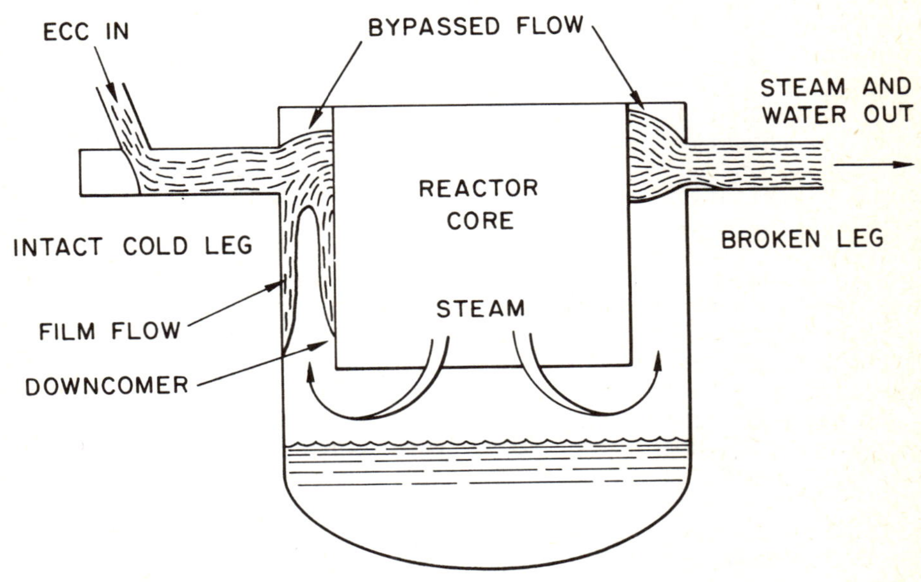

Figure 2. Flow Paths During ECC Injection

TABLE 5. Geometric Dimensi ; used By Segev and Collier [48]

Reference Geometry	1/15 scale	2/15 scale	PWR
Annular Gap, mm	15.	31.	254.
Downcomer length, m	0.41	0.82	5.33
Average Circumference, m	0.918	1.845	12.85

The 1/15 scale model data was suplimented by data from experiments at Creare [2,3,4,5,14]. In addition to steam water experiments, some experiments with air and water were performed for comparison. The resultant data was used to evaluate the effect of phase change. Various length core barrels were used to create different length downcomers.

Ueda and Suzuki [57] studied flooding in annular geometry with the flow on the inside wall of the annulus. The inner diameter of the annulus was 10.0 mm. The outer shroud diameters were 28 mm, 35 mm, and 40 mm which produced annular gaps of 9.0 mm, 12.5 mm, and 15.0 mm, respectively. The test section length was 1 meter. Liquid was supplied through a 80 mm long sintered bronze section at the top of the inner wall. The liquid flow rates varied from 3.3 to 16 ml/s. This corresponded to liquid film Reynolds numbers of 472 to 2360. The air velocity varied from 5.5 to 9 m/s. Measurements were made of the liquid and gas flow rates. The maximum liquid film height was measured using a contact probe at locations 0.3 m, 0.6 m, and 0.9 m below the sinter section. Single phase air velocity distribution was measured using a pitot tube. Visual observations were also made.

Kusuda and Imura [35] investigated flooding in annular geometries by using inserts in their thermosyphon to produce an annular, open, top flooded configuration. Inserts of diameter 12.1 mm, 19.1 mm, and 25.2 mm were inserted into the 28.4 diameter, 1 meter long, tubes to produce annular gaps of 8.2 mm, 4.7 mm, and 1.6 mm, respectively. Inserts were also used in the 0.3 meter long, 10.7 mm ID tubes. Inserts of 6.5 mm and 8.0 mm produced annular gaps of 2.1 and 1.4 mm, respectively.

Crowley, Block, and Cary [14] investigated flooding in a 1/15-scale PWR downcomer geometry. This experiment was aimed at predicting lower plenum filling rates in postulated reactor accidents. As a result, flooding was only one of the effects studied. Other effects included lower plenum voiding, cold leg steam, and superheated downcomer walls. The experimental facility consisted of a thick walled steel vessel. Variable core barrels were used to provide annular gaps of either 12.7 mm or 25.4 mm. The water was injected to simulate an emergency core cooling system (ECCS). Water flow up to a maximum of 6.3 l/s per cold leg was supplied. The maximum total water flow was 18.9 l/s. The total available steam flow was 1.4 kg/s. The water injection temperatures varied from 289 K to 322 K. The steam injection temperature varied from 373 K to 444 K. System operation pressure was 0.86 MPa. Measurements were made of the lower plenum filling rate, pressure drop, and flow rates.

Kusuda and Imura [35] used inserts with a few combinations of other fluids. Ethyl-alcohol was used in the 28.4 mm ID thermosyphon with inserts of 12.1 mm, 19.1 mm and 25.2 mm to produce annular gaps of 8.2 mm, 4.7 mm, and 1.6 mm, respectively. Normal-hexane was used with 6.5 mm and 8.0 mm diameter inserts in the 10.7 mm ID, 0.3 meter long tube to produce annular gaps of 2.1 and 1.4 mm.

Ueda and Suzuki [57] used an aqueous sec.-octyl alcohol solution in annular flooding studies. The tube diameters for these studies were the same as their water-air studies, 10 mm ID, 28 mm, 35 mm and 40 mm OD.

3.3 Simulated annular geometries

Crowley [15] investigated steam and water counter current flow in a 1/30 scale, flat-plate plastic model of a pressurized water reactor downcomer. The simulated annulus consisted of two 305 mm x 457 mm polycarbonate plates separated by a 9.5 mm gap. The water entrances were 43 mm holes that bridged the annulus. The water was supplied at rates up to 940 g/s at temperatures of 286 K, 311 K, and 333 K. The steam flow was in the range of 15 to 105 g/s at atmospheric pressure and a temperature of 375 K to 376 K. The variables measured included steam and water flow rates and steam pressure in the supply barrel up to a maximum of 13 kPa.

3.4 Other geometries

Verschoor [58] investigated the limiting velocity of vapor in packed columns.

Shearer and Davidson [51] investigated the standing wave generated by upward flow of air through an orifice with a rod with downward flowing water film in the center of the orifice. Although not flooding in the classical sense, this counter-current flow situation was used to study the forces on large disturbance waves which are often associated with flooding. A 25.4 mm diameter rod was used. The orifice was 38.1 mm in diameter on the upper side and expanded to 68.3 mm on the lower side of a 4.8 mm thick plate. The air and water rates were varied over a very small range of stability. If the air rate decreased slightly, the wave disappeared. If it increased slightly, droplets were ripped off the crest of the wave. If the water flow rate was reduced slightly, the tube did not wet fully. If it increased slightly, rippling occured and made the wave unstable. Once a stable wave had been produced, it was photographed and a brass model of the wave was constructed and fitted with internal pressure taps. By using the brass replica, it was possible to obtain pressure readings at the interface.

Ueda and Suzuki [57] performed flooding experiments in simulated nuclear reactor rod bundles of three and four rods. All the rods were 10.0 mm in diameter and 1.0 meter in length with a sintered section for fluid entry at the top. It was possible to vary the amount of liquid to each rod independently. The set of three or four rods was contained in one of a set of shrouds of diameters either 28 mm, 35 mm, or 40 mm. This allowed a variety of rod to rod spacings.

Naitoh, Chino, and Kawabe [41] studied the restrictive effect of ascending steam on falling water during top spray emergency core cooling in a rectangular array of fuel pins. The test section consisted of 64 rods, 12.5 mm in diameter, arranged in a square lattice with 16.3 mm pitch. The length of the test section was 4.508 meters. Steam was injected into the lower plenum to simulate generation of steam in a BWR. Water with various degrees of subcooling was injected in the top of the test section. Therefore several effects were studied in addition to flooding. These effects included condensation and type of water injection nozzel. The flow rate of steam supplied to the test section varied from 43 to 83.3 g/s. The spray water rate varied from 117 to 1033 g/s. Temperature of the spray water varied from 300 K to 370 K. Measurements were made of the water and steam flow rates. The amount of condensation was based on a separate set of measurements with steam only and no spray.

4. ANALYTICAL METHODS

The phenomenon of flooding has not yet yielded to a complete analytical explanation. The various attempts at explanation and prediction of flooding fall under three general categories: empirical; semi-empirical or semi-analytic (depending or point of view); and analytical.

4.1 Empirical correlations

Purely empirical correlations for flooding generally have a weak physical basis are no longer in wide use.

English et al. [24] developed a correlation for gas and liquid flow in a vertical upright partial condenser. A total of 56 flooding determinations were plotted as a function of the liquid to gas ratio. The other variables, tube diameter, liquid density, surface tension, gas viscosity, liquid viscosity, and the angle of the tube taper, were combined using fractional exponents to produce minamal scatter on a log-log plot. The best straight line was then fitted through these points and the slope calculated to determine the fraction exponent for the liquid to gas (L/G) ratio. The results were rewritten in terms of the maximum allowable gas superficial mass flow rate. The equation given below is recursive in the gas superficial mass flow rate.

$$j_g = 1550 \left[\frac{D^{0.3}\ \rho_\ell^{\ 0.46}\ \sigma^{0.9}\ \rho_g^{\ 0.5}}{\mu_\ell^{\ 0.14}\ (\cos\theta)^{0.32}\ (j_\ell/j_g)^{0.07}} \right] \quad (1)$$

where

D = tube diameter, inches

θ = taper angle

j_ℓ, j_g = superficial velocity of liquid and gas, ft/s

The weak dependence on G requires a one or two time iteration.

Diehl and Koppany [19] developed a correlation of the form:

$$V_g = F_1\ (j_\ell/j_g)^{-0.25}\ (\sigma/\rho_g)^{0.5} \quad (2)$$

where

$$F_1 = \begin{cases} (80D/\sigma)^{0.4} & 80D/\sigma < 1.0 \\ 1.0 & 80D/\sigma \geq 1.0 \end{cases}$$

V_g = gas velocity, ft/s

The determination of the F1 constant is based on a equation for the critical diameter presented in the paper. This equation, however, is not consistent in units and a mixed form of units persists throughout the paper. Therefore, this correlation is suspect as to the correct units.

4.2 Semi-empirical correlations

These correlations include the Wallis Correlation and the Kutateladze Correlation. Both of these correlations enjoy wide use as predictive methods and are often used without regard to the limits of their data.

The approach used in the Wallis [63,64] correlation is to reduce the superfical velocity of both the gas and the liquid to its respective dimensionless velocity. The dimensionless velocities are:

$$J_g^* = \frac{\rho_g^{\ 1/2}\ j_g}{[gD(\rho_\ell - \rho_g)]^{1/2}} \quad (3)$$

and

$$J_\ell^* = \frac{\rho_\ell^{1/2}\, j_\ell}{[gD(\rho_\ell - \rho_g)]^{1/2}} \qquad (4)$$

The dimensionless velocities are then combined in the form:

$$J_g^{*1/2} + m\, J_\ell^{*1/2} = C \qquad (5)$$

The value of the constants are then fitted to the data. The value of C varies from 0.88 to 1.00 when m=1.

Kutateladze [36] developed a correlation similar to the Wallis correlation. The liquid dimensionless velocity is of the form:

$$K_\ell = \rho_\ell^{1/2} j_\ell / [g g_c \sigma(\rho_\ell - \rho_g)]^{1/4} \qquad (6)$$

The gas dimensionless velocity is:

$$K_g = \rho_g^{1/2} j_g / [g g_c \sigma(\rho_f - \rho_g)]^{1/4} \qquad (7)$$

The final form of the equation is:

$$K_g^{1/2} + m\, K_\ell^{1/2} = C \qquad (8)$$

The Wallis and Kutateladze correlations are not complete solutions of the flooding phenomenon, but rather are usable tools for finding solutions to specific cases without knowing the complete solution. Therefore, it is not surprising that in a given case, one correlation may perform better than the other when compared to experimental data. Such differences should always be understood in the context of the relation of the specific problem under study to the ranges and limitations of each of the correlations.

Pushkina and Sorokin [44] emphasized the differences between the Wallis and the Kutateladze correlations with two sets of experiments. One set was performed in a 309 mm tube as described previously. The other set of experiments was performed in a series of small tubes with internal diameters of 6.2 mm, 8.8 mm, 9.0 mm, 12.2 mm, and 13.1 mm. All of the small tubes used a funnel as a top liquid feed. Both sets of experiments investigated flooding at the lower limit of liquid flow on the flooding curve (i.e. the point just above film breakdown). The analysis focused on determining a critical air velocity for flooding. The results presented show a critical velocity for air of 14 m/sec to 16 m/sec. In the analysis, an arbitrary 0.2 mm correction for film thickness was used for all tubes less than 12.2 mm. The results do show that a decrease in critical velocity occurs even when using the film correction. If the smaller and larger tubes had been corrected the same, there would have a slight slope to the critical velocity versus tube diameter curve. The Wallis correlation predicts a straight line but with a positive slope when applied to the same cases. Pushkina and Sorokin claimed this data that demonstrated the superiority of the Kutateladze correlation to the Wallis correlation.

Wallis and Makkenchery [66] offered a partial explanation of the differences in the Wallis and the Kutateladze correlations regarding the data of Pushkina and Sorokin. They empasized that the Wallis correlation was originally based on data taken in tubes ranging from 12.7 mm to 50.8 mm. They stated that since the Kutateladze correlation contained no characteristic dimension of the apparatus, that it would be best suited for large diameter tubes provided the liquid film remained thin. For small tubes surface tension effects would be important. To support this claim, Wallis and Makkenchery presented data showing the difference in velocity required to support a hanging film compared to that required for flow reversal was a factor of two greater in small tubes compared to large tubes. Wallis and Makkenchery proposed that the range of surface tension effects could be expressed in terms of a dimensionless tube diameter by using the Bond number.

$$D^* = D[g(\rho_\ell - \rho_g)/\sigma]^{1/2} \tag{9}$$

In addition, the critical dimensionless gas velocity used by Wallis [83] and that used by Pushkina and Sorokin [61] could be related as the square root of the Bond number.

An additional factor, not mentioned by either Pushkina and Sorokin or Wallis and Makkenchery was the method of fluid entry used in Pushkina and Sorokin's experiments. The large 309 mm tube used an extended midsection fluid entry consisting of several thousand drilled holes simulating to the porous entries used by many researchers. Flooding in this large tube was best represented by the Kutateladze correlation as expected. The smaller tubes used an overflowing funnel liquid entrance. Thus, there was no rigid wall supporting the reservor of fluid in the event that it was above the lip of the funnel. Any instability created in this region may be transferred down the tube. The Kutateladze correlation is based on a physical geometric parameter representative of a wavelength similar to a Taylor instability rather than a geometric parameter like the Wallis correlation. Therefore, the Kutateladze correlation may have better represented the effects of any instabilties generated. This is a excellent example of considering all of the the varying parameters which may influence the applicability of a specific correlation to a given situation.

Alekseev et al. [1] developed a correlation based on the Kutateladze, Froude, Weber, and Galileo numbers.

$$K_g = 1.13\ Fr^{-0.20}\ We^{-0.21}\ Ga^{-0.09} \tag{10}$$

where

$$Fr = \frac{Q_\ell\ (\rho_\ell - \rho_g)^{3/4}}{g^{1/2}\ \sigma^{3/4}} \tag{11}$$

$$We = \frac{\sigma}{(\rho_\ell - \rho_g)\ D^2} \tag{12}$$

$$Ga = \frac{g\ \sigma^{3/2}}{U_\ell^2\ (\rho_\ell - \rho_g)^{3/2}} \tag{13}$$

Q_ℓ = mass flow/wetted perimeter

Grolmes et al [25] use the Nusselt solution for average film thickness in the falling film without gas flow as a basis to calculate the critical gas velocity for flooding.

$$V_g = 1.15\ \left(\frac{\rho_\ell}{\rho_g}\right) \left[\frac{g\ \delta_N}{0.006 + \dfrac{200\ \delta_N^2}{(\mu_\ell/\mu_R)^{0.44}}}\right]^{1/2} \tag{14}$$

where

μ_ℓ/μ_R = ratio of liquid viscosity to that of water

and δ_N is the Nusselt film thickness, calculated as follows:

$$\delta_N = \left(3\ Q_L\ \frac{\mu_L}{\rho_L g}\right)^{1/3} \tag{15}$$

Imura et al. [28] also produced a correlation based on Nusselt's film thickness solution.

$$\frac{j_g}{j_\ell}\left(1 + \frac{j_\ell}{j_g}\frac{\rho_\ell}{\rho_1}\frac{S_\ell}{S_1}\right) = \frac{S_\ell}{S}\frac{1}{L}\ \left(\frac{\rho_\ell \sigma}{\delta_N}\right)\left(\zeta - \frac{1}{R/\delta_N - 1}\right)^{1/2}. \tag{16}$$

where

$$S = \pi D^2/4 \tag{17}$$

$$S_1 = \pi D \delta_N (1 - \delta_N/D) \tag{18}$$

$$S_g = (\pi D^2/4)(1 - 2\delta_N/D)^2 \quad . \tag{19}$$

$$\zeta = 0.046 \left(\frac{d^2 \rho_1 g}{\sigma_N}\right)^{0.3} \left(\frac{\mu_1}{\mu_g}\right)^{0.12} \quad . \tag{20}$$

Suzuki and Ueda [55] used four nondimensional groups to form a flooding correlation. The Froud number,

$$Fr \equiv [\rho_G (V_G + V_L)^2/\rho_L g y_i]^{1/2} \quad , \tag{21}$$

where $\rho_g(V_g + V_\ell)^2$ represents the kinetic energy of the gas based on relative motion of the phases, is defined interms of the other three nondimensional groups.

$$F_r = C_1 \, \mathrm{Log}_{10} \left[Re_L^{-1/3} \left(\frac{\rho_L g D^2}{\sigma}\right)^{1/4} \left(\frac{\mu_G}{\mu_L}\right)^{2/3} \right] + C_2 \tag{22}$$

The modified surface tension σ' is defined as:

$$\sigma' = \sigma_N + 1.5|\sigma - 0.05| \mathrm{N/m} \quad . \tag{23}$$

Crowley et al. [14] based a correlation on the Wallis correlation but modified it to account for condensation of steam. This represents an extention of the purely hydrodynamic aspects of flooding to cases involving heat transfer and phase change. However, since the steam-water system is of much importance, it is included in this summary. The gas dimensionless velocity is modified as follows:

$$J_g^* = J_{gc}^* - f(T_{sat} - T_f) \left(\frac{c_p}{h_{fg}}\right)\left(\frac{\rho_f}{\rho_g}\right)^{1/2} J_{f,in}^* \, , \tag{24}$$

where J_g^* refers to the Wallis equation, J_{gc}^* is the value with condensation, and f is a measure of the condensation efficiency

$$f = \left(\frac{P_{LP}}{P_{atmos}}\right)^{1/4} \left(\frac{1}{1 + 30 J_{f,in}^*}\right) \quad . \tag{25}$$

The m and C constants in the Wallis equation are then determined experimentally to be:

$$m = \exp\ (-5.6(J^*_{f,in})^{0.6} \tag{26}$$

$$C = 0.32 \tag{27}$$

A number of correlations have been developed for specialized geometries. Many of these geometries apply to those found in either boiling water reactors or pressurized water reactors.

Sun [54] correlated data for both the tieplates and bottom side-entry orifices of boiling water reactors. The Kutateladze correlation was used, but a different set of constants was established for each case. For the tieplates with three different methods of steam injection,the constant m=1, but the constant C depended on method of steam injection. For the capped steam sparger.

$$(K_g)^{1/2} + (K_f)^{1/2} = 1.89 \pm 6\% \tag{28}$$

For the clarienet steam sparger.

$$(K_g)^{1/2} + (K_f)^{1/2} = 2.10 \pm 13\% \tag{29}$$

For the lower tieplate steam sparger.

$$(K_g)^{1/2} + (K_f)^{1/2} = 2.07 \pm 8\% \tag{30}$$

For the bundle side-entry orifices, the constant, m=0.59 and the constant, C, was calculated in terms of the Bond number.

$$K_g^{1/2} + 0.59\ K_f^{1/2} = 2.14 - 0.0080\ D^* \tag{31}$$

Dilber [18] presented a correlation based on counter-current injection of water above a peforated plate. He defined a dimensionless flow which combined features from both the Wallis and Kutateladze correlations. The correlation is in the form:

$$H_g^{*\ 1/2} + H_f^{*\ 1/2} = C \tag{32}$$

where the dimensionless flow, H, is defined as :

$$H^*_{f,g} = [\rho_{f,g}/gw(\rho_f - \rho_g)]^{1/2} j_{f,g} \tag{33}$$

and w is :

$$w = D_h^{(1-\alpha)} [\sigma/g(\rho_f - \rho_g)]^{\beta/2} \tag{34}$$

$$\beta = \tanh \left[\frac{2\pi\ D_h}{t_p}\ (A_h/A_t)\right] \tag{35}$$

A_n/A_t = perforation ratio

D_h = hole diameter

t_p = plate thickness

The use of the hyperbolic tagential function, a, results is a smooth transition between the Wallis (β=0) and Kutateladze (β=1) correlations. The value of C is given in terms of the Bond number.

$$C = 1.07 + 4.33 \times 10^{-3}\ D^* \tag{36}$$

where

$$D^* = n\pi\ D_h[g(\rho_\ell - \rho_g)/\sigma]^{1/2} \tag{37}$$

4.3 Analytical models

Flooding may never yield to a purely analytical approach due to the statistical nature of the turbulence present. However, analytical approaches, even with their often gross simplifications, provide insight and direction into the process.

Richter [45] proposed a correlation which is an extention of the Wallis correlation in an attempt to resolve the differences between the Wallis and the Kutateladze correlations. The approach used by Richter is to set up control volumes for both entire pipe and for the gas bounded by the interface. This allowed him to eliminate the pressure drop by combining the equations for each control volume.

$$\frac{4\ \tau_w}{D} + \frac{4\ \tau_i}{D\sqrt{(\alpha)}} = (\rho_\ell - \rho_G)g(1 - \alpha) \tag{38}$$

τ_w = wall shear

τ_i = interfacial shear

α = void fraction

The three remaining unknowns are the wall shear stress, τ_w, the interfacial shear, τ_i, and the void fraction, α. The wall and interface shear is transformed to wall and interface friction factors, respectively. The wall and interfacial friction factors are then related through Wallis' correction for wavy annular flow [62].

$$f_i + f_w \left(1 + 300 \frac{\delta}{D}\right) \tag{39}$$

f_w = wall friction factor

f_i = interfacial friction factor

δ = film thickness

The above is in terms of the film thickness and pipe diameter. The void fraction can also be expressed in the same terms. Therefore, it is possible to eliminate one variable. Experimental data by Wallis regarding the ratio of d/D for the case of little interfacial shear is used to arrive at a wall friction factor of f_w=0.008. At this point, Richter makes several assumptions concerning the large disturbance waves that have been observed prior to the onset of flooding. First, the wave is assumed to have a semicircular shape. Second, it is assumed that the pressure difference between the base and crest of the wave is due to the dynamic head of the gas flow. Third, the mechanism of flooding is considered to be total entrainment with onset of entrainment coincident with flooding. Using this assumption, stability requirements on the wave require the dynamic head of gas flow to be less than or equal to the ratio of wave heigth to surface tension. The final form of the correlation is of the form:

$$\frac{f_w}{4} D^{*3} J_g^{*6} J_\ell^{*2} - f_w D^* J_g^{*4} + 150 f_w J_g^{*2} = 1 \tag{40}$$

Maron and Dukler [39] present an analysis based on the structure of the film flow. Additional details of the analysis can be found in a report in which Dukler assumes that the liquid inlet is located below the top of the test section so that a liquid may exist both above and below the inlet depending on flow conditions [21]. This analysis assumes the film to be an incompressible Newtonial fluid in laminar motion. The fluid interface is assumed to be smooth for the development of the fluid equations. However, the interface is assumed to be wavy when considering the gas. This allows the introduction of larger pressure gradients and interfacial shear forces. The fluid motion is then solved for one dimensional flow for three limiting cases. The first of these is the free falling film with no interfacial shear. The solution is the well known Nusset profile with a zero velocity gradient at the interface. The second reference case is a falling film with zero velocity at the interface. This represents the limiting condition for complete downflow, since any increase in interfacial shear and hence interfacial velocity will result in upflow in part of the film. The third reference case is a rising film with a zero velocity gradient at the wall. This represents a limiting condition for complete upflow, since any decrease in interfacial shear and hence interfacial velocity will result in downflow in part of the film. The film thickness at each reference case can be related to that at the other two cases. Also the interfacial shear for the second and third cases can be related but the exact value of interfacial shear is unknown except in the first case where it is zero. Therefore, the interfacial shear for any flow condition is defined in terms of a shear parameter, F, which is the ratio of the interfacial shear to that of the second case with zero interfacial velocity.

Rather than using a correlation for interfacial shear, Dukler and Maron define the shear in terms of the interfacial shear for the case of zero interfacial velocity.

$$F = \tau_i/\tau_o = \frac{2\ \tau_i}{\rho_L g \delta_o} \qquad (41)$$

τ_o = interfacial shear for case with zero liquid velocity at interface

δ_o = film thickness with zero velocity at interface

By relating the film thickness to that for the freely falling film, Dukler and Maron solved the one dimensional flow equations for three cases: downflow, split flow, and upflow. The solutions of these respective cases are given below.

$$R_N^3 - 0.75\ \sqrt[3]{4}\ R_N^2\ F - 1 = 0 \qquad (42)$$

$$R_N^3 - \frac{7F}{2}\ \sqrt[3]{0.5}\ R_N^2 + 4F^2\ \sqrt[3]{0.25}\ R_N - \frac{2F^3 + 1}{3} = 0 \qquad (43)$$

$$R_N^3 - 0.75\ \sqrt[3]{4}\ FR_N^2 + 1 = 0 \qquad (44)$$

where

$$R_N = \delta/\delta_N \qquad (45)$$

Figure 3 shows that a single value of film thickness exists for values of $F < 1.0$. This corresponds to a state of total downflow. Two solutions of film thickness are possible for $1 < F < 1.59$. Thus the flow may be either downflow with circulation or split flow. For a value of $1.59 < F < 1.65$, three solutions are possible which correspond to upflow with circulation, split flow, and downflow with circulation. For $F > 1.65$, three solutions are possible. These are upflow with circulation, total upflow, and downflow with circulation. Dukler and Maron use the existence of multiple solutions of film thickness to suggest that a variety of film states are likely at the flooding point and hence the states are by nature unstable. The large disturbance waves observed at the flooding point are explained as transitions between states. Dukler and Maron then speculate that a feedback from film thickness to capillary waves may exist. This would cause small changes in the interfacial shear and thus changes in the flow states. Such a feedback would be a type of limit cycle.

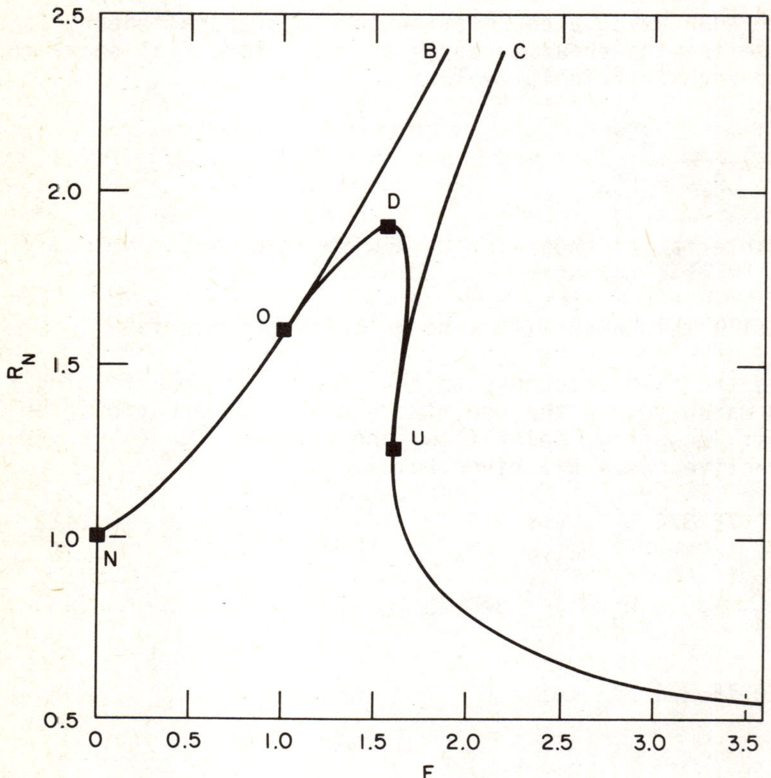

Figure 3. Non-dimensional Friction Factor vs. Non-dimensional Film Thickness (from Reference 21)

5. SUMMARY

Since the early research in the 1930's, the approach to flooding has consisted of several stages. The early papers on flooding are somewhat sparce and confined mainly to chemical engineering until the 1950's. In the late 1950's and early 1960's a large number of expermental studies were performed. The early work seemed to be directed toward empirical relations for packed columns. Analytical attempts to explain packed columns seemed to present an intractable problem and therefore a simpler geometry was sought which would yield to analysis. As a result, a number of studies were conducted in cylindrical tubes. By chance, all of the studies involved tubes in the diameter range from 25 mm (1 inch) to 50 mm (2 inches). A number of other parameters were studied such as water inlet configuration, gas inlet configuration, length to diameter ratio, and phase change effects.

The next major step in understanding the phenomenon of flooding came in the 1960's with the development of semi-empirical correlations which attempted to replace pure data fitting with somewhat of a physical basis. The Wallis correlation and the Kutateladze correlation are the most widely used semiempirical correlations. The correlations, however, are often misused by those attempting to apply them beyond their applicable ranges.

In the 1970's, a large amount of research in the form of large scale modeling was performed. Most of this work was done in support of reactor safety studies by either the nuclear industry or regulatory agencies. This work often was integral in that many effects in addition to flooding were studied in a combined experiment.

In the last few years, more detailed correlations such as those of Richter [44] have attempted to expand the original semi-empirical correlations and provide additional physical basis. Other work, especially by Dukler, has investigated the statistical nature of flooding and examined the structure of both large disturbance waves and small ripples. Dukler and Maron [21,39] have proposed a film model for flooding.

At this point, the authors would like to propose some suggestions for further research and speculate on possible approaches to flooding.

Dukler et al. [21] and Maron and Dukler [39] discuss the importance of the location of the liquid feed entrance relative to the top of the test section. Dukler's analysis assumes flooding to be a film phenomenon versus an entrainment phenomenon with "split flow" as one of the solutions for film thickness at the flooding point. The location of the liquid inlet must be different than the exit for internal recycle to take place in the film. By assuming that flooding is related but not identical to entrainment, it should be possible to reconcile the end effects due to a length of test section above the liquid feed. In this case, the entrained flow is responsible for the internal recycle. For a given distribution of droplets at a given axial location, it should be possible to apply the results of research on droplet deposition to develop a distribution of droplets along the entire downstream section of the tube. The droplet distribution would be a function of gas velocity, droplet size, radial position (in polar geometries), and axial position. In addition, phase change effects may be added. The distribution of any internal recycle due to entrainment then becomes a convolution of a droplet deposition function and an entrainment source function such as given in references 30,33,34. The use of such an approach may require a relaxing of the criteria of no liquid out the top of the tube before flooding since droplet deposition implies only an asymptotic limit once a finite amount of liquid has been entrained.

Levy and Healzer [38] have proposed applying mixing length theory to a wavy, turbulent liquid-gas interface for cocurrent, annular flow. The analysis divides the flow into three regions: gas core, wall film, and wavy gas-liquid transition region. It may be useful to combine such an approach with Dukler's analysis [21,39]. In such an analysis, Dukler's analysis could provide the equations of liquid motion in the film. Work on droplet deposition, discussed above may aid in the analysis of the gas core. The transition region would present the remaining and perhaps most difficult element of the analysis. In this region, entrainment could be based on a model such as that of Ishii and Grolmes [31] modified for counter-current flow. Some statistical data on the shape of the film would then provide the final part of the solution.

In closing, the authors wish to state that, although flooding may never yield to a complete analytical solution, there are several areas of research which should reduce the uncertainty in the present level of knowledge such that accurate predictions may be made. The most useful areas of experimental research appear to be in the development of diagnostic techniques to determine the interface shape and liquid velocity profile within the film. Analytically, this is a good time for reflection to determine if solutions to other fluid problems might be applied to flooding.

ACKNOWLEDGEMENT

This work was performed under the auspices of the U. S. Department of Energy.

NOMENCLATURE

Symbols

A_h	area of holes in perforated plate, m^2
A_t	area of perforated plate, m^2
C, C_1, C_2	experimentally determined constants
C_i	interfacial friction factor
C_w	wall friction factor
C_p	specific heat capacity, J/kg.K
D	tube diameter, m (unless otherwise specified)
D_h	hole diameter (in perforated plate), m
D*	Bond number
F	dimensionless shear factor
Fr	Froude number
g	acceleration of gravity, m/s^2
g_c	gravitational constant
Ga	Galiano number
$h_{\ell g}$	heat of vaporization, J/kg.K
H	dimensionless velocity defined by Richter
j	superficial velocity, m/s
J*	dimensionless velocity defined by Wallis
K	dimensionless velocity defined by Kutateladze
m	experimentally determined constant
P	pressure, Pa
R	ratio of film thickness to Nusselt film thickness
Re	Reynolds number
Q_ℓ	liquid mass flow rate divided by wetted perimeter, kg/m
S, S_ℓ, Sg	flow areas defined by Imura, m^2
T	temperature, K
V	velocity, m/s
We	Weber number
α	void fraction
δ	film thickness, m
δ_N	Nusselt film thickness, m

ζ factor involving experimental constant used by Imura

θ angle of cut of tube entrance, degrees

μ viscosity, Pa.s

ν Kinematic viscosity, m^2/s

ρ density, g/m^3

σ surface tension, N/m

σ^1 modified surface tension used by Suzuki and Ueda, N/m

τ shear, N/m^2

Subscripts

atmos	atmospheric
g	gas
h	hole (used for perforated plate correlations)
i	interface
in	inlet
ℓ	liquid
LP	lower plenum
o	reference state used by Dukler (no interfacial velocity)
N	reference state for falling film with no gas velocity (Nusselt solution)
R	referenced to water
sat	saturation
w	wall

REFERENCES

1. Alekseev, V. P., Poberezkin, A. E., and Gerasimov, P. V. 1972. Determination of flooding rates in regular packings. Heat Transfer - Soviet Research 4,6:pp159-163

2. Bishop, Thomas A., Collier, Robert P., and Kurth, Robert P. 1981. Statistical Analysis of ECC Bypass Data Using a Nonlinear Constrained Maximum Likelihood Estimation Technique. Nuclear Engineering and Design 64:pp87-91

3. Block, James A., and Crowley, Christopher J. 1975. Hot wall experiments in a simulated multiloop PWR geometry. Creare Inc. Technical note TN-202

4. Block, James A., Rothe, Paul H., Fanning, Margaret W., Crowley, Christopher J., and Wallis, Graham B. 1976. Analysis of ECC delivery. _Creare Inc. Technical note TN-231_

5. Block, James A., and Wallis, Graham B. 1974. Effect of hot walls on flow in a simulated PWR downcomer during a LOCA. _Creare Inc. Technical note TN-188_

6. Block, James A., and Wallis, Graham B. 1975. Heat transfer and fluid flows limited by flooding. _Creare Inc. Technical note TN-213_

7. Block, James A., and Wallis, Graham B. 1978. Heat transfer and fluid flows limited by flooding. _American Institute of Chemical Engineers Symposium Series 174 Heat Transfer: Research and Application_ 174:pp73-82

8. Cetinbudaklar, A. J., and Jamerson, G. J. 1969. The mechanism of flooding in vertical countercurrent two-phase flow. _Chemical Engineering Science_ 24:pp1669-1680

9. Chang, F. W., and Dukler, A. E. 1964. The influence of a wavy interface on pressure drop in conduits. _Int. J. Heat Mass Transfer_ 7:pp1395-1404

10. Chu, K. J., and Dukler, A. E. 1974. Statistical characteristics of thin, wavy films: Part II. Studies of the substrate and its wave structure. _A.I.Ch.E. Journal_ 20:pp695-706

11. Chu, K. J., and Dukler, A. E. 1975. Statistacal characteristics of thin, wavy films: Part III. Structure of the large waves and their resistance to gas flow. _A.I.Ch.E. Journal_ 21:pp583-593

12. Cliff, R., Pritchard, C. L., and Nedderman, R. M. 1966. The effect of viscosity on the flooding conditions in wetted wall columns. _Chemical Engineering Science_ 21:pp87-95

13. Cook, D,, Bankoff, S. G., Tankin, R. S., and Yuen, M. C. 1981. Countercurrent Steam-Water Flow in a Vertical Channel. United States Nuclear Regulatory Commission Report NUREG/CR-2056

14. Crowley, Christopher J., Block, James A., and Cary, C. N. 1977. Downcomer effects in a 1/15-scale PWR geometry - experimental data report United States Nuclear Regulatory Commission Report NUREG-0281

15. Crowley, Christopher J. ,Wallis, Graham B., and Ludwing, D. L. 1974. Steam/water interaction an a scaled pressurized water reactor dowmcomer annulus. Dartmouth College Report COO-2294-4

16. Crowley, Christopher J., Wei, S., and Rothe, P. H. 1981. Analysis of Flashing Transient Effects During Refill. United States Nuclear Regulatory Commission Report NUREG/CR-1765 CREARE TN-321

17. Diehl, J. E., and Koppany, C. R. 1965. Flooding Velocity correlation for gas-liquid counterflow in vertical tubes. Chemical Engineering Symposium Series 92/65:pp77-83

18. Dilber, I., Bankoff, S. G., Tankin, R. S., and Yuen, M. C. 1981. Countercurrent Steam/Water Flow Above a Perforated Plate-Vertical Injection of Water. United States Nuclear Regulatory Commission NUREG/CR-2323

19. Dukler, A. E. 1977. The Role of Waves in Two Phase Flow: Some New Understandings. Chemical Engineering Education 1977:pp108-117,138

20. Dukler, A. E., Chopra, A., Maron, D., and Semiat, R. 1979. Two-Phase Interactions in Countercurrent Flow. Annual Report November 1977 - December 1978. United States Nuclear Regulatory Commission NUREG/CR-0669

21. Dukler, A. E. and Smith, L. 1979. Two-Phase Interactions in Countercurrent Flow: Studies of the Flooding Mechanism. Annual Report November 1975 - October 1977. United States Nuclear Regulatory Commission NUREG/CR-0617

22. Dukler, A. E. and Smith, L. 1977. Two-Phase Interactions in Countercurrent Flow Studies of the Flooding Mechanism. Progress Report November 1, 1975 - September 30, 1976. United States Nuclear Regulatory Commission NUREG-0214

23. Dukler, A. E., and Smith, L. 1976. Two phase interactions in countercurrent flow. Studies of the flooding mechanism. Summary report no. 1 under United States Nuclear Regulatory Contract AT(49-24)0194

24. English, K. G., Jones, W. T., Spillers, R. C. and Orr, V. 1963. Flooding in a vertical updraft partial condenser. Chemical Engineering Progress 59,7:pp51-53

25. Grolmes, M. A., Lambert, G. A., and Fauske, H. K. 1974. Flooding in vertical tubes. Symposium on Multi-Phase Flow Systems, University of Strathclyde, Glasgow, Scotland, April 2-4, 1974 A4:pp1-15

26. Hewitt, G. F., and Lacey, P. M. C. 1965. Transistions in film flow in a vertical tube. United Kingdom Atomic Energy Authority Report AERE - R 4614

27. Hewitt, G. F., and Wallis, Graham B. 1963. Flooding and associated phenomena in falling film flow in a vertical tube. United Kingdom Atomic Energy Authority Report AERE - R - 4022

28. Imura, H., Kusuda, H., and Funatsu, S. 1977. Flooding velocity in a counter-current annular two-phase flow. Chemical Engineering Science 32:pp79-87

29. Imura, H., Kusuda, H., Ogata, J., Miyazaki, T., and Sakamoto, N. 1979. Heat Transfer in Two-Phase Closed-Type Thermosyphons. HEAT TRANSFER - Japanese Reasearch 8,2:pp41-53

30. Ishii, Mamoru and Mishima, Kaichiro 1981. Correlation for Liquid Entrainment in Annular Two-Phase Flow of low viscous fluid. Angonne National Laboratory ANL/RAS/LWR 81-2

31. Ishii, M., and Grolmes, M. A. 1975. Inception criteria for droplet entrainment in two-phase concurrent film flow. A.I.Ch.E. Journal 21:pp308-318

32. Jamerson, G. J., and Cetinbudaklar, A. 1968. Wave inceeption by air flow over a liquid film. Proceedings of the International Symposium on Research in Co-current Gas Liquid Flow University of Waterloo

33. Kataoka, I., Ishii, M., and Mishima, K. 1981. Generation and Size Distribution of Droplet in Gas-Liquid Annular Two-Phase Flow. Argonne National Laboratory - Reactor Analysis and Safety Division Report ANL/RAS/LWR 81-3

34. Koestel A., Gido, R. G., and Gilbert, J. S. 1980. Film Entrainment and Drop Deposition for Two-Phase Flow. United States Nuclear Regulatory Commission - NUREG/CR-1634, LA-8475-MS

35. Kusuda, W., and Imura, H. 1974. Stability of a liquid film in a counter-current annular two-phase flow (Mainly on a critical heat flux in a two-phase thermosyphon). Bull. JSME 17:pp1613-1618 [original Trans. JSME 40:pp1082-1088]

36. Kutateladze, S. S. 1972. Elements of the hydrodynamics of gas-liquid systems. Fluid Mechanics - Soviet Research 4:pp29-103

37. Kuzay, T. M., Chen, W. L., and Ishii, M. 1979. Evaluation of Incoherency Effect on Cladding Motion in LMFBR Loss-of-Flow Accident Experiments Based an a Multichannel Model. Proceedings of the Multi-Phase Flow and Heat Transfer Symposium-Workshop Miami Beach, Flordia, edited by T. Nejat Veziroglu 1519-1546

38. Levy, S., and Healzer, J. M. 1981. Application of Mixing Length Theory to Wavy Turbulent Liquid-Gas Interface. ASME Journal of Heat Transfer 103:pp492-500

39. Maron, D. Moalem and Dukler, A. E. 1981. New Concepts on the Mechanisms of Flooding and Flow Reversal Phenomena. Letters in Heat and Mass Transfer 8:pp453-463

40. Naff, S. A., and Whitbeck, J. F. 1973. Steady State Investigation of Entrainment and Countercurrent Flow in Small Vessels. Proceedings ANS Topical Meeting on Water Reactor Safety, Salt Lake City, Utah

41. Naitoh, M. 1979. Heat Removal by Top Spray Emergency Core Cooling. Proceedings of the Multi-Phase Flow and Heat Transfer Symposium-Workshop Miami Beach, Flordia, edited by T. Nejat Veziroglu 1695-1712

42. Naitoh, M., Chino, K., and Kawabe, R. 1978. Restrictive effect of ascending steam on falling water during top spray emergency core cooling. J. Nuclear Science and Technology 15:pp806-815

43. Nedderman, R. M. and Shearer, C. J. 1963. The motion and frequency of large disturbance waves in annular two-phase flow of air-water mixtures. Chemical Engineering Science 18:pp661-670

44. Pushkina, O. L., and Sorokin, Yu. L. 1969. Breakdown of liquid film motion in vertical tubes. Heat Transfer - Soviet Research 1,5:pp56-64

45. Richter, H. J. 1981b. Flooding in Tubes and Annuli. International Journal of Multiphase Flow 7,6:pp647-658

46. Salazar, Ronald P., and Marschall, Ekkehard 1978a. Time-average local thickness measurement in falling liquid film flow. Int. J. Multiphase Flow 4:pp405-412

47. Salazar, Ronald P., and Marschall, Ekkehard 1978b. Three-dimensional surface characteristics of a falling liquid film. Int. J. Multiphase Flow 4:pp487-496

48. Segev, A., and Collier, R. P. 1980a. A Mechanistic Model for Countercurrent Steam-Water Flow. A.S.M.E. Journal of Heat Transfer 102:pp688-700

49. Seki, N., Fukusako, S., and Koguchi, K. 1981. An Experimental Investigation of Boiling Heat Transfer of Flourocarbon R-11 Refrigerant for Concentric-Tube Thermosyphon. _ASME Journal of Heat Transfer_ 103:pp472-477

50. Sekoguchi, K., Hori, K., Nakazatomi, M., Nakano, K., and Nishikawa, K. 1977. On ripple of annular two-phase flow - 1. Statistical characteristics of ripples. _Bull. JSME_ 20:pp844-851 [original _Trans. JSME_ 42:pp3551-3559 (1976)]

51. Shearer, C. J., and Davidson, J. F. 1965. The investigation of a standing wave due to gas blowing upwards over a liquid film; its relation to flooding in wall-wetted columns. _J. Fluid Mech._ 22:pp321-335 Shuler, P. J., and Krantz, W. B. 1977. Spatially growing three-dimensional waves on falling film flow. _Int. J. Multiphase Flow_ 3:pp606-614

52. Stainthorp, F. P. 1965. The development of ripples on the surface of a liquid film flowing inside a vertical tube. _Trans. Instn. Chem. Engrs._ 43:ppT85-T91

53. Stainthorp, F. P. 1967. The effect of co-current and counter-current air flow on the wave properties of falling liquid films. _Trans. Instn. Chem. Engrs._ 45:ppT372-T382

54. Sun, K. H. 1979. Flooding Correlations for BWR Bundle Upper Tieplates and Bottom Side-Entry Orifices. _Proceedings of the Multi-Phase Flow and Heat Transfer Symposium-Workshop_ Miami Beach, Flordia, edited by T. Nejat Veziroglu 1615-1635

55. Suzuki, S., and Ueda, T. 1977. Behavior of liquid films and flooding in counter-current two-phase flow. Part 1. Flow in circular tubes. _Int. J. Multiphase Flow_ 3:pp517-532

56. Telles, A. S., and Dukler A. E. 1970. Statistical characteristics of thin, vertical, wavy, liquid films. _Ind. Eng. Chem. Fundam._ 3:pp412-421

57. Ueda, T. and Suzuki, S. 1978. Behavior of liquid films and flooding in counter-current two-phase flow. Part 2. Flow in annuli and rod bundles. _Int. J. Multiphase Flow_ 4:pp157-170

58. Verschoor, H. 1938. Limiting vapour velocity in packed columns. _Trans. Instn. Chem. Engrs._ 16:pp66-76

59. Wallis, Graham B. 1961. Flooding Velocities for Air and Water in Vertical Tubes. United Kingdom Atomic Energy Authority Report AEER-R123

60. Wallis, Graham B. 1962. The transition from flooding to upwards cocurrent annular in a vertical pipe. United Kingdom Atomic Energy Authority Report AEEW - R 142

61. Wallis, Graham B. 1968. Phenomena of liquid transfer in two-phase dispersed annular flow. _Int. J. Heat Mass Transfer_ 11:pp783-785

62. Wallis, Graham B. 1969. _One Dimensional Two-Phase Flow_ McGraw-Hill, New York

63. Wallis, Graham B. 1970a. Annular two-phase flow. Part 1: A simple theory. _ASME J. Basic Engineering_ 92:pp59-72

64. Wallis, Graham B. 1970b. Annular two-phase flow. Part 2: Additional effects. _ASME J. Basic Engineering_ 92:pp73-82

65. Wallis, Graham B., deSieyes, D. C., Rosselli, R. J., and Lacombe, J. 1980. Countercurrent annular flow regimes for steam and subcooled water in a vertical tube. Electric Power Research Institute Report EPRI-NP-1336

66. Wallis, Graham B., and Makkenchery, S. 1974. The hanging film phenomenon in vertical annular two-phase flow. _J. Fluids Engineering_ 96:pp297-298

67. Wallis, Graham B., and Kuo, Jing Tzong 1976. The behavior of gas-liquid interfaces in vertical tubes. _Int. J. Multiphase Flow_ 2:pp521-536

EXPERIMENTAL STUDIES OF TRANSIENT FLOW REVERSAL IN AN ATMOSPHERIC PRESSURE WATER RIG

J.E. Byrne, H.S.Ergur and M.L. Ding

Simon Engineering Laboratories
University of Manchester, Manchester, England.

ABSTRACT

Experimental results are presented for heat transfer during flow reversal in an annular channel. The reversal consisted of a linear velocity change from upflow to downflow in the test section. The electrically heated test section power input was gradually increased in a series of experiments to produce a crisis in the apparatus. Results indicate that such a crisis would lead to eventual failure of the test section. Plots of the power and rate of velocity change required to produce a crisis against inlet subcooling are presented for both an annulus and tubes of different wall thickness.

1. INTRODUCTION

Transient critical heat flux (CHF) has received much attention recently particularly when associated with the safety analysis of water reactors. The transients are of many types. Power excursions, depressurisation, loss of fluid, pump coast down, flow reversal etc. have all received considerable attention. An excellent literature survey has been conducted by Leung(1) listing over 100 references. Griffith et al (2), (3), (4) have carried out extensive experimental work using Freon 113, and their theoretical analysis based on steady state CHF correlations and pool boiling dryout produced conservative predictions. Leung et al (5) carried out similar experiments using Freon 11 and high speed photography. Their results are in qualitative agreement with those presented here.

In general, the prediction of CHF during transients is a combination of a computer code to calculate local conditions combined with some form of local steady state CHF correlation. Most calculations produce conservative results when compared with experimental data. The present study is aimed at obtaining results for a well defined excursion. Only flow reversal is considered here in an attempt to understand the mechanisms involved without the complication of depressurisation or other forms of transient superimposed on the results. Atmospheric pressure water was chosen for the obvious reasons of cheapness and convenience. The flow reversal selected was of the simplest type, a linear velocity change from up flow to downflow. The reversal deceleration was variable between 0.5 m/s^2 and 2.0 m/s^2.

2. EXPERIMENTAL APPARATUS

FIG 1 shows the arrangement of the loop. The test section was an annulus consisting of an electrically heated stainless steel tube 9.5 mm diameter and wall thickness 1.6 mm surrounded by a transparent circular section of inner diameter 19 mm. The heated length of the test section was 1.5m and the inner tube contained 40 thermocouples welded on the inside 25.4mm apart. The rig consisted of an upper and lower plenum. Water was pumped from the lower plenum through a preheater which controlled the subcooling and then flowed up through the test section to the upper plenum. A motor driven valve was used to initiate the flow reversal. On opening at a controlled rate the pump delivery bypassed to the lower plenum and the upper plenum discharged through the test section, preheater and turbine meter to the the lower plenum. Sufficient liquid in the upper plenum ensured that downflow at the required velocity could be maintained for 5 minutes. Tubes of inner diameter 9.5mm and 9.8mm and wall thickness of 0.81mm 3.05mm respectively were used as well.

2.1 Instrumentation

Since the thermocouples were directly connected to the A.C. heated test section, the leads were subjected to the local A.C. voltage. Also in the annular section, the difficulty of welding the thermocouples to the inside of the tube meant that some A.C. voltage difference appeared across the leads. Although only a matter of millivolts, this was sufficient to swamp the temperature signals. An integrating type of digital voltmeter would suffice here for steady state measurements, but for transient work, a scanning system of this type was far too slow.

Consequently, the thermocouple signals were fed through a unity gain differential amplifier with 106dB common mode rejection ratio to eliminate common signals on both leads and then to a four stage 8 pole filer with gains selected to amplify the required signal. The overall performance of the filter produced uniform

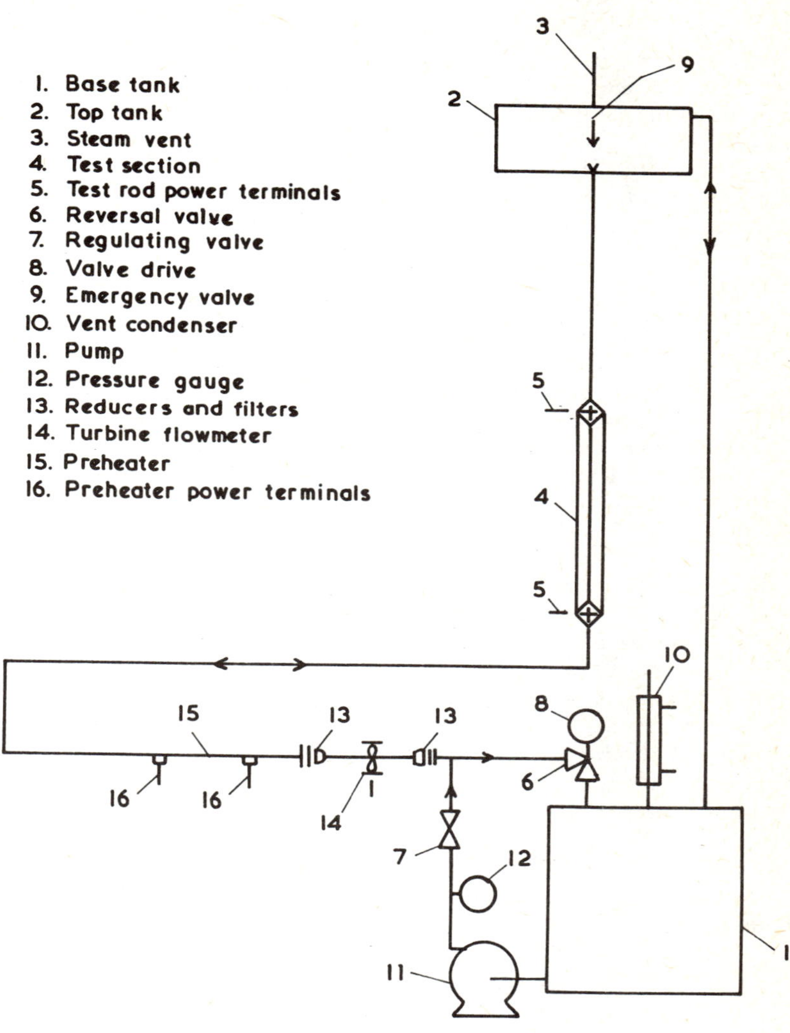

Fig. 1 Schematic diagram of flow loop.

amplification from D.C. to 20 Hz but gave a reduction of -48dB at mains frequency of 50 Hz below the amplified D.C. level. This signal was then fed to a digital computer with A to D convertors capable of scanning 16 channels in less than 0.5 ms with the installed software. Logging was controlled by a real time clock working on an interrupt system and if time permitted was capable of transferring the data directly to floppy disk giving a very large storage capability. Logging rate was timed to occur at even divisions of mains frequency and hence samples could be averaged over mains cycles to produce further rejection of mains frequency. In all, the system performed very well with no measurable 50HZ component. The cost of having one amplifier with filtering per channel with standard components proved very reasonable.

Flow rate was measured using a 16 mm diameter turbine meter. The output signal of the meter was A.C. with a frequency proportional to volume flow rate. To make this compatible with the logging system, a frequency to voltage conversion circuit was constructed. A pressure transducer was used to measure pressure drop across the test section, and in the transparent annular test section, conditions could be studied using cine photography.

2.2 Experimental Procedure

Seconds before initiating the reversal, all channels to be logged were calibrated. This was part of the software handling procedure and was necessary because of slow drift in the filtering amplifier circuits. The experiment was started with a set power level on the test section and a stable inlet subcooling. The motor drive to the reversal valve was set for a particular rate of opening. Throwing a switch started the logging system, the motor drive and, in some instances, the cine camera. Power was maintained on the test section for the duration of the measurements before tripping. Most experiments lasted for 20s or 40s and data was collected every 0.04s or 0.08s.

3. RESULTS

These can be divided into two categories.

(i) Experiments where flow reversal was successfully achieved in some cases with barely detected boiling, in others with some dry out patches forming in the bottom half of the test section.

(ii) Experiments where void formation was to such an extent that the flow stagnated over considerable periods and would eventually lead to test section failure.

It was difficult to separate these cases precisely as only a small increase in power level was required to move from success to crisis. Conditions were not exactly repeatable due to small errors in valve opening rates at the same drive setting.

FIG 2 shows typical results. In 2(a) the wall temperature rises by 10K to 15K during the reversal and decays slowly back to the steady state temperature of the reversed flow. FIG 2(b) shows the flow behaviour with a linear decay from about 1.6 m/s to -1.6 m/s although the circuit characteristics produce some tail off from linearity before full reverse flow is achieved. In FIG 2(c) however the power level has been increased by about 5% but the results are very different. The temperatures are very similar for the first 5 s. Vigorous boiling with dryout occurs in the bottom half of the section for another 3s and then the wall temperature starts rising to a peak at 13s, then a sudden quench which produces more vapour and a further rise to a higher peak. This "sawtooth" waveform with increasing amplitude continues but power is tripped to avoid damage to the section after about 30s. The steep wall temperature rise agrees with results for adiabatic heating at this power level, and the sudden decrease during quench is typical of cooling with a two phase flow heat transfer coefficient. FIG 2(d) indicates the flow behaviour. Instead of reversing normally, the flow is opposed by void formation which results in near stagnation for some period of time before a sudden downflow caused the quenching. Note also that the top of the test section (thermocouple 40) temperature excursion is relatively modest, showing that the crisis occurs towards the bottom. In general the top end remains relatively cool.

A considerable number of experiments of this type have been conducted with varying rates of velocity change, power supplied and subcooling. It appears that the formation of a dry patch in the test section is not indicative of a crisis on its own. Indeed if the section flow rate in up and downflow is such that CHF will not occur, then it is the void formation during the reversal which prevents the flow reversing that will cause a crisis in the rig. Successful reversals have been accomplished with dry patches forming for short period of time before quench. Thus the phenomonen must be influenced by the flow characteristics of the apparatus.

4. ANALYSIS

A simple analysis of the flow looks at a lump of fluid which enters the test section, slows down during the reversal to reach zero velocity at the top of the section before reversing through the section. This lump is subjected to more heat input than any

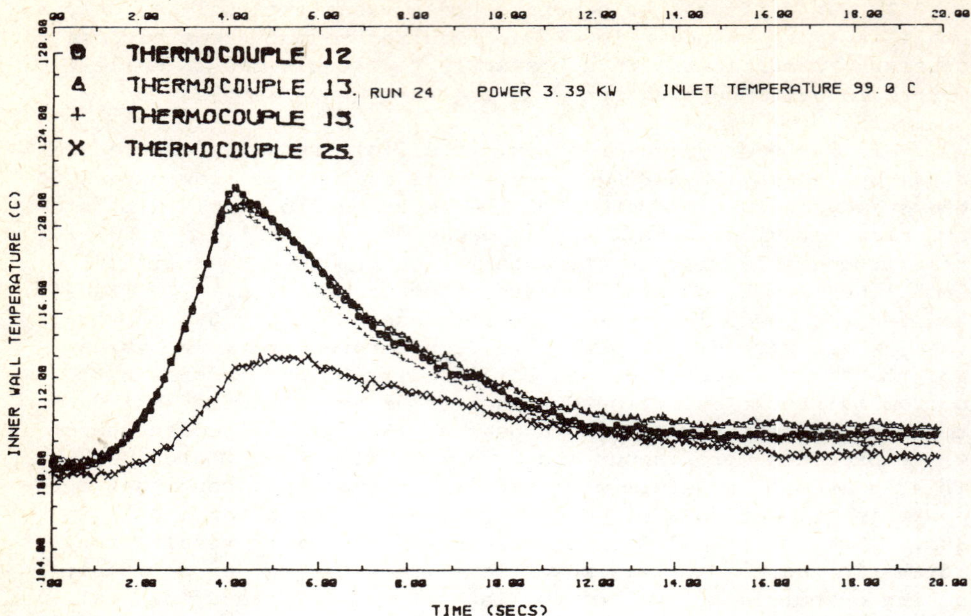

Fig. 2a Inner wall temperature distribution as a function of time.

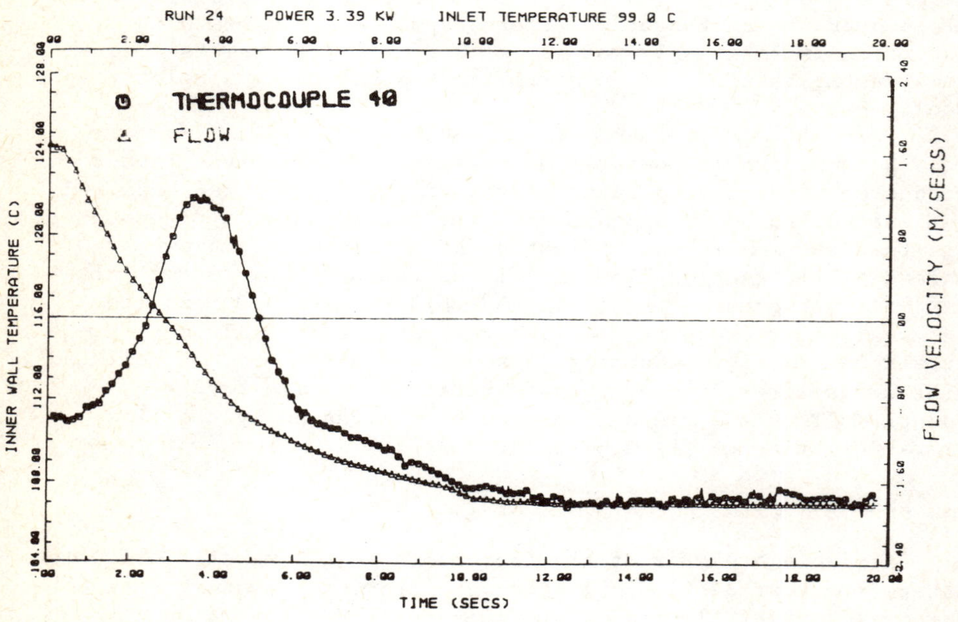

Fig. 2b Inner wall temperature distribution as a function of time.

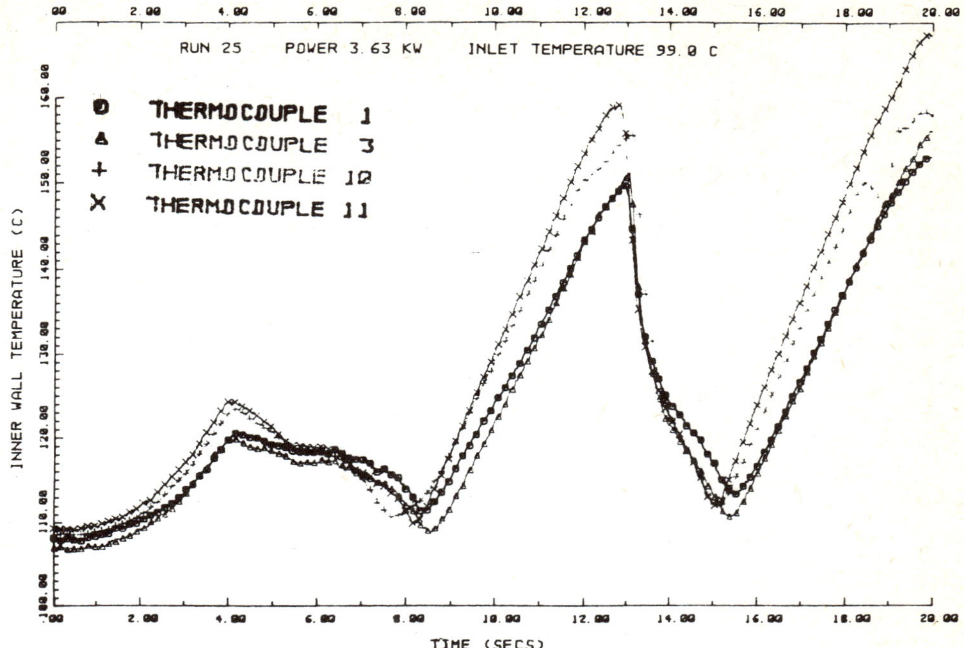

Fig. 2c Inner wall temperature distribution as a function of time.

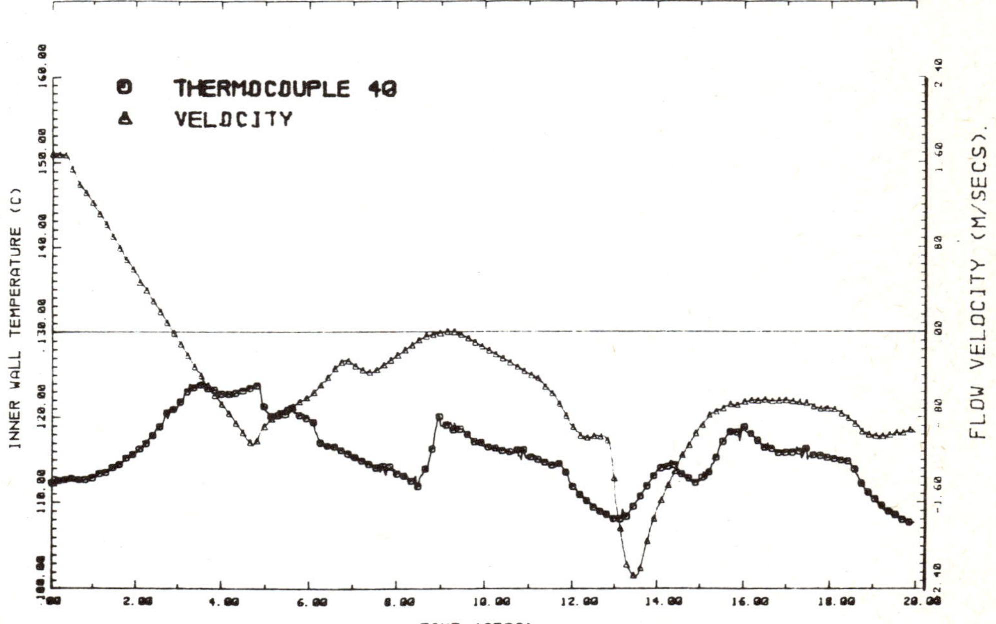

Fig. 2d Inner wall temperature distribution as a function of time.

other since its transit time in the section is maximum. The velocity change with time is (see nomenclature)

$$w = w_o - at$$

and if reaches the top of the section at time

$$t_c = \frac{w_o}{a}$$

travelling a distance $$L = w_o t_c - a \frac{t_c^2}{2} .$$

hence the lump remains in the test section for a total time of

$$2 \sqrt{\frac{2L}{a}}$$

If all the power generated goes into the fluid then the temperature rise of the lump is

$$\Delta T = \frac{P}{\rho_f V c_p} 2t_c = \frac{P}{m c_p} 2 \sqrt{\frac{2L}{a}}$$

Hence plotting the energy generated in the test section during the transit time of the lump per unit volume of liquid in the section against inlet subcooling will indicate whether subcooling can be removed in this worse case lump. Thermal inertia of the heater will influence this somewhat since the heat transfer coefficient is bound to alter during the reversal. Note also that we should expect critical conditions to occur near the bottom of the test section.

FIG 3(a) summarises the results for the annular section. Energy input to the critical lump during reversal is plotted against initial subcooling neglecting any thermal inertia in the heater. The solid line indicates the energy required to remove subcooling. However the actual energy input requires further analysis of the single phase flow.

4.1 Transient Single Phase Flow

The approach adopted in many codes to analyse single phase transient flow is to use the usual steady state correlations with the local Reynolds number during the transient.

For a thin section heater (assumed plane) insulated on one side conduction through the wall with internal heat generation is governed by the equation

$$\frac{\partial T_m}{\partial t} = \frac{k}{\rho_m c_m} \frac{\partial^2 T_m}{\partial x^2} + \frac{Q_m}{\rho_m c_m} \qquad (1)$$

If the Biot number $\frac{hl_m}{k}$ is small then it can be treated as a lumped system

$$\frac{\partial T_m}{\partial t} = \frac{Q_m}{\rho_m c_m} - \frac{h}{\rho_m c_m l_m} (T_m - T_f) \qquad (2)$$

and for the fluid flow

$$\frac{\partial T_f}{\partial t} + w \frac{\partial T_f}{\partial z} = \frac{h}{\rho_f c_p l_f} (T_m - T_f) \qquad (3)$$

where l_f for an annulus is the equivalent diameter.

For the annulus case the Biot number ≃ 1 at the start of the transient and hence to cope with this problem, define a temperature for the metal $\theta_m = T_m - T_{mo}$ where T_{mo} is the initial metal temperature before the transient.

Let $\theta_m = b_o + b_2 x^2$ where b_o and b_2 are functions of time only, then the thermocouple temperatures on the insulated face θ_m^* are given by

$$\frac{\partial \theta_m^*}{\partial t} = \frac{\frac{Q_m}{\rho_m c_m}\left(1 - \frac{h}{h_o}\right) - \frac{h}{\rho_m c_m l_m}(\theta_m^* - \theta)}{1 + \frac{hl_m}{2k}}$$

where $\theta = T_f - T_o$

Using the dimensionless parameters

$$\Phi_m = \frac{\theta_m^* \rho_m c_m a}{Q_m w_o} \qquad \Phi = \frac{\theta \rho_m c_m a}{Q_m w_o} \qquad \tau = \frac{a}{w_o} t \qquad X = \frac{2a}{w_o^2} z$$

$$R = \frac{\rho_m c_m}{\rho_f c_p} \frac{l_m}{l_f} \quad \text{and} \quad K = \frac{h w_o}{\rho_m c_m a l_m}$$

we obtain the equations

$$\frac{\partial \Phi_m}{\partial \tau} = \frac{(1 - \frac{h}{h_o}) - K(\Phi_m - \Phi)}{1 + \frac{hl_m}{2k}}$$

$$\frac{\partial \Phi}{\partial \tau} + 2(1 - \tau)\frac{\partial \Phi}{\partial X} = R\left(\frac{\tau + K(\Phi_m - \Phi) - (1 - \frac{h}{h_o})}{1 + \frac{hl_m}{2k}}\right)$$

given the velocity $w = w_o - a\,t$

These equations are hyperbolic with one characteristic being X = const. ie. normal to the heater wall whilst the other one is $X - \tau(2 - \tau)$ = const. following the motion of the fluid.

These equation were solved on a computer and results are compared with the first part of the temperature transient FIG 4(b). The comparison is poor, and indicates that pseudo steady state correlations are not very accurate in transient flow.

The problem was reformulated. Given the experimental profiles of temperature, what value of h should be chosen to produce agreement. The computer program was altered to adjust h to obtain a match with a specified temperature at a specified axial position. Results are presented in FIG 4(a). During deceleration of the flow the heat transfer coefficient lags behind the steady state value, does not quite go to zero and takes some time to recover during the acceleration, producing a higher peak temperature than the calculated one.

The energy received by the worse case lump was also calculated and was found to be 70% of the generated energy for the annular case. Hence in FIG 3(a) it is clear that in some cases a crisis has occured even though bulk subcooling has not been removed from the worst case lump. FIG 3(b) and FIG 3(c) show results for tube sections. As yet full analysis of the tube case has not been carried out.

4.2 Two Phase Flow

Following the fluid lump the energy equation is

$$\rho \frac{di}{dt} = \frac{dp}{dt} + Q_f$$

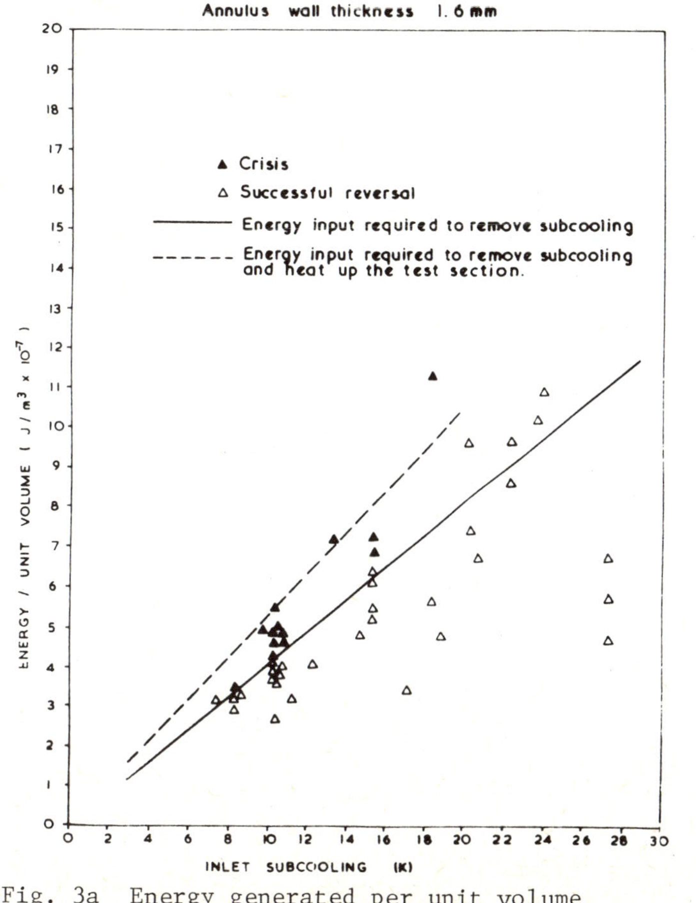

Fig. 3a Energy generated per unit volume during the transient versus inlet cooling.

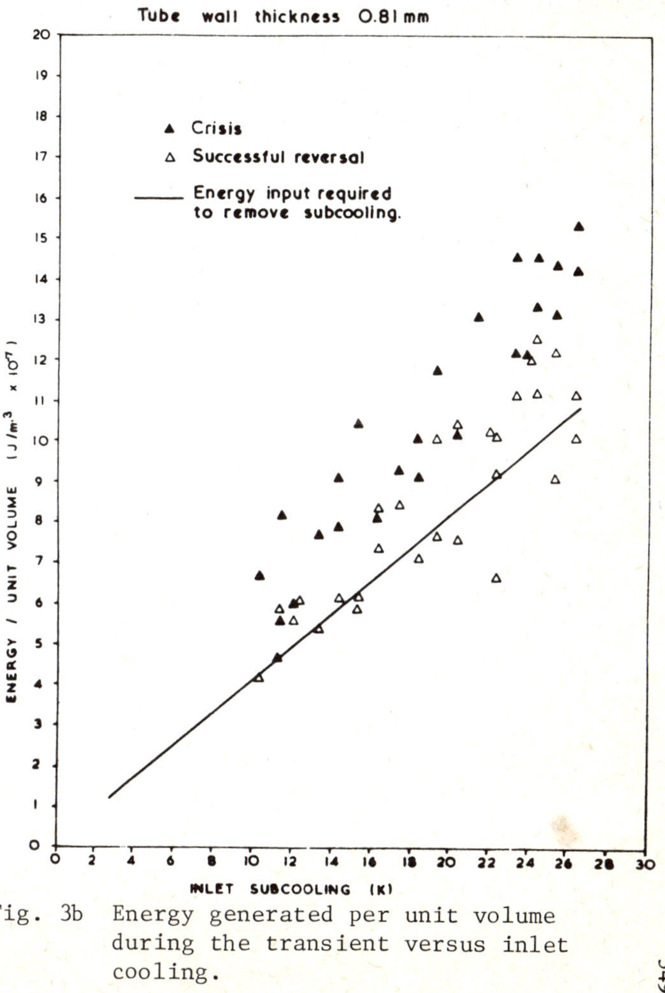

Fig. 3b Energy generated per unit volume during the transient versus inlet cooling.

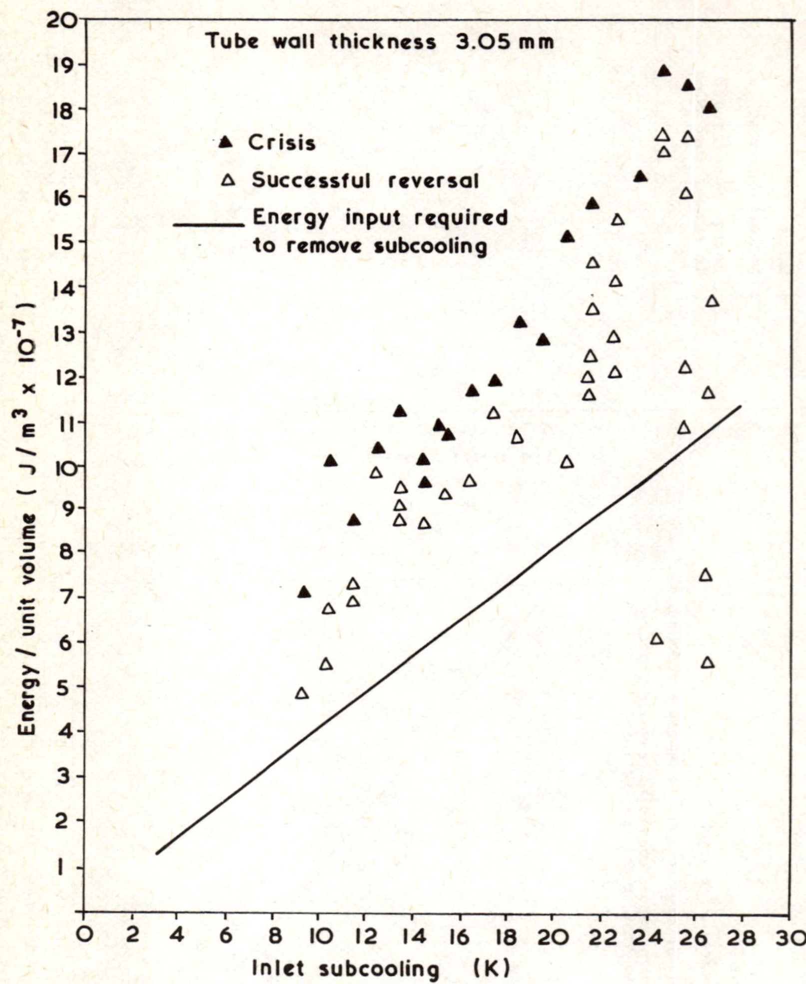

Fig. 3c Energy generated per unit volume during the transient v. inlet cooling.

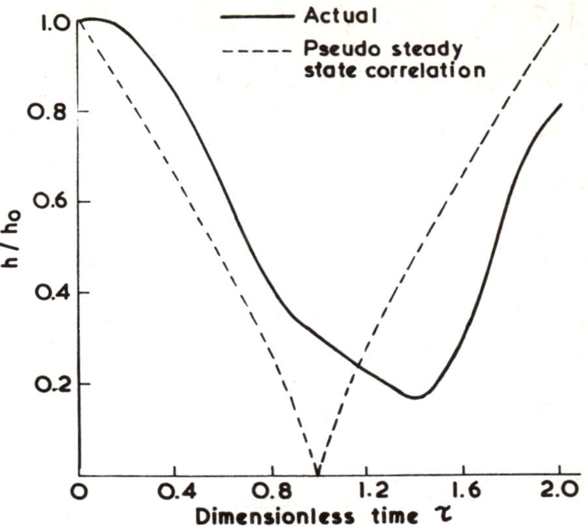

Fig. 4a Heat transfer coefficient v. time during single phase flow reversal.

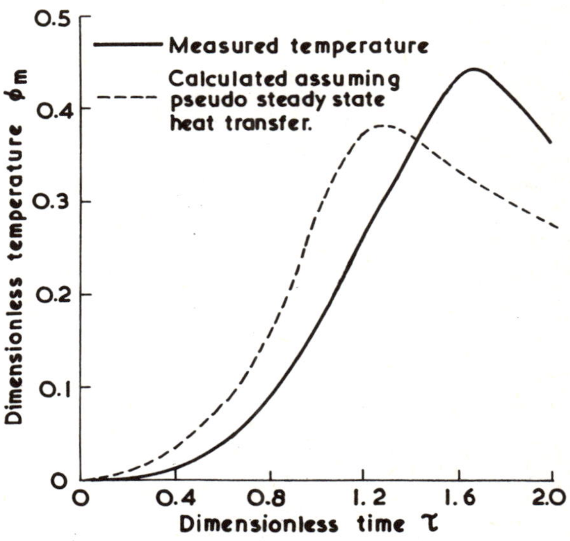

Fig. 4b Temperature v. time during single phase flow reversal at Z/L = 0.25.

In terms of α and p as independent variables the equation becomes

$$\rho_g \rho_f \frac{i_{fg}}{\rho} \frac{d\alpha}{dt} = \frac{dp}{dt} \left(1 - \rho \frac{di}{dp}\right) + Q_f$$

Now $\rho \frac{di}{dp}$ is $O(10^3)$ for $\alpha \to 0$ and is less than unity for $\alpha \to 1$ in atmopheric pressure water, giving a pressure term which is important during the initial phase of boiling.

Solving the above equation we get

$$\frac{v_g}{v_{fg}} - \alpha = \frac{\exp\left(-\frac{Q_f v_{fg}}{i_{fg}} t\right)}{\frac{v_{fg}}{v_g} - \frac{v_{fg}}{i_{fg}} (\rho_f - \rho_g) \int_o^t \frac{\partial i}{\partial p} \frac{dp}{dt} \exp\left(-\frac{Q_f v_{fg}}{i_{fg}} t\right) dt}$$

Define an average $\frac{\partial i}{\partial p} \frac{dp}{dt} = \lambda$ and hence for the boiling time t

$$\frac{Q_f v_{fg}}{i_{fg}} t = -\mathrm{Log}_e \left(\frac{1 - \frac{\alpha v_{fg}}{v_g} - \left(\frac{v_g}{v_{fg}} - \alpha\right) \frac{\lambda (\rho_f - \rho_g)}{Q_f}}{1 - \left(\frac{v_g}{v_{fg}} - \alpha\right) \frac{\lambda (\rho_f - \rho_g)}{Q_f}} \right)$$

For the annular case the time taken to achieve a void fraction of 0.8 is at the most 1.26/P(kW) whilst to remove subcooling takes a time of 1.29 ΔT/P(kW). Thus except at very low subcoolings, most of the time in the test section is spent removing subcooling, and boiling takes place fairly rapidly in 0.15 s to 0.4 s for the power range used. This probably explains the sensitivity of the experiments near the crisis point.

An important point to stress is that the flow is not quite stagnant during the boiling phase since up to 3s may elapse before adiabatic heating of the test section, in excess of the time required to boil all the test section liquid off. Hence there is certainly liquid in the bottom half of the test section for more than the boiling time.

5. CONCLUSIONS

The occurence of CHF during a flow reversal does not necessarily lead to a crisis. The vapour generation in the test section as a consequence can seriously impede the flow causing near stagnation conditions and a serious crisis.

Heat transfer even during the single phase flow part of the transient is not accurately predicted using a local condition steady state correlation.

ACKNOWLEDGEMENTS

Thanks are due to the Nuclear Installations Inspectorate who funded the construction of the loop and to Mr. M. Laithwaite who was responsible for the original construction.

NOMENCLATURE

a	acceleration of the fluid, m/s^2
b_1 , b_2	functions of time only
c	specific heat, J/kg K
h	heat transfer coefficient, W/m^2K
i	specific enthalpy, J/kg
k	metal thermal conductivity, W/m K
K	dimensionless heat transfer coefficient
l	length scale, m
L	length of test section, m
m	mass of liquid in test section, kg
p	pressure, N/m^2
P	power generated, W
Q	heat input per unit volume, W/m^3
R	dimensionless ratio of properties and scales
t	time, s
T	temperature, K
ΔT	temperature difference, K
v	specific volume, m^3/kg
V	volume of fluid in heated test section, m^3
w	velocity of fluid, m/s
x	distance normal to the test section axis, m
X	dimensionless distance
z	distance along test section axis, m
α	void fraction
Θ	temperature minus initial temperature, K
Φ	dimensionless temperature
ρ	density, kg/m^3
τ	dimensionless time

Suffix

c	transit time
f	liquid
g	vapour
m	metal
o	conditions at start of transient
p	liquid constant pressure

Superfix

*	insulated wall of test section

REFERENCES

1. Leung J.C.M. "Critical heat flux under transient conditions: a literature survey." Argonne National Laboratory ANL-78-39

2. Griffith P., Pearson, J.F. and R.J. Lepkowski. "Critical heat flux during loss of coolant accident." Nuclear Safety 18 3 (1977) 298.

3. Smith R.A., Price, F.A. and P. Griffith. "An analysis of CHF in flow reversal transients." Trans ASME, Journal of Heat Transfer 98 (1976), 153-158.

4. Price F.A. and P. Griffith. "Transient critical heat flux during flow reversal." EPRI Project 292-2-1, NP-149.

5. Leung J.C.M., Bankoff S.G., Henry, and O.C. Jones. "Occurance of critical heat flux during blowdown with flow reversal." Nuclear Eng. and Design 52 (1979) 303-324.

UNBOUNDED CRITICAL FLOWS AND JET FORCES

M. Cumo

E.N.E.A.: Italian National Committee
for Nuclear and Alternative Energies
Rome, Italy

ABSTRACT

After a survey of jet flows and jet forces equations and of current design formulae, blow dow tests and critical flows through nozzles and channels of different geometry are briefly described, with reference to recent advances both theoretical and experimental [1-5].

Unbounded flow fields are subsequently examined [6-8], with special attention to experimental investigations.

1. DESIGN CORRELATION

As it is well known, in nuclear power plants pipes with "high energy" fluids are considered as potential sources of different types of impulsive forces, due to the remote possibility of a break in the pressure boundary: reaction forces on the pipes themselves, whips and jet forces [1]. Following to boundary breaks of area A_R, the reaction force F due to the emerging flow of fluid at initial pressure p_o may be conventionally computed, in "steady state" regime, as:

$$F = K_1 K_2 p_0 A_R \tag{1}$$

where K_1 is a thrust coefficient equal to 1.26 for saturated water, and K_2 an "amplification factor" ($1 \leqslant K_2 \leqslant 1.2$) which accounts for the dynamic reactions due to the impingement against hard restraint supports. These forces may be considered stepwise and constant, or time dependent.

The jet forces R_A exerted by the emerging flows on external targets may be, in a very schematic and rough way, conventionally computed by means of few geometric variables (fig. 1) like the cross area projection A of the target in a plane normal to the jet axis, the asymptotic transversal section of the jet, A_∞, the distance between the break section and the target, L, the diameter D of the broken pipe and the break area A_R. Usual computations lie on the following relationships:

$$R_A = F \quad \text{for } L/D < 5 \text{ and } A_R \leq A$$
$$\text{or } L/D \geq 5 \text{ and } A_\infty \leq A$$

$$R_A = FA/A_R \quad \text{for } L/D < 5 \text{ and } A_R > A$$

$$R_A = FA/A_\infty \quad \text{for } L/D > 5 \text{ and } A_\infty > A$$

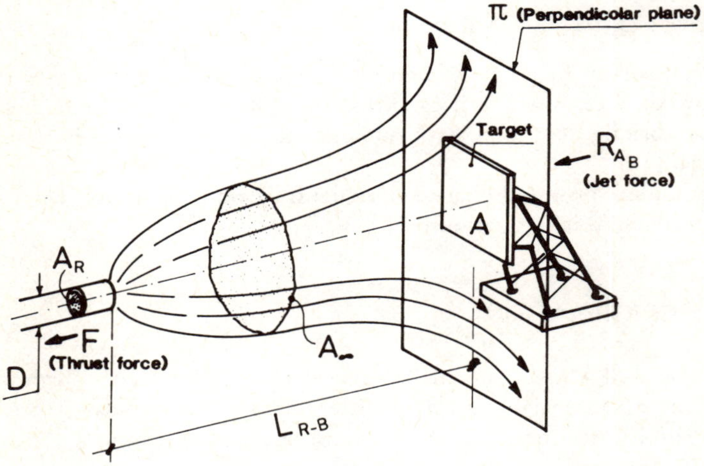

Fig. 1 – Sketch of the jet geometric parameters

2. JET THRUST AND IMPINGEMENT FORCES

Let's consider now the very simple case of a jet emerging from a discharge channel (guillotine break) and impinging against a normal, plane target (DE, fig. 2) in a quasi-steady regime (constant physical parameters, short time interval).

The control volume ABCDE has the "lateral" walls CD and EF sufficiently far from the jet axis. Let's suppose negligible, in comparison with the jet dynamic pressure, the ambient pressure and the hydrostatic height. The impingement force R exerted by the jet on the target and defined by:

$$\vec{R} = \iint_{DE} \vec{p}\, dS \tag{2}$$

in measured by:

$$\vec{R} = \vec{F} - \vec{M}$$

where $\vec{F}$, the jet thrust at the channel outlet, is given by:

$$\vec{F} = \iint_{A_R} (\vec{p}\, dS + \rho \vec{v}\, \vec{v}\, dS) \tag{3}$$

and $\vec{M}$ is the momentum, per unit time, of the jet flow through the lateral walls DC and EF:

$$\vec{M} = \iint_{DC,EF} \rho \vec{v}\, \vec{v}\, dS \tag{4}$$

Obviously $\vec{F}$ remains constant in every transversal section of the expanding jet (momentum conservation, negligible interaction with the ambient air).

Usually $\vec{F}$ is related to the area A (break area of the discharging channel) and to the total pressure in it, p_o, by means of a thrust coefficient ζ_T:

$$\xi_T = F/(p_0 A) \tag{5}$$

Neglecting the wall friction, as said above, one has $1 < \zeta_T < 2$.
For instance, for saturated mixtures up to 150 bar: $1 < \zeta_T < 1.25$.
At increasing subcoolings ζ_T ranges up to 2.

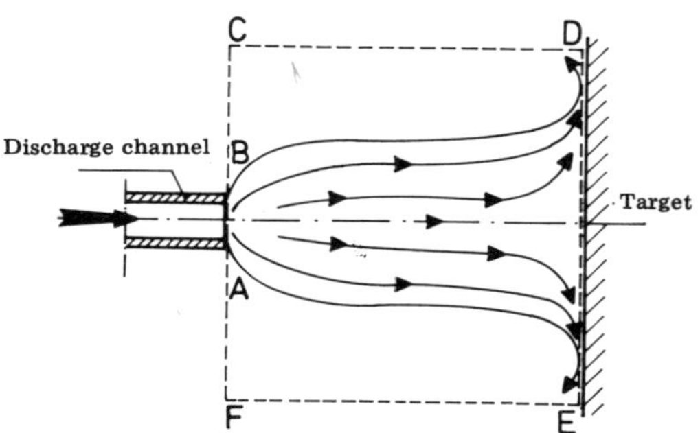

Fig. 2 – Sketch of the jet impinging on a flat, normal target (DE).
The dashed lines indicate the control volume.

Particular tests, recently performed by Centi, Favarelli, Gaspari and Lo Nigro with the FOG test facility of the AMN (Genova) (Fig. 3), have confirmed that the thrust coefficient ζ_T for two-phase flows and for discharge tubes characterized by $0.1 \leqslant f\ L/D \leqslant 15$ is within the range $1 \div 1.26$, as evidenced, for instance, by the diagram in fig. 3 bis.

Fig. 3 – Jet impingement test configuration of AMN FOG facility (Genova).

The reaction force of the emerging fluid on the pipe walls $\vec{R}$ (t) , may be computed as the integral vector for the pressure forces and frictions on the walls themselves through the momentum conservation equation. For instance, with reference to the simple sketch of fig. 4, let's consider the pipe extension OAA'B from the vessel 0 to the break area (B), computing the reaction force $\vec{R}$ (t) on the pipe segment AA'; Let's define a control volume (V_c) which contains with its surface (S_c) the fluid of the segment AA'; the surface S_c is delimited by the normal sections in A and A' - to which are respectively referred the indexes i and j; and by the internal wall of the pipe. Neglecting the gravitational term (high speed flows), the momentum conservation provides:

$$\vec{R}(t) = -\iint_{S_C} \vec{v}\rho\vec{v}\, dS - \iint (\vec{p}_i dS_i + \vec{p}_j dS_j) - \frac{\delta}{\delta t} \iiint_{V_C} \rho\vec{v}\, d\Omega \qquad (6)$$

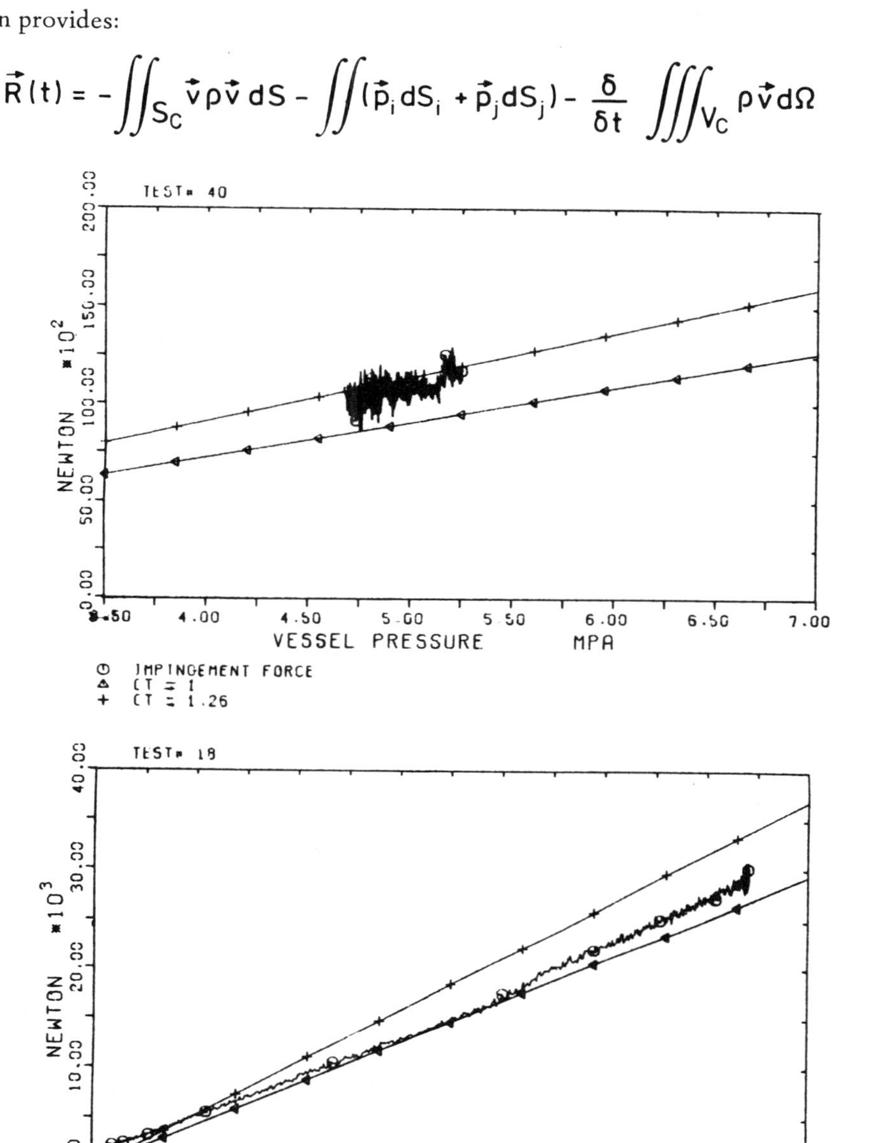

Fig. 3 bis – Predicted and measured impingement (steam pipe break) or thrust (saturated water pipe break) forces. (FOG tests, AMN Genova).

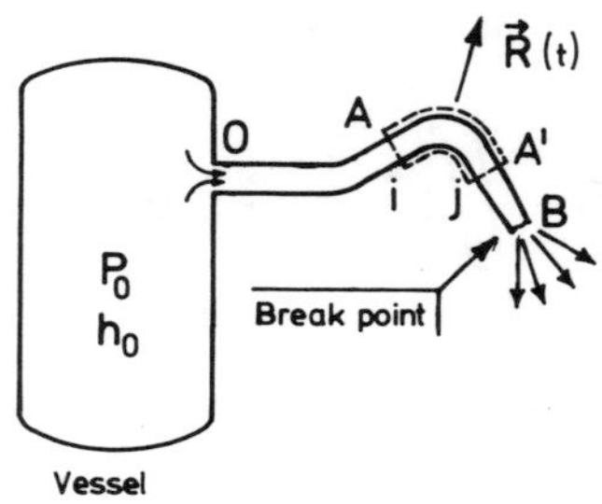

Fig. 4 – Sketch of the control volume in a pipe length subjected by the reaction force R (t).

In a guillotine-type breack (typical L.O.C.A.) with a high pressure two-phase flow the reaction force R (t) increases monotonou up to a limit value, R (∞), corresponding to a steady value of the flowrate, in a time interval which is a little multiple of the transit time (in the pipe OB) of a sonic wave. In the examined example:

$$-\vec{R}(\infty) = \vec{F}_i + \vec{F}_j \tag{7}$$

thrust functions in A and A' defined by:

$$\vec{F}_i = \iint_A (\vec{p}dS + \rho\vec{v}\vec{v}\,dS)\text{, ect.} \tag{8}$$

For a two-phase homogeneous flow, for instance,

$$F_i = \left[\rho_{g_i} v_{g_i}^2 \alpha_i + \rho_l v_{l_i}^2 (1-\alpha) + p_i\right] A_i \quad \text{ect.}$$

3. BLOW DOWN TESTS AND CRITICAL FLOWS THROUGH NOZZLES AND CHANNELS

The very simple examples (figs. 2 and 4) so far considered evidence the need of typical tests consisting in the blowdowh n of pressure vessels through discharge pipes of different geometries (L, D). There is a need of a correct prediction of the blow-down transient in the vessel (pressure and temperature within the vessel, discharge flowrate in the critical flow, etc.) for different conditions of the fluid within the vessel itself (subcooled and saturated water, mixtures, steam, etc.)

A very good data source is the recent "Workshop on Jet Impingement and Pipe Whip" sponsored by CNEN and AMN in Genova, in July 1981 [2].

The major experiment is due to the Marviken Project at Studsvik. In these tests, with discharge pipe diameters of 500 mm, the shortest blowdown time is of the order of 50 seconds during which nearly 300 tons of steam and water are expelled into the jet test section with a maximum flowrate of the order of 14.000 kg/s.

O. Sandervag [3] has revealed three distinguishable periods: a "subcooled period" characterized by a gradually decreasing depressurization rate (fig. 5), a "saturated liquid period" with a constant depressurization rate and a "steam period" with a much quicker depressurization. Following test initiation, the pressure decreases rapidly due to the imbalance between the volumetric flow out of the vessel and the vapour generation rate in the liquid (delayed nucleation, metastibility state). When flashing starts, the pressure recovers and a stable depressurization rate is established.

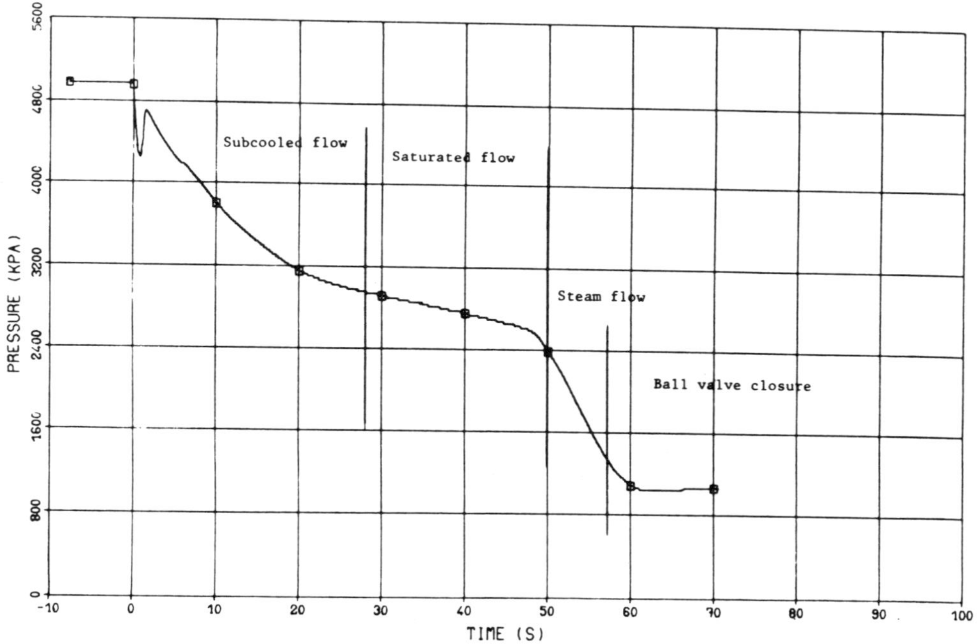

Fig. 5 – Typical history of the vessel pressure in a Marviken test.

The Marviken free jet test data reveal that the static pressure decreases rapidly downstream of the nozzle. At about 1000 mm downstream of the nozzle, the jet over-expands to pressures less than the ambient containment pressure.

At further distances downstream, the static pressure recovers to the containment pressure.

In a smaller scale, another typical blowdown experiment has been performed in Pisa, Italy [4] (fig. 6).

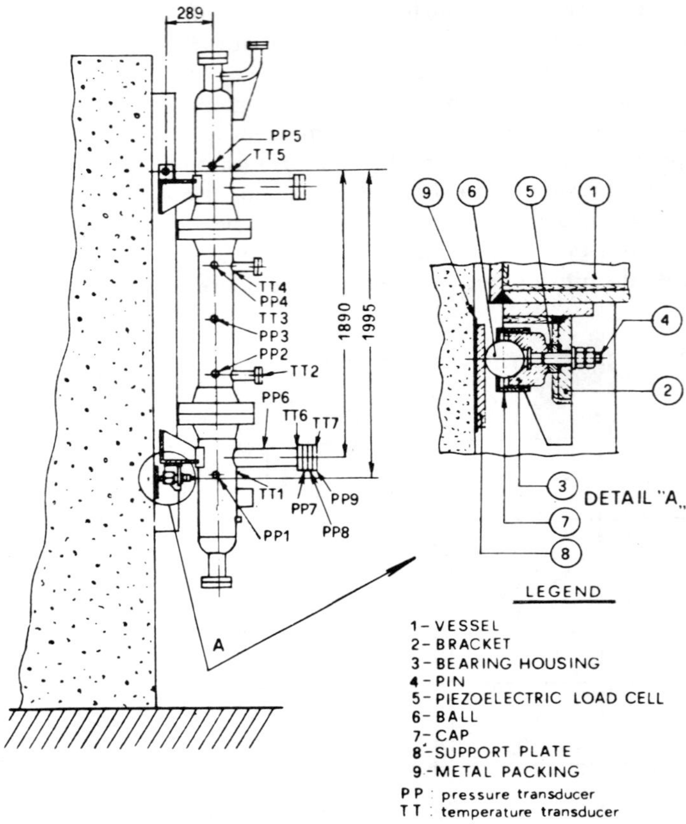

Fig. 6 – PIPER vessel for blowdown experiments in Pisa University.

Following to these tests, a procedure has been developed to obtain all the blowdow related quantities in the reservoir and at the exit of the broken pipe when pressure in the vessel and at the throat and jet reaction thrust histories are known from the experiments. This is very convenient and simple experimental set-up. The performed analysis consists in writing the complete expression of the thrust for stationary flow as a function of local exit pressure.

$$F = \left\{ \left[\alpha \rho_g S v_l + (1-\alpha) \rho_l v_l \right]^2 \left[\frac{(1-X_e)^2}{\rho(1-\alpha)} + \frac{X_e^2}{\rho_g \alpha} \right] + (p_e - p_a) \right\} A \tag{9}$$

The unknowns being the outlet void fraction a, the liquid and vapour densities ρ_l and ρ_g the exit slip ratio $S = v_g/v_l$, the exit quality X_e and the exit liquid velocity v_1.

The further main assumptions are:

- homogeneous mixture in the pressure vessel;
- thermodynamic equilibrium at the exit section.

The main conclusions deriving from the application of the aforesaid method to the experimental data are, according to D'Auria and Vigni:

- the suggested experimental appoach seems a good one to obtain informations about many important variables which cannot be, otherwise, easily measured (two-phase critical flow, steam and water exit velocities, reservoir quality, liquid level in it);

- in evaluationg the two-phase critical flow the homogeneity assumption at pipe exit seems more appropriate than the selection of a slip correlation

(like Moody $\left[S=\left(\frac{\rho_L}{\rho_g}\right)^{1/3}\right]$ or Fauske $\left[S=\left(\frac{\rho_L}{\rho_g}\right)^{1/2}\right]$).

- RELAP 4 - Mod. 6 calculations, while leading to a sufficient agreement between calculated pressure transients and experimental ones, are not equally satisfactory concerning to other variables (specially those at the channel exit); the common practice to adopt flowrate multipliers leads to values for the variables appearing in the flowrate expression which are physically unacceptable (figs. 7 and 8).

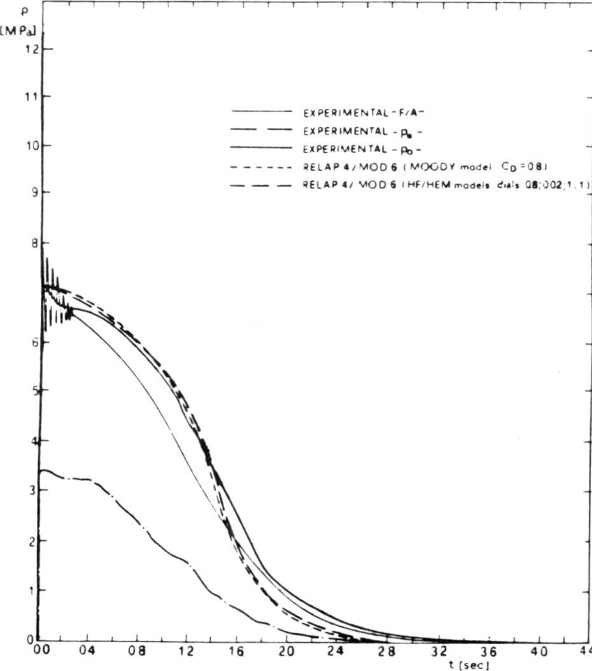

Fig. 7 – Pressures and thrust as recorded with the experimental device of fig. 6.

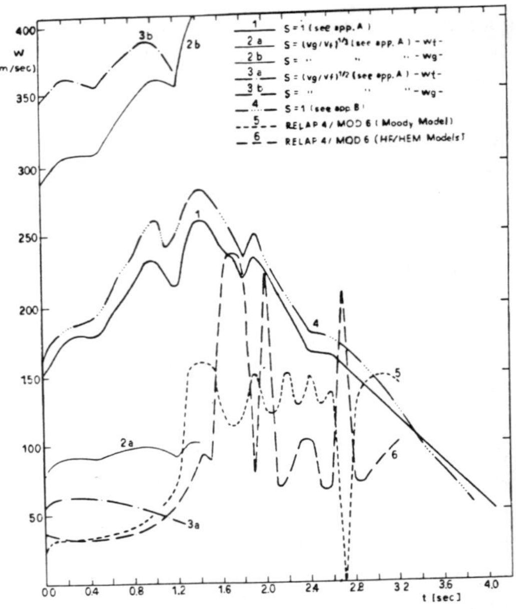

Fig. 8 – Exit velocities following different hypothesis and models.

Critical flows in steady state conditions with subcooled and saturated water and with saturated steam through discharge channels of different lenght - to - diameter ratios (fig. 9) have been systematically measured in the CNEN laboratories at Rome Casaccia Nuclear Center [5]. These tests have provided the critical flowrates as well as the static pressure profiles along the channels (fig. 9 bis, 9 ter).

The knowledge of the experimental value of all these variable has allowed the reproduction of pressure profiles leading to the value of the critical pressure in the outlet section of the channel when the assumed value of the flowrate is exactly equal to the critical one, employing well known pressure drop correlations. Obviously, in the case of critical flows of saturated steam through the employ of the condition that in the outlet section the fluid velocity is equal to the velocity of sound it is possibile to obtain the critical outlet pressure, $p_{c, out}$, by means of a very simple relation:

$$\frac{P_{c,out}}{P_{in}} = M_{in} \sqrt{\frac{(\gamma-1) M_{in}^2 + 2}{\gamma+1}} \qquad (10)$$

where M denotes the Mach number and M_{in} its value in the inlet section:

$$M_{in} = G V_{in} / \sqrt{\gamma P_{in} v_{in}} \qquad (11)$$

being γ the ratio c_p / c_v.

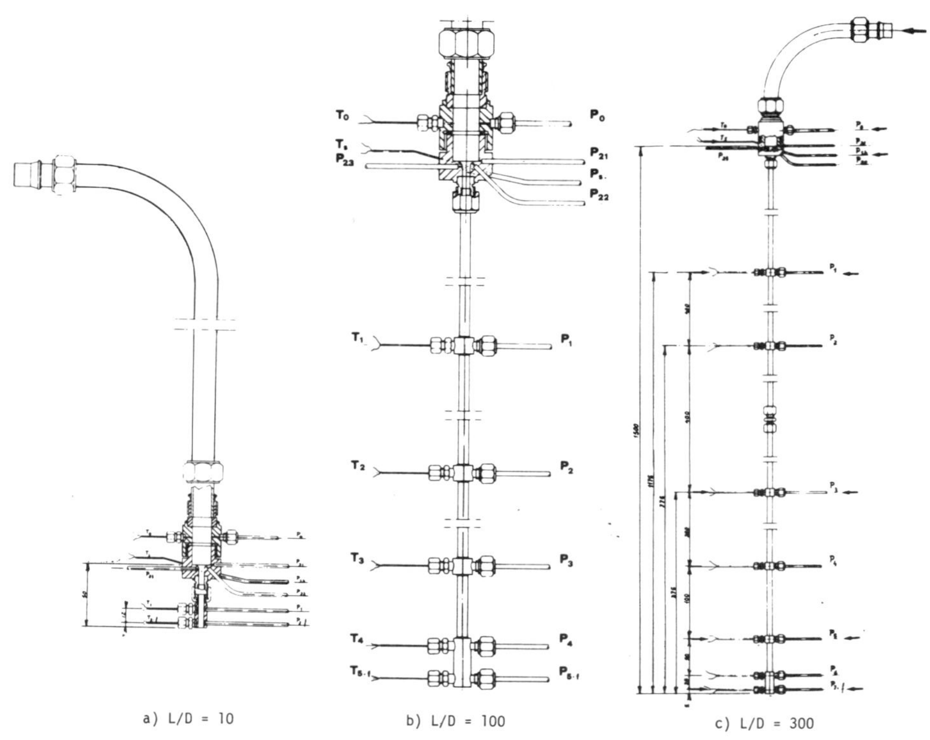

Fig. 9 – Typical test sections for critical flows in CNEN experiments.

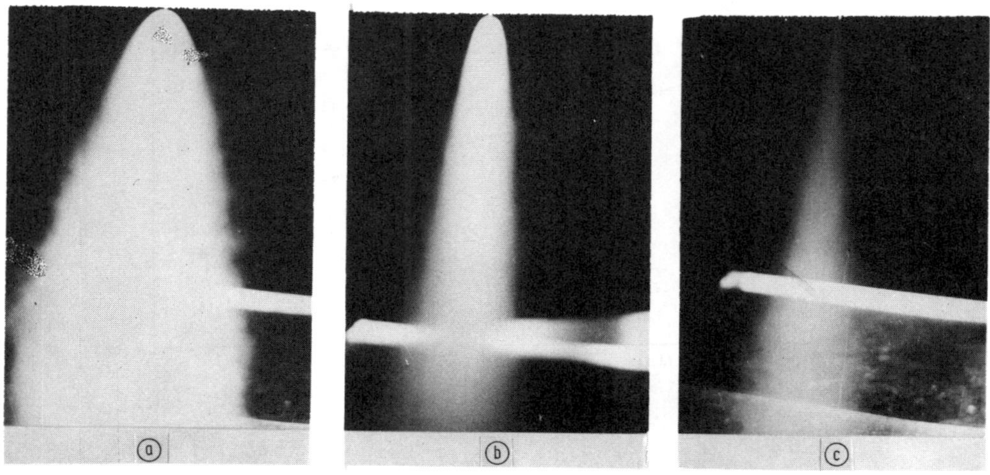

Fig. 9 bis – Three typical aspects of jet flow configurations:
a) subcooled liquid $\Delta T = 30$ °C; b) saturated liquid; c) saturated steam;

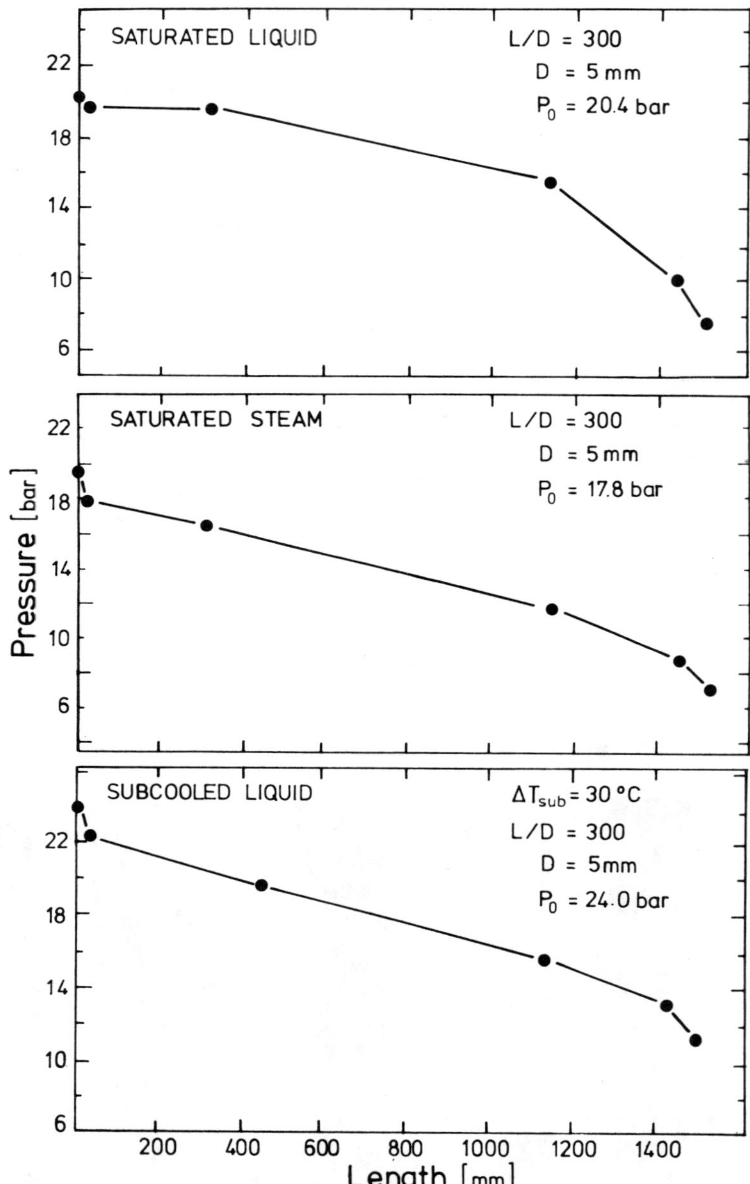

Fig. 9 ter – Typical pressure drop trends along the L/D = 300 test section.

These tests have shown that the knowledge of the critical pressure (in the outlet section of the channel) is equivalent to the knowledge of the critical mass flowrate, independently from the fluid inlet state, with the simple employ of the well known two-phase pressure drop correlations. Predicted and measured values are within the experimental error ranges [9].

As far as critical flow of subcooled liquids is concerned, a very simple computation procedure for the critical mass flowrate has been suggested [8]. Looking through all the experimental data it's possible to deduce that the critical mass flowrate, G_c, may be simply calculated by means of the well known single phase pressure drop correlations $\Delta T_{sub} \geqslant 20$ °C)

$$G_C = \sqrt{\frac{2\,\Delta p}{K_R v_0}} \qquad (12)$$

Where:

- Δ_P is equal to $P_o - P_{sat}(T_o)$;
- K_R is the overall pressure drop coefficient, $K_i + fL/D$
- v_o is the liquid specific volume at stagnation conditions.

Of course, K_R values may be obtained both from tables and diagrams of hydraulic engineering handbooks and, in very special cases (for instance very short discharge channels) by means of ad hoc cold water hydraulic tests.

In fig. 10 a comparison between computed (G_{calc}) and measured (G_{exp}) values of the critical mass flowrate is available.

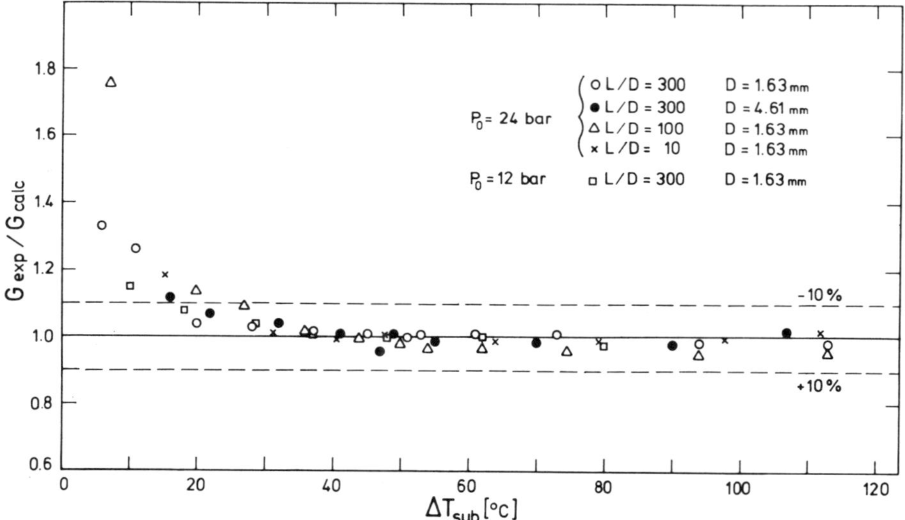

Fig. 10 – Comparison between measured (G_{exp}) and computed (G_{calc}), by means of equation (12), values of the critical mass flowrate versus inlet subcooling.

4. UNBOUNDED FLOW FIELD

If, through the discharging channel, the critical mass flowrate and the critical pressure, at the channel outlet section, are known, the jet flow characteristics are determined. Pressure and temperature profiles, both axial and radial are the objet of this kind of investigation, with the particular aim to deduce jet forces against external obstacles (targets). It is a common opinion that current models used for two-phase jet load calculations (above shown) seem to be overly conservative, in many situations. Let's consider, with reference to fig. 11, a simple model for two-phaşe jet loading on flat axisymmetric targets, proposed by Tomasko, Weigand and Thompson of the SANDIA Laboratories [6].

They have selected a two-dimensional, inviscid system of homogeneous equilibrium equations. Obviously, the governing equations are the conservation of mass, momentum and energy:

$$\frac{\delta \bar{U}}{\delta t} + \frac{\delta \bar{M}}{\delta r} + \frac{\delta \bar{N}}{\delta z} - \bar{Q} = 0 \tag{13}$$

where the vector are:

$$\bar{U} = r \begin{Bmatrix} \rho \\ \rho V_r \\ \rho V_z \\ \rho E \end{Bmatrix} \quad \bar{M} = r \begin{Bmatrix} \rho v_r \\ \rho v_e^2 + p \\ \rho v_r v_z \\ \rho v_r (E + P/\rho) \end{Bmatrix} \quad \bar{N} = r \begin{Bmatrix} \rho v_z \\ \rho v_z v_r \\ \rho v_z + p \\ \rho v_z (E + P/\rho) \end{Bmatrix} \quad \bar{Q} = \begin{Bmatrix} 0 \\ -p \\ 0 \\ 0 \end{Bmatrix} \tag{14}$$

being ρ the density, r and z the radial and axial coordinates, E the internal energy, p the pressure and V_r and V_z the radial and axial velocities.

The break diameter D can be introduced into equation [13] as a length scale:

$$x' = x/D \qquad \text{and} \qquad y' = y/D.$$

This substitution has no effect on the steady state character of the equations; thus the diameter dependence has been scaled from the problem. The above system is combined with an equation of state to form a closed system of equations.

In the above system the steady solution does not have energy exchange with the surroundings.

Additionally, the above relations assume that heat transport within the fluid is not important; therefore, one of the equation in [13] could be eliminated in favour of the statement that the entropy at a fluid fixed location is constant. Of course, this only applies where there are no shocks in the field; when shocks appear, the Rankine-Hugoniot relations must be applied. There are two phenomena, choked flow at the break and a standing shock at the target, which need some definition and discussion. One-dimensional homogeneous equilibrium choked flow (HEM) is assumed as the initial condition for the computational model. For 1-D HEM the assumptions are:

$$H_0 = H + \frac{V^2}{2} \quad ; \quad S = S_0 \quad ; \quad G = \rho v$$

(G is the mass flux).

The choking may be defined as the state the fluid flow reaches where the downstream properties do not influence the mass flowrate; the fluid at this state is supersonic, thus choked flow occurs when $\left|\frac{V}{c_s}\right| < 1$ switches to $\left|\frac{V}{c_s}\right| \geqslant 1$ where c_s is the HEM sound speed. For most HEM flow conditions, the SANDIA 2-D computer model predicted the presence of a stationary shock wave immediately upstream of the target. The existence of this standing (almost normal) shock wave at the target was not unexpected; its existence and strength are well documented in single phase gas flows.

At a shock, kinetic energy is exchanged for thermal energy; consequently, there is a stagnation pressure decrease across the shock (fig. 12).

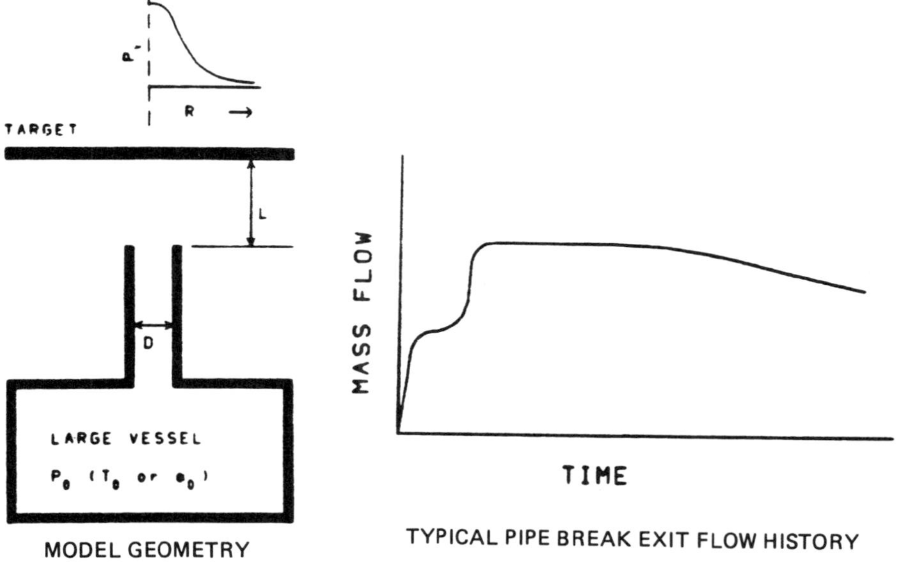

Fig. 11 – Model geometry assumed in the SANDIA model.

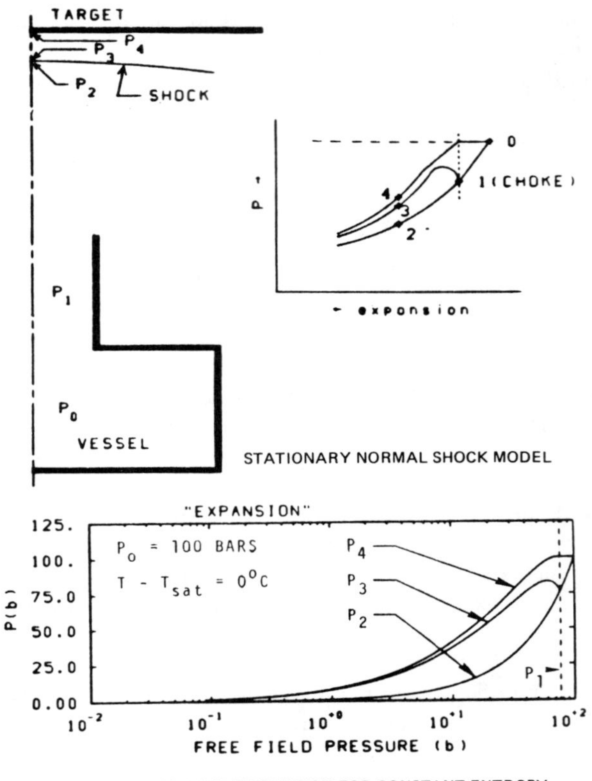

Fig. 12 – Examples of solutions found by Tomasko, Weigland and Thompson.

5. JET FLOW TESTS

The behaviour of two-phase jets, especially those of initially subcooled liquid, is rather complex and undoubtely needs preliminary experimental investigation. The radial expansion of the two-phase jet depends strongly on the degree of thermodynamic disequilibrium that there is in the liquid jet core. The delay times of thermodynamic disequilibrium play a central role. "Ad hoc" tests performed in CNEN with critical flows of subcooled liquids through narrow orifices (see the scheme of fig. 13) have permitted the measure of the extension of the meta-stable liquid jet L_m before the inception of flashing [7], knowing the flowrate and the outlet liquid velocity, u_{cr}, the measure of the meta-stability time $\Delta\tau = L_m/u_{cr}$, it's possible. For specific mass flow-rates of the order of 30.000 kg/m^2 s the meta-stability time is of the order of 10^{-4} (compared with the 10^{-3} s value often adopted in critical flow codes). While the degree of subcooling does not influence greatly this delay time, it has a deeper influence on the shape of the two-phase expanding jet at the orifice outlet (fig. 14).

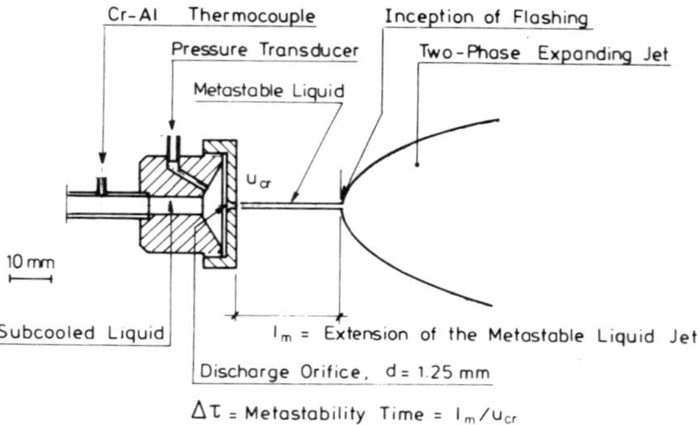

Fig. 13 – Test section and idealization of the flow model for the measure of the metastability time in the subcooled liquid jet.

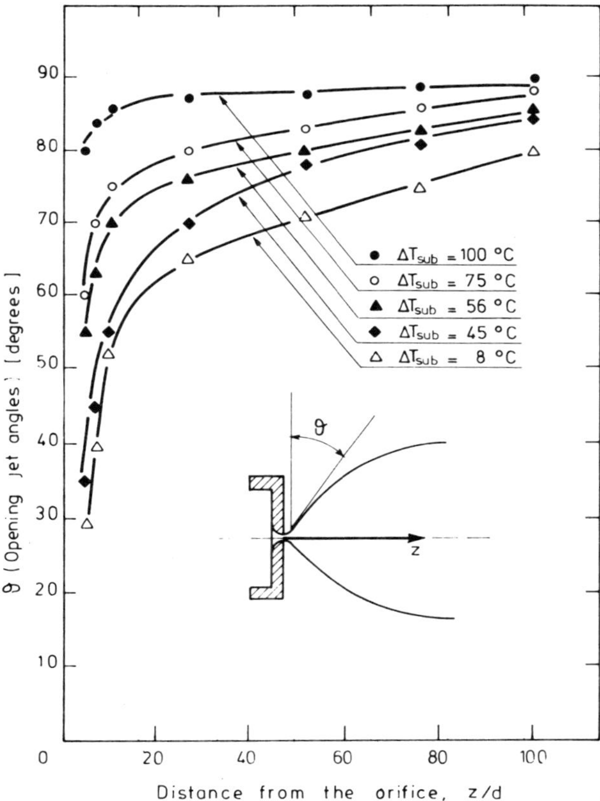

Fig. 14 – Opening angles of the expanding jets at different inlet subcooling, at the orifice outlet.

Similar observations on the jet shape in dependence of the degree of inlet subcooling have been performed at CNEN also for critical flows through channels [8]. Some of these tests have, for instance, outlined that the opening of the jet increases with subcooling up to almost 50 °C (with respect to the saturated critical flow) to decrease again at higher subcoolings, narrowing more than the saturated case (fig. 15).

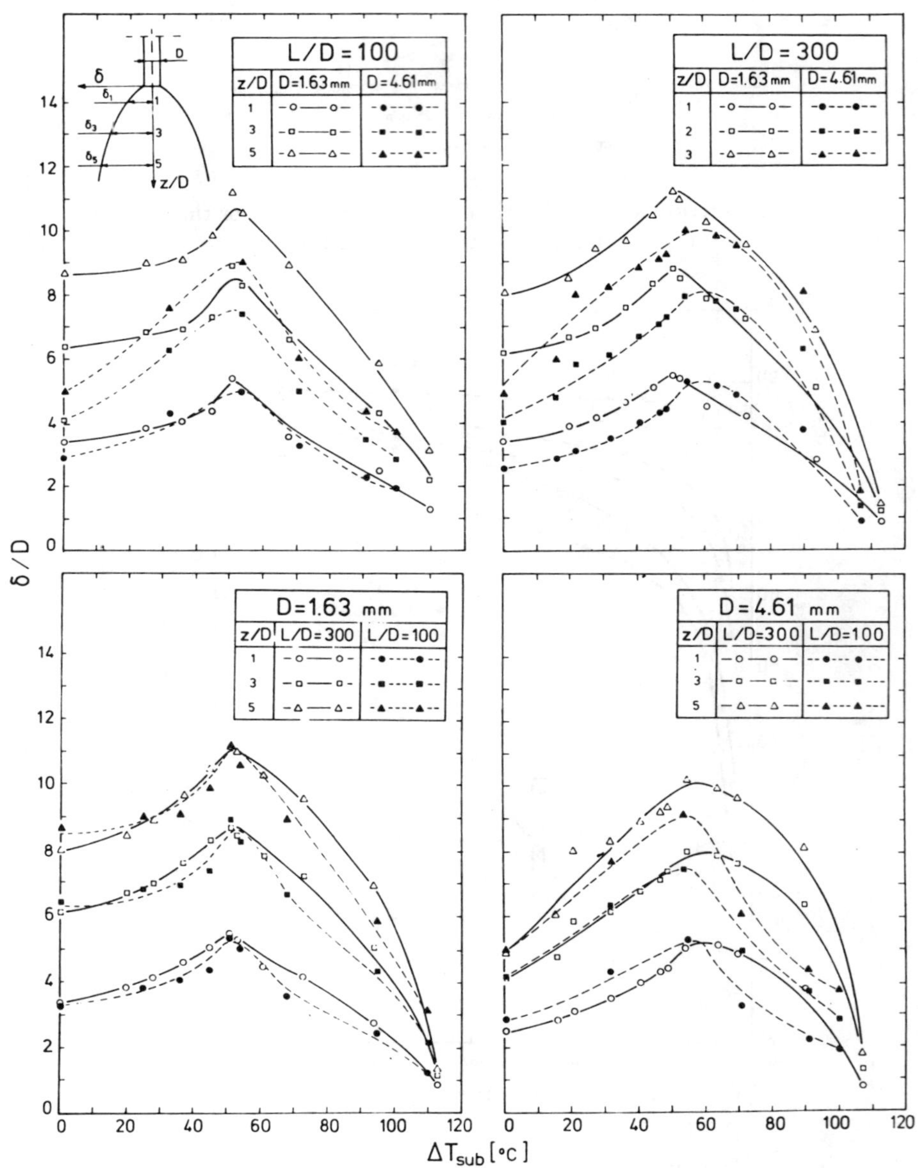

Fig. 15 – Jet external width versus inlet subcooling.

Extensive measurements have been performed on the axial and radial pressure and temperature profiles within the jet for different stagnation conditions in the pressure vessel (saturated and subcooled water, saturated steam) and for different discharging channels' geometries. For all these tests the external shape of the jet has been photographed, to deduce general shape-geometry correlations. For subcooled critical flows the radial pressure profiles show the existence of central peaks which reveal the presence of liquid cores flashing superficially all around. These liquid cores, of reduced diameter, progressively evaporate at increasing distances from the outlet of the channel (figs. 16 and 17). The flashing doesn't behave as a bulk phenomenon, but as a surface phenomenon.

The few examples show the complexity of the two-phase flow within the jets. Analitical models for pressures and temperatures distributions need, before all, a complete understanding of the phase distribution. Further investigations, in progress at CNEN with the employ of X-rays devices, will provide hopefully these informations for different fluid conditions.

To limit the extent of the extrapolations of theoretical models, derived from single phase models, a lot of experimental informations are simultaneously required. Among them, the critical mass flowrate, the critical pressure in the outlet section of the discharge channel, the phase distribution within the jets at different distances from the outlet section, pressure and temperature radial and axial profiles within the jets, jets' shape. As it is well known, two-phase flows exhibit different flow regimes and have a complicated behaviour even within constant section channels. When these two-phase flows are flashing and unbounded, without external constraints (the interaction with the ambient air is generally negligible) a deeper insight is necessary for a preliminary understanding of the flow regimes which must be assumed in theoretical models. The formulation of viable two-phase flow models seems the necessary, preliminary step for the final assessment or reliable and general computation codes.

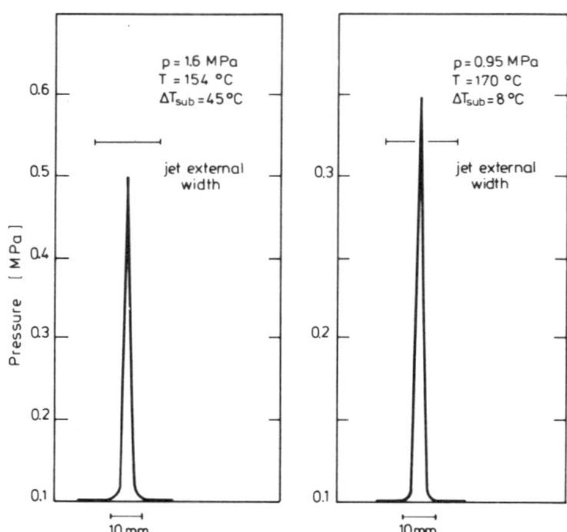

Fig. 16 – Radial pressure profiles within the jet showing the existence of a liquid core.

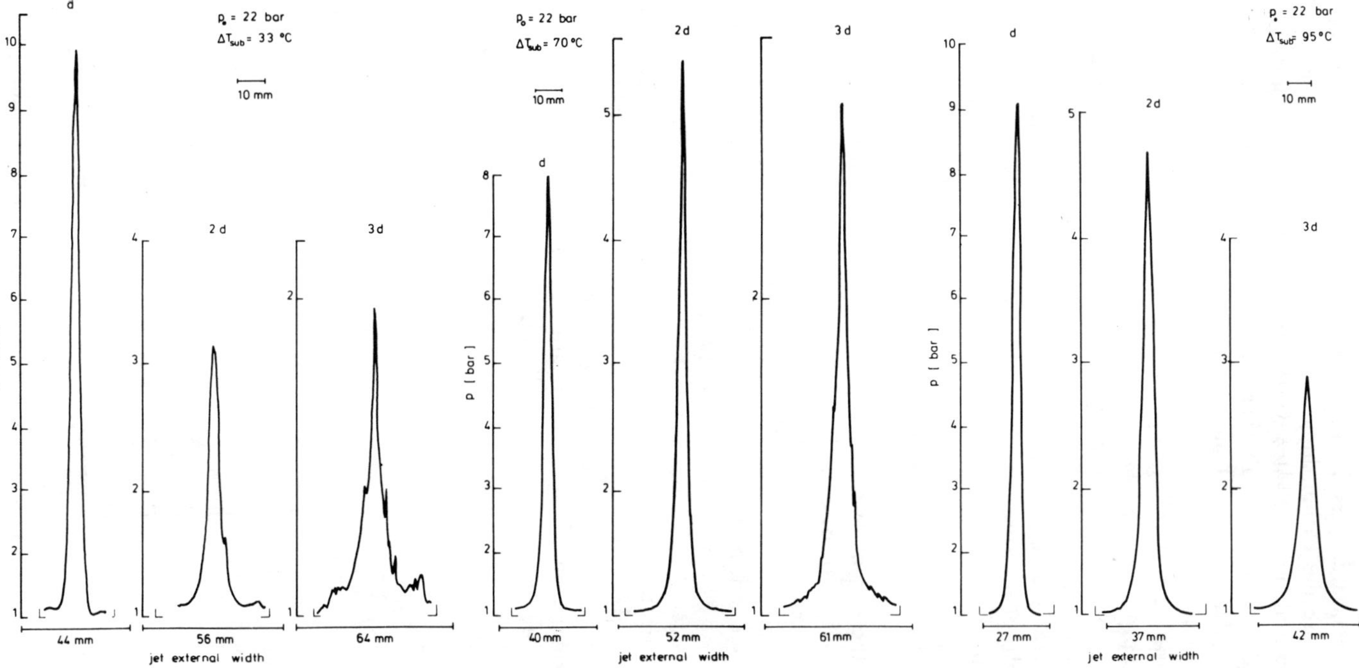

Fig. 17 – Pressure distribution within the jet at different distances from the outlet section and different inlet subcooling.

NOMENCLATURE

A	area
c_p	specific heat at constant pressure
c_v	specific heat at constant volume
c_s	velocity of sound
D	diameter (with special reference to the nozzle or to the discharge channel)
E	internal energy
F	force
$K_{1,2}$	coefficients
P_a	ambient pressure
G	critical mass flowrate
L	generic distance or lenght of the discharge channel
M	Mach number (v/C_s)
M	momentum force
r	radial
R	reaction force
S	surface, area
S	slip ratio
t	time
u	velocity
v	velocity
V	specific volume
X	quality
x	coordinate
y	coordinate
z	axial coordinate
α	void fraction
γ	c_p/c_v
ζ	coefficient
τ	time
ρ	density

INDEXES

B	target
c	control, critical condition
e	exit, outlet
g	vapour phase
i, j	related to two different surfaces in inlet
j	jet
l	liquid phase
o	related to stagnation conditions or initial conditions
out	exit, outlet
R	break
T	Thrust.

REFERENCES

1. Lahey, R., Moody, F.J. "BWR Thermohydraulics" (Ed. A.N.S.)

2. "Workshop on jet impingement and pipe whip" sponsored by CNEN and AMN, Genova, June (29-July 1, 1981).

3. Sandervag, O. "The Marviken full scale jet impingement tests". Workshop on jet impingement and pipe whip, Genova June 29 - July 1, 1981).

4. D'Auria, F., Vigni, P. "Fluid dynamic anlysis of steam water flow from a pressure vessel". Workshop on jet impingement and pipe whip, Genova (June, 29 - July 1, 1981).

5. Farello G.C. "Reduced scale investigation on critical flow and jet forces". Workshop on jet impingement and pipe whip, Genova (june 29 - July 1, 1981)

6. Tomasko, D., Weigand, G.G. and Thompson, S.L. "A model for two-phase jet loading". Workshop on jet impingement and pipe whip, Genova (june 29 - July 1, 1981)

7. Celata, G.P., Cumo M., Farello G.E., and Incalcaterra, P.C. "On the critical flow of subcooled liquids". 7th International Heat Transfer Conference, September 6 - 10, Munich (1982)

8. Celata, G.P., Cumo M. Farello G.E. and Incalcaterra, P.C. "Critical flows and jet forces". European Two-Phase Flow Group Meeting, June 2 - 4, Paris (1982).

9. Celata G.P., Cumo M., Farello G.E., Incalcaterra, P.C. and Naviglio A. "Critical flow of saturated steam". XXXVI ATI Conference, Viareggio, October 5 - 9 (1981).

A REVIEW OF TWO-PHASE FLOW INSTABILITIES

S. Kakaç and T. N. Veziroglu

University of Miami
Coral Gables, Florida, U.S.A.

ABSTRACT

The main objective of this paper is to sum up the experimental and theoretical work carried out by the authors and their groups over a period of several years, demonstrating and explaining three different types of two-phase flow oscillations, namely, density-wave type, pressure-drop type oscillations and thermal oscillations, encountered in various boiling flow channel systems.

Classification of two-phase flow instabilities is summarized first. Three distinct types of two-phase flow oscillations and their mechanisms are explained. Since 1950's with the beginning of commercialization of nuclear reactors, the interest in two-phase flow instability studies started to grow in the western countries followed by the Soviet Union and China. Hence, in the literature review, the most available work to our knowledge is cited. The results of our experimental work on transient and sustained instabilities in single, double channel, cross-connected double channel, four parallel channel and four cross-connected parallel channel upflow systems are presented. The effects of heat flux variations, inlet subcooling, flow rate, inlet and exit restrictions are indicated. The effect of heat transfer augmentation on two-phase flow instabilities in a vertical single channel is also included. Numerical models which are based on the assumption of homogeneous two-phase flow and thermodynamic equilibrium of the two phases to predict both the steady state and transient behavior of forced convection boiling upflow two-phase flow in a single channel are summarized and some of the results of these solutions are presented. Two different two-phase flow models, namely a constant-property homogeneous flow model and a variable property-flux model which are used for the prediction of the pressure-drop and density-wave instability thresholds in a single boiling channel upflow system are mentioned and some of the solution results are presented.

1. INTRODUCTION

In recent years, considerable interest has been expressed concerning the phenomena of flow instability in two-phase systems. The phenomenon of thermally induced two-phase flow instability is of interest for design and operation of many industrial systems and equipment, such as steam generators, thermosiphon reboilers, refrigeration plants, and other chemical process units. Oscillations of flow rate and system pressure are undesirable as they can cause mechanical vibrations, problems of system control, and in extreme circumstances, so disturb the heat transfer characteristics that a heat transfer surface may burn out. In a recirculating plant, where burnout is to be avoided, flow oscillations can lead to transient burnout. Under certain circumstances, large flow oscillations can lead to tube failure due solely to the increase in wall temperature, but a more likely cause of failure would be thermal fatigue with a continual cycling of the wall temperature; the thermal stresses set up in the wall and the cladding material in the case of nuclear reactor, fuel element can be broken down by thermal fatigue. Furthermore, the movement of the burnout boundary, due to flow oscillations, may also lead to thermal fatigue of the tube wall, especially in "once-through" boilers.

It is clear from these problems that flow instabilities are to be avoided, and every effort needs to be made to ensure that any plant has an adequate safety margin against them.

Therefore, the understanding of two-phase flow instability is extremely important to the design, control and performance prediction of any system having such phenomena. It is necessary to establish the influence of pressure, flow rate, temperature, quality, inlet and exit restrictions, the effect of property variations and heat input on two-phase flow oscillations. However, there are other operational and design parameters as well, which influence the thresholds of two-phase oscillations, such as the variation of thermal flux during the oscillations, variation of friction coefficient during transients and with flow regime.

The understanding of most instability mechanisms is fairly complete now, but the analytical modeling of these phenomena remains limited.

1.1 Types of Instabilities

Most steady boiling flow processes involve small-scale

instabilities: bubbles form and collapse; a given point in the flow may be occupied by the liquid phase in one instant of time, and by vapor in the next instant, and then by liquid again; these are called microscopic instabilities that occur locally at the liquid-gas interface. Microscopic instabilities can be treated by methods of classical fluid mechanics. On the other hand macroscopic instabilities involve the entire two-phase flow system. Microscopic oscillations do not lead to sustained oscillations of considerable amplitude by themselves. But microscopic oscillations may trigger the actual system oscillations if there are some other macroscopic instability mechanisms at work. If the small-scale local instabilities do not cause any sustained oscillations, such flows are called as stable flows. In other words, a flow is stable if when momentarily disturbed, its new operating conditions tend asymptotically towards the initial operating point. Sometimes another steady-state point may not be possible in the vicinity of the original point. Perturbation may grow and a new operating point is obtained. These instabilities are called static instabilities, since the cause of the phenomenon lies in the steady-state laws. The threshold for static instabilities can only be predicted from steady-state considerations. Known types of static instabilities include flow excursions (Ledinegg instability), flow pattern transitions, nucleation instabilities, bumping, chugging and geysering.

A flow is said to be subject to a dynamic instability when the inertia and feedback effects have an essential part in the process. The most common dynamic instabilities are characterized as (1) density-wave type oscillations, (2) pressure-drop type oscillations, and (3) acoustic oscillations.

The mechanisms for the dynamic instabilities can be explained by propagation time lags and feedback phenomena present in any two-phase flow system. The momentary disturbance takes some time, which is proportional to the propagation wave speed, in reaching other points along the system; e.g. a disturbance in the inlet fluid velocity is felt at the system exit only after a certain time interval has passed. These delayed disturbances along the system are then reflected back to the initial point of disturbance creating a new disturbance, and so on. When certain conditions are satisfied, this process can go on indefinitely by feeding on itself, thus giving rise to sustained flow oscillations. The period of the oscillations is related to the time required by a perturbation wave to travel along the system.

Pressure perturbations travel at the sonic velocity of the mixture, therefore the oscillations associated with time-delay and feedback effects of pressure wave propagation are necessarily of very high frequencies (10-100 Hz) and in most cases they are experienced along with a shrill sound; that is why they are called acoustic oscillations. In the experimental studies that the authors carried out jointly at the University of Miami and at the Middle East Technical University, acoustic oscillations have not been observed. They often seem to be triggered by or coupled with heat transfer mechanisms such as subcooled boiling (collapse of bubbles) and film boiling (variation of the film thickness and vapor generation rate owing to changes in pressure). In literature, the acoustic oscillations are reported to have been observed at various systems with various fluids which will be mentioned in the following section. Acoustic oscillations do not constitute an important category of the two-phase flow instabilities. They are more important in combustion stability studies for rocket engines.

In our experimental studies we observed three identifiable types of dynamic oscillations, namely density-wave type, pressure-drop type, and temperature oscillations, which may be encountered in a boiling flow, each one being associated with a specified mode of operation, depending on mass flow rate, variation of heat input, and system pressure.

Density-wave oscillations. The density-wave oscillations are related to kinematic wave propagation phenomena. In two-phase flow systems fluid waves of alternatively higher and lower density mixtures travel across the system. The compressibility of the vapor is not an important factor in generating density-wave oscillations. The origin and mechanism of the density-wave oscillations can be most clearly understood by considering an evaporator and duct (Fig. 1) followed by an exit restriction at the end. The pressure at the inlet and exit reservoirs is kept constant so that the total pressure drop across the system is constant at all times. We assume that the rate of vapor generation in the evaporator (heater) is constant. Now, suppose that at time, t = o, the exit restriction pressure drop undergoes a sudden infinitesimal drop from its

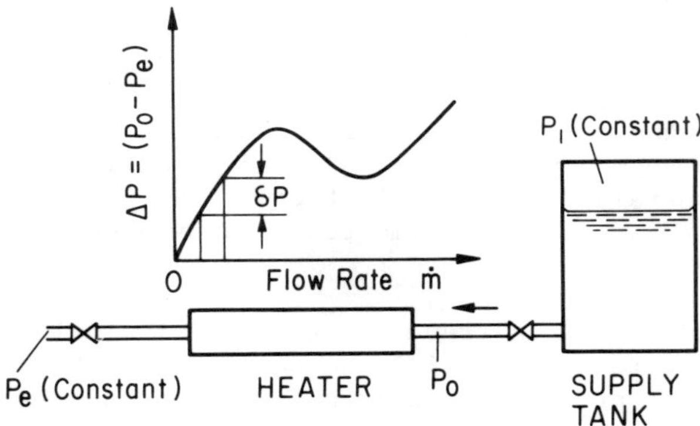

Fig. 1 Simplified system for density-wave type oscillations.

steady-state value. Such a momentary disturbance may result in two-phase flows due to the effect of the microscopic instabilities, i.e., a low density two-phase mixture may pass through the exit restriction. The pressure drop through the exit restriction will decrease and this decrease in the exit pressure drop ($p_o - P_e$) is propagated to the inlet almost instantaneously at the speed of sound and P_o will decrease and ($p_1 - p_o$) will increase since P_e = constant. As a result, inlet velocity is increased infinitesimally.

Therefore the enthalpy wave with a higher density fluid originated at t = o at the inlet reaches the boiling boundary at time t_1 and a wave of higher-density mixture is set to travel along the heater. At time t = t_2 the density wave reaches from the boiling boundary to the heater exit. At time t = t_3, the density wave reaches from the heater exit to the exit restriction. The sum of these three time intervals, $\Delta t = t_1 + t_2 + t_3$, constitutes the transit time of a fluid particle through the system. When this higher density wave reaches the exit restriction after Δt time of its origination the exit pressure drop will increase infinitesimally (see Fig. 1) and as a result ($p_1 - p_o$), which is the driving force, will decrease. Therefore inlet velocity will be decreased infinitesimally and a lower density fluid is set to travel along the heater. But it takes another time interval of Δt, for this adjustment to be felt at the exit restriction. When the new density wave (lower density) arrives at the exit restriction, the pressure drop is given this time a negative perturbation and practically at the same time the inlet velocity increases accordingly. It takes one high and one low wave to make one cycle so it can be concluded that the periods of the density-wave oscillations are roughly equal to twice the transit time of a fluid particle through the system. It is assumed that the pressure drop is concentrated at the exit restriction. In reality, the pressure drop is distributed in the system, then in actual cases, the periods may not be equal to twice the transit time of a fluid particle in the system. It is of particular interest to observe that compressibility of the vapor is not an important factor in generating density-wave oscillations. The essential feature is that two components with different densities are present, and may be mixed in different proportions.

In the above discussion it is assumed that the pressure waves are assumed to propagate at infinite speeds through the system for purposes of analyzing the density-wave oscillations, since the pressure waves travel at sonic speeds which are incomparably larger than the mixture velocities encountered in two-phase flow systems.

Pressure-drop type oscillations. It is found that the pressure-drop type oscillations occur only when the pressure drop across the test section decreases with increasing flow rate and their periods were governed by the volume and the compressibility of the vapor in the system, including the compressibility introduced at the upstream

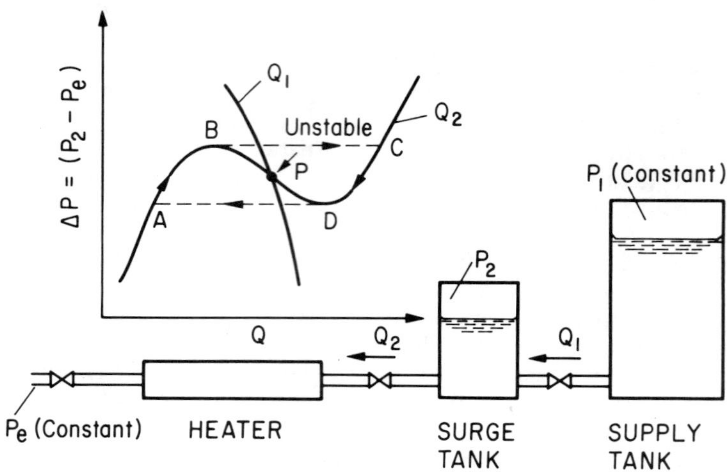

Fig. 2 Schematic of flow diagram and limit cycle of pressure-drop type oscillations.

of the heater through a surge tank which is partially filled with air.

The mechanisms of the pressure drop oscillations can be understood from the steady-state pressure drop relations (Figure 2).

$$(p_1 - p_2)_s = k_1 Q_1^2 \tag{1}$$

$$(p_2 - p_e)_s = f(Q_2) \tag{2}$$

where p_1 and p_e are constant pressures at the supply tank and at the exit, respectively; p_2 is the surge tank pressure, Q_1 and Q_2 are the surge tank inlet and exit flow rates, K_1 is a constant representing the frictional pressure drop between the supply tank and the surge tank, f is a function relating the system pressure drop to the heater inlet flow rate Q_2. For any steady-state operating point, the flow rate, $Q = Q_1 = Q_2$ must satisfy both of these equations.

With p_1 and p_e constant, two curves for p_2 versus Q can be drawn; one curve represents p_2 as a function of Q_1 from Eq. (1) (flow valve characteristics), the other represents p_2 as a function of Q_2 from Eq. (2) (system steady-state characteristics). Any steady-state operating point is an intersection of these two curves (Fig. 2). If two curves intersect (p) in a region where $d(p_2 - p_e)_s/dQ_o<0$, a small increase in p_2 will cause Q_2 to decrease more than Q_1. As Q_1 exceeds Q_2, the liquid level in the surge tank will rise and p_2 will increase further. The system is then statistically unstable and the liquid accumulates in the surge tank increasing the surge tank pressure, the operating point will move along the boiling curve from point P to point B. There is no stable intercept of the two curves. The process cannot stop there because of the imbalance between the surge tank inlet and exit flow rates and a flow excursion occurs taking the operating point to C. At C, Q_2 is greater than Q_1. The increased flow rate through the heater now works in evacuating the surge tank. Due to the decompression of the air-vapor mixture in the surge tank, the pressure, p_2, decreases and the operating point moves along the curve from C to D, another flow excursion takes the operating point to A and the process repeats itself along the limit cycle ABCDA.

This model explains the general phenomenon of pressure-drop oscillations and gives a good estimate of the period of the oscillations. It also explains the appearance of the density-wave oscillations on the rising pressure portion of the pressure-drop oscillations. Since the rising pressure portion corresponds to operation on the branch AB, where density-wave oscillations are observed, one would expect to see the density-wave oscillations in this portion of the cycle.

The compressible volume essential for the occurrence of the pressure-drop type oscillations can be introduced through a surge tank in the upstream of the heater. However, the internal compressibility in the long test sections can be sufficiently large for purposes of generating pressure-drop oscillations.

The periods of the pressure-drop oscillations are determined by the time constant associated with the amount of compressible volume but not by the heater residence time of a fluid particle as was the case for the density-wave oscillations. The periods of these oscillations are generally larger than those experienced in density-wave oscillations. However, as the amount of compressibility is reduced, the frequencies can be comparable with those of the density-wave oscillations which may make it difficult to distinguish them from each other in such cases.

Acoustic oscillations, density-wave type oscillations and pressure-drop type oscillations are the most common type of the dynamic two-phase flow instabilities. However, there are many other kinds of instabilities associated with two-phase flow systems. One of them is thermal oscillations. These oscillations appear to be associated with the instability of the liquid film, and are accompanied by large fluctuations in wall temperatures in a constant heat flux system. The flow oscillates between transition boiling and film boiling at a given point, the wall superheat fluctuates accordingly to accommodate the constant heat generation in the wall as the heat transfer coefficient of the two-phase flow regime oscillates between the wet and dry wall conditions.

2. LITERATURE REVIEW

Since 1950's with the beginning of commercialization of nuclear reactors, the interest in two-phase flow and heat transfer studies started to grow in the Western countries, followed by the Soviet Union and China. Hence, in the literature survey on two-phase flow instabilities the "Western countries" and "Soviet Union and China" will be covered separately for convenience.

2.1 Two-Phase Flow Instability Research in the West

The earliest research in the field of two-phase flow is ascribed to a German engineer, Lorentz, who investigated the hydrodynamics of two-phase flows in pipes in 1909 by treating the gas-liquid mixture as a single homogeneous fluid [1]. Studies on the unstable two-phase flows started much later. The first successful analysis is by Ledinegg in 1938 [2]. Ledinegg instability is a classical example of static two-phase flow instabilities. While studying the conditions leading to flow instabilities in heated parallel boiler tubes, Ledinegg observed that, under certain conditions, the steady-state curve of the system pressure drop versus flow rate had a negative slope. Under the circumstances, a flow excursion to another steady-state operating point may occur because the flow rate is not a single-valued function of the pressure drop. This criterion has been employed by a number of other investigators [3-6] with variations in methods for determining the slope.

In one of the earlier studies, Wissler, Isbin and Amundsen [7] investigated the oscillatory behavior of a natural circulation loop. They solved the homogeneous model equations by an analog computer. Qualitatively, they obtained reasonable results; however, the quantitative differences between the theoretical and the experimental cycles were too large due to the simplicity of their assumptions regarding the density distribution along the channel and the frictional pressure drop.

Using a Lagrangian description of flow, Wallis and Heasley [8] integrated the energy and the continuity equations for a channel subjected to a cyclic variation of the inlet flow rate.

Quandt [9] published data on the stability of a high-pressure (up to 1600 psia) boiling water channel operating in parallel with a large by-pass to produce constant pressure drop, approximating experimentally a boiler with many parallel channels. His results show that for a given heat flux and flow rate, flow oscillations occur over a certain range of inlet subcooling. Bouré [10] reported results of a number of European studies displaying the same behavior.

Meyer and Rose [11] solved the conservation equations in the time domain with a digital computer using a finite difference scheme for a heated boiling channel. They tried to predict the oscillatory tendencies of the channel by observing the behavior of the solution in the time domain. This study was one of the first in a string of computer codes attempting to solve the two-phase flow equations directly.

Gouse and Andrysiak [12] investigated the flow oscillations in three transparent parallel vertical heated channels. They verified that the essential mechanism for the oscillations is not due to flow patterns but due to boiling, after observing that flow oscillations would occur for various flow regimes when there was boiling but would not occur for similar flow regimes of air-water mixtures. They studied a small Freon-113 system operating at a pressure close to the atmospheric and showed that for a given heat flux and flow rate, the flow oscillations would occur over a range of inlet subcoolings except at very high or very low values.

During their experiments with Freon-11, Stenning and Veziroglu [13] identified three distinct two-phase flow instability modes. They classified them as density-wave, pressure-drop and thermal oscillations. Zuber [14] referred to the density-wave oscillations as thermally induced oscillations.

Zuber [6] formulated the flow instability problem in terms of the drift-flux model in order to take into account the effect of slip between the phases. He used a small perturbation analysis to obtain a characteristic equation with two time delays.

Villeneuve [15] reported the two-phase flow instability research in Grenoble during the early sixties, proceeding in three directions. First, an attempt was made to explain the oscillations by the effects of the finite propagation velocities of the disturbances through the two-phase mixtures, the wave propagation velocity in the homogeneous mixture being assumed to be equal to the fluid velocity. The second one, a "density-effect" model, assumed the fluid to be of one phase but with the specific properties varying linearly with the mixture enthalpy. A third model was also reported to be under development based on the assumption that the instabilities were generated by a variation of friction in connection with a change in the flow patterns.

The "density-effect" model was used by Bouré [16] and Bouré and Mikaila [17] to explain the oscillatory behavior of heated channels. Under conditions of thermodynamic equilibrium, Bouré assumed the fluid density to be a function of only the enthalpy.

Various reviews on two-phase flow instabilities have been published, the most notable ones being done by Bouré et al. [18],

Ishii [19], Bergles [20] and Yadigaroglu [21]. Acoustic oscillations in single-channel boiling systems have been analyzed by Bergles et al. [22].

The most common instability mechanism in two-phase flows is associated with kinematic wave propagation effects. Such oscillations related to kinematic wave propagation phenomena are called density-wave oscillations. The well-accepted terminology, density-wave oscillations, is from Stenning and Veziroglu [13] illustrating the fact that fluid waves of alternatively higher and lower density mixtures travel across the system.

The second major kind of dynamic instability is named as pressure-drop oscillations following the terminology set by Stenning and Veziroglu [13]. They found that the pressure-drop oscillations occurred only when the pressure drop across the test section decreased with increasing flow rate, and their periods were governed by the volume and the compressibility of the vapor in the system.

The compressible volume essential for the occurrence of the pressure-drop oscillations was introduced through a surge tank in the upstream of the heater by Stenning and Veziroglu [13] and in various studies afterwards [23-34]. However, as confirmed by Maulbetsch and Griffith [35], the internal compressibility in the long test sections can be sufficiently large for purposes of generating pressure-drop oscillations.

Veziroglu [24] and Veziroglu and Lee [25] investigated pressure-drop and density-wave instabilities in a single vertical boiling channel using Freon-11. Both modes of oscillations were compared with those encountered in horizontal flows. They observed that the vertical upward flows were more stable than the horizontal flows. Later, Veziroglu and Lee [26] performed experiments with distilled water. They found that, within the range of those experiments, the pressure-drop and the density-wave oscillations always occurred together.

Veziroglu and Lee [26, 27] also studied the instabilities in two parallel vertical channels using Freon-11 as the test fluid. They observed that the density-wave oscillations occurred simultaneously in both of the parallel channels, whereas the pressure-drop oscillations could occur in either one or both of the channels. The density-wave oscillations in the two parallel channels were out of phase by 180°, while the pressure-drop oscillations, whenever they occurred simultaneously, were in phase. Later they studied the effect of cross-connections between the channels on the instability [28]. In contrast to the case without cross-connections, the density-wave oscillations in cross-connected channels were in phase.

Akagawa et al. [36] and Hayama [37] studied the distribution of flow rates and flow instabilities in long parallel tubes of steam generators.

Later Veziroglu and Lee [29] investigated sustained instabilities in cross-connected channels by using distilled water as the working fluid. They also performed some transient instability experiments with Freon-11. The response of a cross-connected parallel channel system to sudden changes in the flow rate and the heat input was investigated. Kakac, Veziroglu, Akyuzlu and Berkol [30] studied the sustained and the transient boiling flow instabilities in a cross-connected four-parallel channel upflow system.

The pressure-drop type oscillations are compound dynamic instabilities generated by the operation of static excursive effects in cooperation with the effects of compressible volume inertia and system delayed feedbacks. Consequently, pressure-drop type oscillations might be expected to exist even in certain single-phase systems where there are similar effects in operation. Indeed, a 1956 study by Marris [38] deals with the experimental and analytical investigation of the hydraulic oscillations occurring in a simple surge tank operating under the condition of constant hydraulic power input to the turbine.

There are other types of oscillations associated with two-phase flow systems. One of them is thermal oscillations reported by Stenning and Veziroglu [13]. As identified by them, these oscillations appear to be associated with the instability of the liquid film, and are accompanied by large fluctuations in wall temperatures in a constant heat flux system. Two-phase flow instabilities may occur due to thermodynamic nonequilibrium effects connected with subcooled boiling as reported by Jeglic and Grace [39]. Berenson [40] and Schuster and Berenson [41] studied the flow oscillations in a system of five parallel channels. D'Arcy [42] investigated the boiling flow instabilities in three parallel internally heated annular channels. Davies and Potter [43] performed a linearized perturbation analysis of a boiling channel under constant pressure drop condition as in parallel channels.

Friedly et al. [44] investigated the linear stability of a supercritical fluid flowing through a horizontal channel. A flow rate-to-pressure drop transfer function was obtained for the linearized system and the stability of the characteristic equation was tested by means of Nyquist plots. Density-wave oscillations can also appear in fluids operating in the near-critical and the supercritical thermodynamic region. These were investigated by Zuber [45].

Carver et al. [46] studied the density-wave instabilities and their effect on dryout conditions in electrically heated 19-element bundle of a forced-circulation water loop. They found that the dryout patterns were affected significantly by the occurrence of the oscillations.

Crowley et al. [47] studied the density-wave type oscillations in three parallel vertical transparent heated channels with a constant pressure drop maintained across the ends, using Freon-113 as the test fluid. They mapped the boundaries between the stable and the unstable regions and also recorded both the periods and the amplitudes of the oscillations.

Crowley and Bergles [48] studied the problem of the fluid-to-fluid modeling of density-wave instabilities, particularly the case where a low-pressure fluorochemical flow system is used to model a high-pressure water system. Application of the scaling parameters obtained from the analytical formulations led them to conclude that good similarity can be achieved by using full-size models.

Collins et al. [49] investigated the parallel-channel instability of a full-scale simulated nuclear reactor channel operating in vertical high-pressure boiling water upflow system, connected in parallel with an unheated by-pass between common inlet and outlet headers.

Matsui [50, 51] investigated the effects of the heat input, inlet subcooling, drag coefficient, inlet velocity on boiling flow instabilities. The periods of the oscillations increased when the heat input or the drag coefficient was decreased or when the inlet subcooling was increased. Yadigaroglu and Bergles [52] investigated the effects of the wall heat storage and the pressure dependence of the saturation enthalpy. They observed that the phase of the oscillation of the boiling boundary depends in most cases uniquely on the phase of the enthalpy perturbation at the steady-state location of the boiling boundary. Friedly [53] studied the stability behavior of a once-through liquid nitrogen evaporator under conditions when the system pressure drop was dominated by the pressure drops across the inlet and the exit valves. He observed that the amplitude of the oscillations consistently increased as the ratio of the exit pressure drop to the inlet pressure drop was increased. Ishii and Zuber [54] developed a linearized stability analysis using the drift-flux model of Zuber and Findlay [55]. The two-phase friction factor and the vapor drift velocity were assumed to be constant, the wall heat storage was neglected and thermal equilibrium conditions were assumed. Saha et al. [56] discussed the incorporation of subcooled boiling into the drift-flux model of Zuber and Findlay [55]. They reported that, at low subcoolings, a non-equilibrium model including the effects of subcooled boiling gives better results. At higher subcoolings, the experimental system is more stable than the predictions of both the equilibrium and the non-equilibrium models.

Density-wave oscillations can be manifested in all kinds of two-phase flow systems including BWRs [57] and in reboilers and other chemical process equipment [58].

Fukuda and Kobori [59] investigated the stability of parallel channels in forced and natural circulation, using water. They encountered pressure-drop oscillations at very low steam qualities and density-wave oscillations at higher qualities. Fukuda and Hasegawa [60] developed a method for the evaluation of the oscillation modes and instability criteria in parallel multichannels. Takitani [61, 62] studied the boiling flow instabilities in a vertical channel using Freon-11. He analyzed the density-wave oscillations by using a simplified homogeneous model which treated the heater in three distinct regions, the flow rate being uniform through each region. Nakanishi et al. [63] analyzed the density-wave oscillations in a boiling channel by using a linearized drift-flux model. Ozawa et al. [64] studied the instabilities in an air-water two-phase flow system with upstream compressible volume. The flow patterns during oscillations and the effects of the compressible volume and the average superficial velocity on the limit cycles were investigated.

Ünal [65] investigated the inception of the density-wave oscillations in sodium-heated once-through steam generator tubes and correlated the experimental threshold data. A study by Romberg and Rees [66] uses a linearized drift-flux formulation with methods of multivariate spectral analysis to assess the hydrodynamic behavior of boiling channels near the instability threshold. Romberg and Rees [67] also compared the linearized instability results of a distributed homogeneous model, a lumped-parameter model and an empirical model.

In addition to the boiling flow instabilities, the instabilities associated with condensation flow processes are an important area of the general two-phase flow instability field.

In a relatively recent review, Bankoff [68] surveyed the research on condensation rates involving steam in contact with fairly thick layers of cold water. Current understanding of the problem is apparently limited and there is need for more research. Bhatt and Wedekind [69] generated a simple linearized model for the prediction of the oscillation thresholds in condensing flows through a horizontal tube with inlet and outlet throttling under conditions of complete condensation. The "steam mean void fraction model" originally introduced by Wedekind et al. for both evaporating and condensing flows [70] is apparently less useful for the evaporating flows due to the effect of the time-varying quality condition at the system exit.

Pitts [71] conducted a quasi-steady-state analysis of steam chugging phenomena in a vertical condensation pipe connecting a drywell reservoir (subjected to a time-decaying steam supply simulating LOCA, initially filled with air) to a half-filled reservoir (wetwell). Reasonable agreement with the results of the two large-scale experiments was reported.

During their experiments on the subcooled decompression of water in a vertical tube under imposed temperature gradients, Takeda and Toda [72] observed that violent oscillations on the pressure occurred only in systems with large temperature gradients and only in low-temperature regions of these systems.

Henry et al. [73] showed the flow regime dependence of the propagation velocity, concluding that the homogeneous solutions for the wave propagation velocity might be in considerable error depending on the flow regime. Later, Henry [74] developed an empirical relationship giving the ratio of the actual wave velocity to the homogeneous isothermal solution. Mori et al. [75] studied the pressure waves of finite amplitude propagating in a two-component (air-water or nitrogen-water) two-phase medium or in liquid in an elastic tube.

Liquid-solid and gas-solid two-phase flows have been drawing increasing attention in recent years due to growing importance of applications such as pipeline transportation of solids. Zenz [76] examined the minimum air velocity (saltation velocity) required to transport the solids without saltation. Particulate slug formations are observed when the air velocity is below this critical value. Canada et al. [77] investigated the void fraction distribution in gas-fluidized beds of different sizes and observed different flow regime limits. The period of the pressure-drop oscillation characteristic of the bed was related to the bubble or slug residence time in the bed. Tomita et al. [78] investigated the wavelike motion of particulate slugs in a horizontal pneumatic pipeline.

Cumo et al. [79] made an experimental investigation of the density-wave type oscillations with different heat flux profiles in parallel channels. Nakanishi [80] presented some results of experimental work on the modes of flow oscillation in multi-channel two-phase flow systems which were conducted on the test loops with Freon-113 and with air-water mixtures. Akagawa [81] presented recent research work on two-phase flow in universities and colleges in Japan. He tabulated Japanese research work on flow instabilities. The state-of-the-art of Japanese research on two-phase instabilities has also been presented by Nakanishi [82]. The extensive activities on two-phase flow dynamics in both countries, US and Japan, are presented in two-phase flow dynamics. The volume represents both reviews and original contributions [83].

2.2 Two-Phase Flow Instability Research in the Soviet Union and China

The investigation of two-phase flow instabilities has been conducted in the Soviet Union since 1940. At that time in Once Through Boiler Bureau, an experimental installation consisting of a heating coil, which was heated by an electric current passing through external coils, was installed. Test pressures ranged from 10 to 20 kg/cm^2. Later in the same laboratory, a large steam-heated pulsation circuit which had five parallel coils that were connected by means of inlet and outlet headers, was installed. Test pressures ranged up to 90 kg/cm^2. On these installations, Davidov [84, 85] and Davidov and Shenin [86] conducted a series of experiments and theoretical studies. Davidov found that the three following determining criteria are of great importance about two-phase flow oscillation processes: (a) the Reynolds number referred to the parameters of boiling water; (b) the pressure in the loop; (c) the length of the economizer, referred to the total length of the heated loop. Semenovkel [87] conducted a series of experimental work using four parallel coils, forced-feed circulation, with a test pressure up to 90 kg/cm^2. He found the factors which enhanced the occurrence of oscillations in the region of small steam qualities were the following: A drop in water velocity in the tube, an increase in heating intensity, and an increase in tube diameter.

Armand [88] identified bubbly, slug and annular flow regimes for air-water flow through a vertical tube. He associated pressure "pulses" and instances of flow reversals with the slug flow regime. Petrov [89] studied both the instabilities in parallel channel upward flow systems and horizontal flow systems. Morozov [90] by using a simplified model studied the boiling flow in a coil of a once-through boiler. Klaziakova and Gloskel [91] investigated the instability of boiling flows in vertical tubes. Test pressures ranged from 140 to 230 kg/cm^2.

Treshev [92] conducted experimental studies of steam-water flow oscillations in heated tubes. The study showed that if the flow resistance before boiling point was large enough, entering fluid temperature was saturation temperature or below it, and mass flow rate was high, then the oscillations did not occur or were small. Celov, Smilnov and Zerkov [93] carried out an experimental study on the flow stability boundaries of steam-generating tubes connected in parallel with uneven heat inputs. Dolgov and Sudnitsyn [94] studied experimentally the hydrodynamic instability in boiling water reactors. They found out that the amplitude of oscillations could be reduced by increasing inlet pressure drop, reducing inlet subcooling temperature or raising the pressure.

According to the accumulated experimental data, the Standard of Hydraulic Calculations for Steam Boilers (1961) [95] proposed two parameters for determining oscillations in horizontal tubes: One parameter expresses the relationship between the ratio of pressure drops and mass velocity, the other parameter expresses the relationship between the ratio of pressure drops and the pressure.

Baldina and Kalinin [96] and Baldina et al. [97] experimentally studied the boiling flow instabilities in four parallel horizontal tubes and three snake type tubes. Test pressures ranged from 21 to 235 kg/cm^2. Test medium was water. The stability-instability boundary curves were obtained.

Habenski, Baldina and Betilson [98-101] carried out an analysis of discharge oscillations in a system of parallel boiling tubes. By numerical integration of the governing equations describing unsteady heat transfer and fluid mechanics in a vapor generating channel, a diagram for determining the smallest mass velocity that could prevent the flow from oscillating was obtained. In their results, the minimum mass velocity was a function of the enthalpy at the inlet (subcooling), the flow resistance at the entrance of the tube, the tube diameter, the heat flux, the tube length and the pressure.

Koshelev, Sulnov and Nikitina [102] reported the test results for three parallel heated tubes. The phase shifts in flow oscillations occurred even under supercritical pressure conditions. Leleev [103] studied the oscillation phenomena occurring when a medium passed through the tube of a steam generator under supercritical pressure. Schevalz, Kozmin and Glagolev [104] studied the flow stability experimentally in "U" type tube and inverse "U" type tube elements at supercritical pressure. They also studied the influence of heat input and hydraulic unevenness on flow stabilities.

Nikmatolin [105] analytically studied the hydrodynamic crisis phenomena of a two-phase critical flow with dispersed-annular structure. Drizius et al. [106] studied the oscillations of wall temperature of heated tubes in the subcooled boiling region. They found out that whenever there were pressure oscillations, the inner wall temperature oscillations had the same frequency as that of the pressure oscillations. Drizius et al. investigated [107] the pressure oscillations in heated channels in the subcooled boiling region.

Kalinin [108] studied the oscillation problem in a trans-sonic two-phase flow. Miloslafski [109] studied the vibration problem of large pipes in power plants. He predicted the flow instability in liquids and liquids undergoing phase change in pipelines. Glikman et al. [110, 111] studied the forced oscillations of two-phase two-component flows in single channels when there were small losses at the channel exit. Nesis et al. [112] investigated the physical phenomena that caused high frequency oscillations in tubes in subcooled boiling regions. Using a small moveable hydrophone placed in a tube, relationships between acoustic pressure, its frequency, amplitude and heat flux were studied. Thermo-acoustic oscillations in containers [113-117] and in channels [118-121] have also been studied.

In China, Chen and Chen [122] studied the flow characteristics in boiling flows for both upward and downward flows. Bi [123] experimentally investigated the hydrodynamic instabilities in once-through evaporators. Fan [124] experimentally studied the two-phase flow instabilities in a low-pressure natural circulating boiling loop. Wang and Wang [125] studied the multivalue problem of the hydrodynamic characteristics of a La-Mont Boiler. The experiments and improvements on large once-through boilers have also been reported [126-128].

Lin [129] investigated the vapor quality and flow rate in two-phase flows through sharp edged circular orifices. Chen [130] studied the circulation characteristics and internal processes in large capacity boilers. Wu [131] investigated the heat transfer between horizontal finned tubes and the condensing Freon-12 and Freon-113.

2.3 Mathematical Modeling

On the issue of the proper formulation of two-phase flow equations, there are several proposals in the literature [55 and 132-160] along with a long list of computer codes which must be updated quite often. Various analyses accomplished so far can be collected under two general headings, namely, the homogeneous flow model and the slip-flow model. The homogeneous flow model considers the two-phases flowing as a single phase possessing mean fluid properties. The realization that the density of the homogeneous two-phase mixture can be treated as a function of only the mixture enthalpy but not of both enthalpy and pressure was a breakthrough for the analytical investigation of the two-phase flow instabilities. Serov in 1953 was apparently the first to realize that such an approximation was possible [133]. After about ten years, Serov [137, 138] integrated the momentum equation and obtained the characteristic equation for a distributed parameter system. This analysis takes into account the variation of the inlet flow rate and the heat transfer coefficient but neglects the effects of the boiling boundary displacement.

The slip-flow model considers the phases to be artificially segregated into two streams--one of liquid and one of vapor. In this model's simplest form, the phases are in thermodynamic equilibrium, and each phase is assumed to travel at a mean velocity. In the more general case of unequal phase velocities, an interphase constitutive relationship is needed. This can be in the form of a slip correlation relating the slip ratio, $S = u_v/u_\ell$, to the quality and pressure, a drift velocity, e.g. Zuber and Findlay [55], or a void fraction correlation, e.g. Martinelli and Nelson [161].

The drift-flux model of Zuber and Findlay [55] is relatively simple to use and realistic enough for most purposes. In Wallis' [141] drift-flux formulation, the void fraction and volumetric flux are assumed to be uniformly distributed over the channel cross-section. Flow regime effects are not accounted for in the Wallis' model.

In all of its different formulations, the drift-flux model is an approximation of the more rigorous two-fluid models. It is presumably most valid for cases in which the drift velocity is significant compared with the volumetric flux. This limits its usefulness to the bubbly, slug and churn flow patterns.

In its most general form, the two-fluid model is formulated by considering each phase as a separate stream flowing together in cocurrent or countercurrent flow. Thus the model is expressed in terms of two sets of conservation equations governing the balance of mass, momentum and energy in each phase. In addition to these six field equations, there are seven constitutive relations. Simplifications can be made on the two-fluid models by some kinds of averaging over a cross-sectional area [142] or over a time interval [147].

There are several ways to average and, as Ishii [147] notes, the choice of averaging and instrumentation are closely coupled, since in general measured quantities represent certain mean values. An excellent treatise on the merits and weaknesses of two-fluid models has been provided by Ishii [147].

Gidaspow developed a model [151, 152] using the methods of non-equilibrium thermodynamics and Fick's diffusion law for transient one-dimensional flow.

In a study, Ishii [158] proposed new equations for the drift velocity of dispersed phase in dispersed flow (bubble and droplet) and the vapor drift velocity in annular flow. Ishii reports the steady-state predictions of this correlation to be within ± 30% of the available experimental data for annular two-phase flows.

For steady flows all two-phase flow models produce somewhat different but reasonable results. So the choice of the right model is especially important in analyzing the two-phase flow instabilities.

Yeşin [162] and Yüncü [163] studied two-phase flow instabilities in a horizontal boiling channel by the use of a homogeneous equilibrium model.

Dogan [164] developed a numerical model based on the assumption of homogeneous two-phase flow and thermodynamic equilibrium of the two-phases. The model was used to predict the pressure-drop and density-wave type oscillations and to study the effect of inlet subcooling.

Akyüzlü [165] studied two-phase flow instabilities by solving a set of non-linear hyperbolic partial differential equations by explicit finite-difference technique. The model is based on the assumptions of homogeneous two-phase flow and thermodynamic equilibrium of the phases. The numerical results have been compared with experimental findings.

Lahey [166] presented a state-of-the-art analytical model which can be used to determine the linear stability of boiling channels of boiling water nuclear reactors. The model takes phasic slip, nonuniform heat flux and the wall dynamics into account to analyze the density wave oscillations.

Selected bibliography on density waves and modeling of these oscillations is presented by Yadigaroglu [167].

This brief survey has given an overview of the broad spectrum of research activity in the area of two-phase flow instabilities.

The following is the summary of the experimental and theoretical work accomplished on two-phase flow instabilities by the late Dr. A. H. Stenning, the authors and their group at the University of Miami and at the Middle-East Technical University, over a period of several years, demonstrating and explaining three different types of two-phase flow oscillations encountered in forced convection boiling systems [23-34].

3. EXPERIMENTAL SYSTEM AND EXPERIMENTATION FOR A SINGLE CHANNEL SYSTEM

The single-channel experimental system is shown schematically in Fig. 3. This experimental setup was modified extensively for the purposes of studying two-phase flow oscillations in multi-channel systems.

3.1 General Description

Test fluid Freon-11 is supplied from a main tank pressurized by nitrogen gas, typically at 60 psig. A thermostatically controlled immersion heater in the main tank and a cooling unit before the test section provide an inlet temperature range of -40°F to 130°F with a control accuracy of ±1.0°F. Following the electrically heated test section is a recovery system consisting of a condenser and a collector tank. Mixture of saturated liquid and vapor is led through the condenser coil, which itself is cooled by refrigerated brine at 32°F. Condensed liquid is then stored in a recovery tank which is maintained at constant pressure. This arrangement ensures a constant level of exit pressure.

All the tubing except the recovery section is made of 0.295 in. ID Nichrome pipe, and the connections are made by Swagelok-type threaded joints. In the recovery section, tubing of much larger diameter is used to minimize pressure losses, thus maintaining a nearly constant level of exit pressure from the test section.

A 3"-long transparent piece of tubing is incorporated into the loop right after the heater. This part provides a very clear view of the two-phase flow regimes inside, and makes it possible to record the two-phase flow oscillations in their various modes by means of a high speed camera.

Two fine filters, one installed at the main tank exit and a second one after the surge tank, keep the test section and sensitive devices such as the rotameter and valves free from entrained particles. Basically, the experimental setup is an open-loop forced

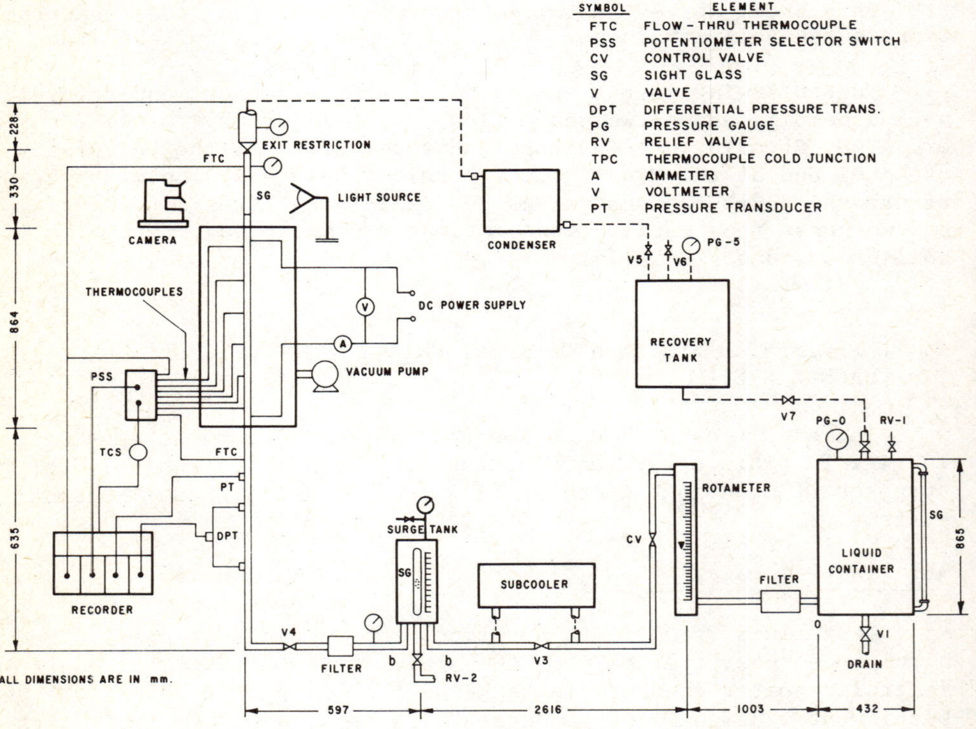

Fig. 3 Schematic drawing of the experimental system.

convection boiling flow system operating between constant inlet and exit pressures. The loop can be examined as being comprised of three major segments which are, in the order of arrangement, the fluid supply section, the test section and the recovery section.

3.2 Fluid Supply System

This part comprises a main tank, a filter, a rotameter and a subcooling controller installed in that order. A pressurized nitrogen supply connected to the main tank provides the necessary pressure to pump the liquid through the system. A pressure regulating valve is fitted into the high pressure nitrogen supply section to maintain a constant tank pressure and a continuous flow of fluid during each run. A bourdon type pressure gauge and a relief valve are installed on the tank for safety purposes. Also included in the tank is a thermostatically controlled immersion heater to provide above ambient test temperatures. A needle valve and a rotameter provide control and measurement of flow rate.

The final component in the line is the subcooling controller. The subcooling controller has a rated cooling capacity of 14,000 Btu/hr and, together with the immersion heater in the main tank, provides a combined temperature range of 0°F to 150°F available at the heater inlet.

3.3 Test Section

Test section includes a surge tank, a heater and a riser followed by an exit restriction. The main body of instrumentation is also clustered in that section.

The surge tank, acting as a capacitance, is an important dynamic component of the loop, especially during pressure-drop oscillations. The surge tank is connected to the lab's compressed air system to provide control for the liquid level in it. Through a valve arrangement, enough air is admitted into the tank to maintain an average level of liquid during the whole course of a run. A sight glass attached to the tank is used to check the initial liquid level and to observe the variations of that level during oscillations.

A 24-inch portion of the loop also functions as a heater. The heater is interconnected to the loop by special couplings which need no axial clearance for dismantling and reinstallation of the heater section. Thus it is possible to change and replace the heater section by different tubes with different wall characteristics or internal configurations, as many times as required and without disturbing anything else in the rest of the system. The heater is resistively heated by controlled D.C. supplied from a transformer-rectifier unit. This unit is connected to the lab's 120-volt line, has a capacity of 20 kW and is equipped with a continuous current setting. The electrical insulation of the heater assembly from the rest of the system is provided by two teflon bulkhead unions at both ends.

To minimize heat losses, a radiation guard is built around the heated tube and the whole assembly is then housed in a vacuum jacket. A vacuum pump connected to this jacket maintains a vacuum of less than 1.0 mm Hg absolute. The test section is terminated by an exit restriction. The exit restriction is basically a sharp edged orifice with an inner diameter of 1/16 inch.

A needle valve is included in the test section between the surge tank and the heater. The valve is used to stabilize the system by affecting the inlet resistance. However, it is used infrequently and only as necessary; otherwise, it is kept fully open while recording data.

3.4 Recovery System

Following the test section is a recovery system consisting of a condenser and a recovery tank. The mixture of saturated liquid and vapor discharging from the exit restriction is led through the condenser coil cooled by refrigerated brine. The condensed liquid is then collected in the recovery tank, which is maintained at constant pressure to ensure a constant level of exit pressure at the test section. When the liquid level in the main tank falls below a certain value, the tank pressure is released and the test liquid collected in the recovery tank is pumped into it by pressurizing the latter using compressed air.

3.5 Instrumentation

Flow rate and heat input, along with pressure and temperature at appropriate localities are measured during each run. Temperature measurements are made by standard copper constantan thermocouples. Five of these, fixed to the outer surface of the heater at 5-inch intervals, are used for heater wall temperature measurements. To avoid floating voltage effects, thermocouple ends have been electrically insulated from the wall surface by 0.4 mm thick mica flakes. Two flow-through thermocouples at the exit side and a third one at the inlet side of the heater are used for fluid temperature measurements at these points.

Pressures are measured at five different points by bourdon type Heise Pressure gauges. In addition to that, a strain gauge type pressure transducer (Dynisco Model TCTP-25-1C) is used to record pressure variations at the heater inlet. A calibrated rotameter (AMETEK Model 20-4120 Size 2) is used for flow rate measurements. However, the slow response rate of the rotameter makes it unsuitable for measurement of rapid flow variations. Due to the cushioning effect of the surge tank, oscillations are either much slower or not transmitted at all to the rotameter side. Therefore, the rotameter reading is taken to indicate the average flow rate into the test section. For this reason also, a differential pressure transducer is used to detect and to measure the rapid flow rate oscillations.

Outputs of the pressure transducers are fed directly to a four-channel amplifier-recorder unit (Sanborn Model 854-5460). A variable speed drive for the recorder paper movement and a time marker, combined with separate gain control for each channel, enables detailed recordings of signal histories. Thermocouple outputs are fed selectively, either to a digital voltmeter (Weston Mode 1242) or to the remaining two channels of the recorder.

Heat input is obtained by measuring the current and the voltage drop across the heated section. Conventional instruments are used for this purpose.

3.6 Photographic System

In order to determine the actual flow regime in two-phase flows, it is essential to have good visual studies. For this purpose, a high-speed photographic investigation is carried out. In this investigation, a continuous film is taken during the steady and unsteady operation, and near the stability-instability boundary of the system. A HYCAM 40-0007 high-speed camera is used for recording the flow through the transparent section of the test loop. During the filming, two flood-type light generators are used for proper lighting of the transparent section.

The camera has a speed of up to 3000 pictures per second and a built-in timing light generator with 100 Hz and 1000 Hz frequency.

Analysis of the films uses a 16-mm Lafayette Analyzer, Model 00200. Data from the film speed, timing marks and projector speed are combined to obtain the pertinent information from the recorded film.

3.7 Experiments

Employing the above-described experimental system, experiments were carried out to investigate the basic modes of two-phase flow instabilities. Special emphasis was given to sustained density-wave type and pressure-drop type oscillations. The effects of heat transfer variations were studied at different heat inputs. Inlet temperature and heat flux were chosen as test parameters, while the mass flow rate was varied in small decrements. Stability boundaries and oscillatory characteristics were determined for each individual test.

Depending upon the particular conditions of the flow during the test, the pressure in the heated section varied. The supply tank pressure was held constant at a pressure of 75 psia and the exit pressure maintained around 25 psia.

Two different series of experiments were carried out for the experimental investigation, utilizing Freon-11 and Freon-113 as test fluids. In the first series of these experiments, sustained boiling flow instabilities were investigated. The second series of experiments are transient experiments, and will determine the transient behavior of the boiling flow system in response to a step change, either of flow rate or heat flux.

3.8 Experimental Procedure

For sustained stability experiments, the following experimental procedure was followed:

For a given heater test section, different sets of experiments corresponding to various heat inputs were conducted. Each set was composed of a sufficient number of tests to cover the available subcooling range, and stability boundaries were determined in each case. Oscillations were identified by the cyclic variation of test section pressure and flow rate, and by observing the transparent section, the pressure gauge pointers and the recorder. In defining the stability boundary, short life transients were disregarded and only sustained oscillations were considered.

A test preparation procedure was followed before each experiment: the test liquid was transferred to the main tank, either from the recovery tank or from the original containers, and the immersion heater switched on with the thermostat set for a predetermined test temperature. For low temperature tests, the subcooling controller was set for the proper temperature. All the instrumentation and mechanical components were checked for faultless operation. An ice bath prepared in a thermos bottle served as the reference junction temperature for the thermocouples. The recorder was switched on half an hour before the experiment. This allowed the amplifiers and transducers to warm up, as suggested by the manufacturer. The amplifier settings were checked periodically as well, in order to ensure proper calibration.

The actual test procedure can be outlined as follows:

1. With enough liquid in the main tank and the inlet temperature set, the tank was pressurized by nitrogen gas.

2. Flow rate and heat input were increased gradually to the desired starting point, and the system was allowed to become steady, as indicated by system pressure, temperature

and flow rate.

3. Measurements of temperature, pressure, flow rate and heat input were taken and critical observations noted.

4. The mass flow rate was reduced by a small amount using the inlet control valve and the system behavior was observed.

Following each adjustment, the system was allowed to stabilize, and the procedure was continued, starting from Step 3.

The above-described procedure was then repeated for different heater surface conditions, heat inputs and inlet temperatures [33, 34].

3.9 Typical Experimental Results

Different sets of experiments were run with Freon-11. In most of the experiments, two basic modes of oscillations, namely the pressure-drop type and density-wave type oscillations, were observed in the experiments. The stability-instability boundaries for each type were determined experimentally and they were plotted on system pressure drop vs. mass flow rate curves. As a typical example, the

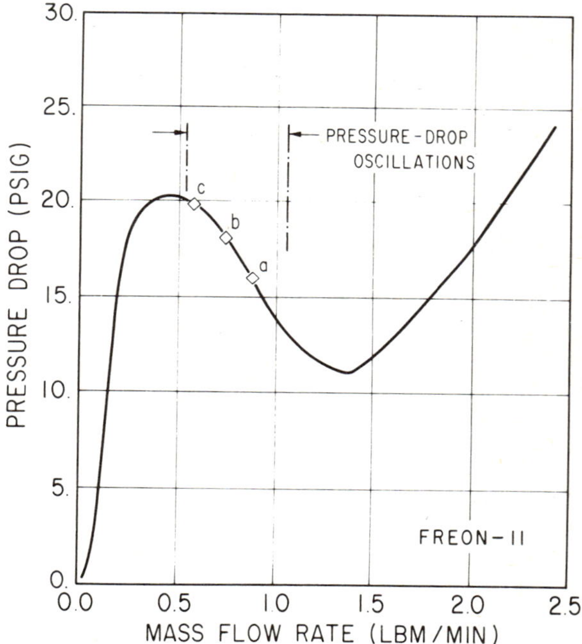

Fig. 4 Stability boundary for pressure-drop oscillations (Heat input: 1023 Btu/hr, Inlet temperature: 25°F) [32].

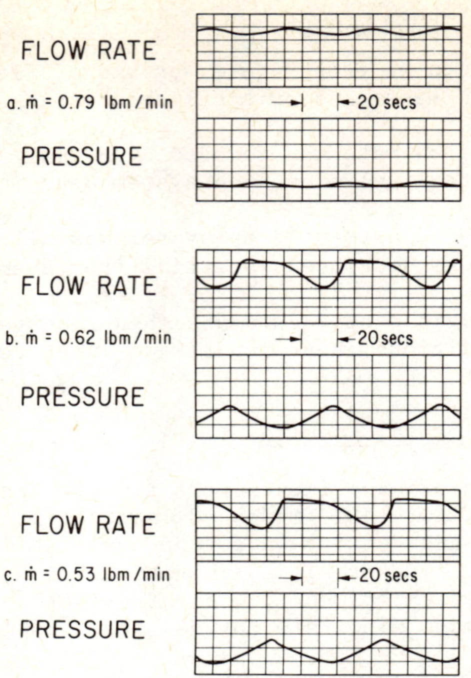

Fig. 5 Typical recordings of pressure-drop oscillations [32].

results of the tests corresponding to 1023 Btu/h heat input and 25°F inlet temperature are given below:

Figure 4 shows a typical plot of pressure drop versus mass flow rate for a single channel upflow boiling system; pressure drop being the pressure difference between the surge tank and the system exit and the mass flow rate being the mean mass flow of the fluid entering into the system per minute. As the flow rate is reduced, the pressure drop across the system also reduces until a flow rate of 1.32 lbm/min is reached at which time the liquid leaving the heater is almost saturated. Further reduction in flow rate is accompanied by an increase in pressure drop as bulk boiling commences. For flow rates between 1.32 lbm/min and 0.45 lbm/min, the slope of the pressure drop vs. mass flow rate is negative. At a flow rate of 0.90 lbm/min sinusoidal flow rate and pressure oscillations of very small amplitude are observed with a period of 50 seconds. As the flow rate is reduced, the amplitude of the oscillations grows. For mean flow rates of 0.79, 0.62 and 0.53 lbm/min, the flow rate and pressure recordings of the oscillations at the heater inlet are reproduced in figures 5a, 5b and 5c, respectively. The oscillations continue as the flow rate is reduced and the slope remains negative. No oscillations are observed near the crest of the curve. At a mean flow

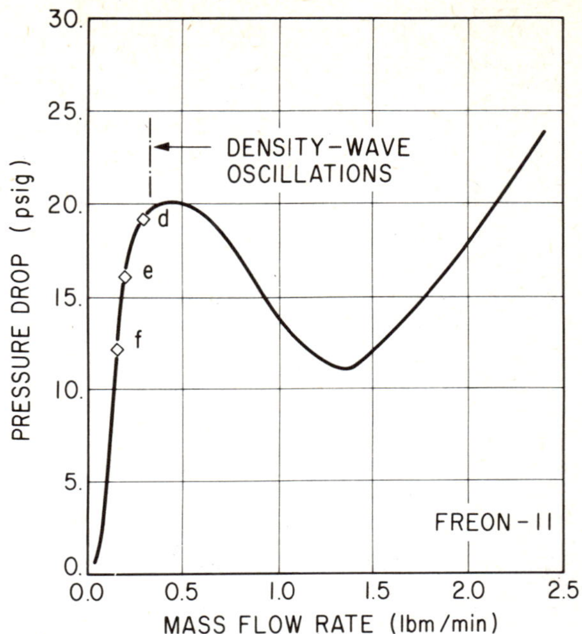

Fig. 6 Stability-instability boundary for the density-wave type oscillations (Heat input: 1023 Btu/hr, Inlet temperature: 25°F) [32].

rate of 0.36 lbm/min, figure 6, the high frequency oscillations with periods around 2 seconds and small amplitudes commence. Their periods decrease with a decrease in flow rate. Typical recordings of these high frequency oscillations are shown in figures 7e, 7f, and 7g for mean mass flow rates of 0.24, 0.19, and 0.15 lbm/min, respectively. The first mode of oscillations, which occur at the negative slope region of the curve, is identified to be pressure-drop type oscillations. The second mode, which has relatively high frequencies, is identified to be density-wave type oscillations.

In some of the experiments with film boiling a completely different type of oscillations was also observed [31]. In this type of oscillations the wall temperature fluctuated periodically between 262°F and 396°F (figure 8). During the temperature oscillations it was observed that the part of the heater that these oscillations occurred, the heater was operating in the film boiling zone. These slow oscillations were called thermal oscillations because of the large fluctuations in wall temperature not encountered in other types. It appeared that a relatively rare combination of flow and

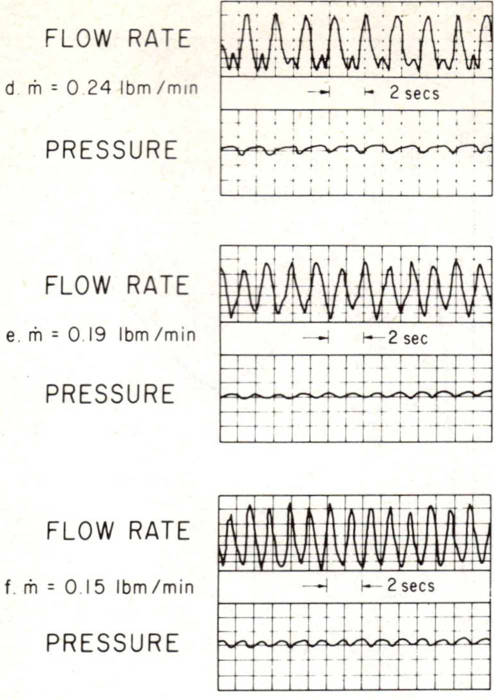

Fig. 7 Typical recordings of density-wave oscillations [32, 177].

heat transfer characteristics may be necessary to permit thermal oscillations to occur.

In the above-described experiments, the density-wave type and the pressure-drop type oscillations were encountered.

The steady-state data were plotted as curves of the inlet plenum to system exit pressure drop versus mass flow rate for various heater power inputs, and stable and unstable regions with respect to density-wave and pressure-drop type oscillations were obtained. The dotted line at the bottom represents the additional pressure drop from the surge tank to the inlet plenum when the heater inlet valve is fully open (Fig. 9). These are steady-state operating curves for the system. The region to the left of the black dots in the figure represents the region where the density-wave oscillations were encountered. As it can be seen from the figure, at sufficiently high flow rates where all liquid flow persists throughout the system, the relation is roughly parabolic corresponding to turbulent liquid flow. As the flow rate is decreased and net vapor generation starts, a decrease in flow rate is accompanied by an increase in pressure drop. Negative slope region

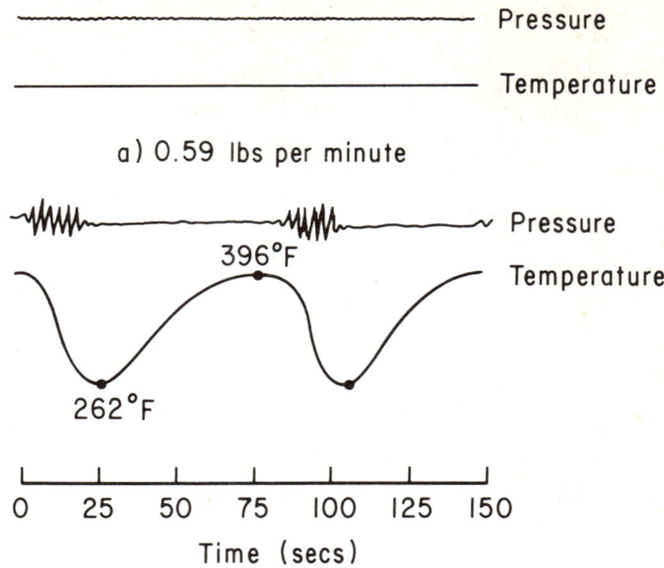

Fig. 8 Pressure and temperature oscillations at 700 watts heat input [31].

widens and becomes steeper at increasingly higher heat inputs.

The main conclusions for three different modes of two-phase flow oscillations encountered in earlier studies in single-channel systems (vertical and horizontal) can be conveniently summarized in three parts [23-34]:

Conclusions about density-wave oscillations:

(a) Density-wave oscillations occur on the boiling positive slope branch of the pressure drop versus flow rate curves.
(b) Increase in mass flow rate increases stability.
(c) Increase in heater power decreases stability.
(d) Increase in overall density ratio decreases stability.
(e) Increase in heat fraction expended in removing subcooling decreases stability.
(f) Increase in inlet pressure drop increases stability.
(g) Period of density-wave oscillations is of the order of the residence time of fluid from heater inlet to exit plenum.
(h) Boiling upward flows are more stable than horizontal flows.

Conclusions about pressure-drop oscillations:

(a) Pressure-drop oscillations occur on the negative slope branch of the pressure drop versus flow rate curves.

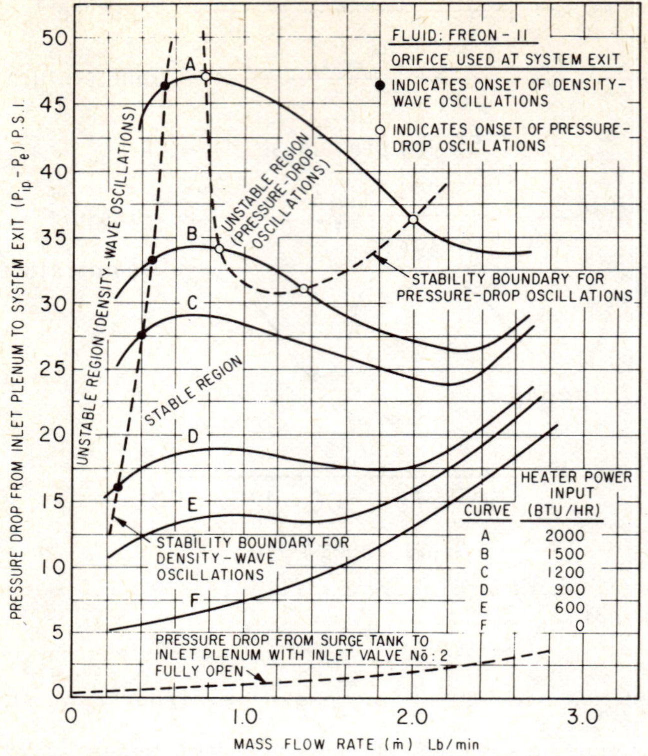

Fig. 9 Pressure drop from the inlet plenum to system exit versus mass flow rate steady-state relationships [31].

(b) Increase in negative slope decreases stability.
(c) Increase in heater power decreases stability.
(d) Increase in exit pressure drop decreases stability.
(e) Increase in inlet pressure drop increases stability.
(f) Period of pressure-drop oscillations increases with increase in flow rate.
(g) When minimum flow rate of a pressure-drop oscillation cycle falls into density-wave oscillations region, superimposed density-wave oscillations appear on low flow part of the cycle.
(h) Boiling upward flows are more stable than horizontal flows.

Conclusions about thermal oscillations:

(a) Thermal oscillations occur in a region between density-wave and pressure-drop oscillations regions when there is film boiling.
(b) During part of the cycle there are always superimposed density-wave oscillations.
(c) Increase in heat input and/or decrease in flow rate decreases

stability.

(d) Increase in inlet pressure drop increases stability.

(e) Period of thermal oscillations is greater than those of pressure-drop and density-wave oscillations.

4. BOILING FLOW INSTABILITIES IN PARALLEL CHANNEL SYSTEMS

4.1 Two-Parallel Channel Systems

The objective of the experiments was to study the modes of oscillations in a two parallel channel boiling upflow system--two parallel channels containing a heater each, installed between an inlet plenum and an exit plenum and fed with test fluid from a pressurized tank [27-29].

The experimental setup was the same as described in figure 3. Two heater geometries used in these experiments are shown in figure 10.

Freon-11 was used as the test fluid as it has the advantages of low boiling point and low burn-out heat flux. These permit operations up into the film boiling regime without causing excessive heater-wall temperatures. Various experiments were carried out as described in section 3.8. In each set of the tests, the inlet temperature of the test liquid (Freon-11) and the power inputs into the heaters were fixed. The flow rate through the system was then varied over the range allowed by the system characteristics. In

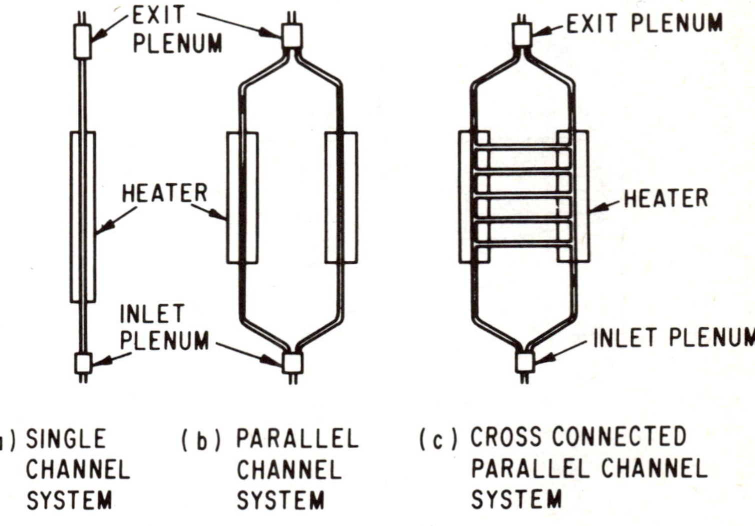

Fig. 10 Schematic diagrams of various heater geometries.

general, the system would be stable within the portions of the flow range and unstable in other portions. In the unstable region, two different modes of oscillations were observed. They were the same as those observed in single channel upward boiling flows: Density-wave type flow and pressure oscillations were encountered for all the combinations of heat inputs whenever the flow rate through one of the channels was low enough to result in a high inlet to exit density ratio for that channel. These oscillations did not produce any noticeable temperature changes in the heater-wall temperatures.

Whenever the density-wave type (flow and pressure) oscillations started, they would be present in all the channels of the system. The oscillations in the two heater channels were out of phase by 180°, and the oscillations in the common inlet channel represented the algebraic sum of the oscillations in the individual heater channels and as such their amplitudes were relatively small as compared to the density-wave oscillation amplitudes in the individual channels (figure 11).

The density-wave type oscillations could be stopped by introducing a pressure drop between the inlet plenum and any one of the heaters, but not by a pressure drop between the surge tank and inlet plenum. Introduction of a pressure drop at the inlet to the higher heat input channel would produce a damping effect for the oscillations in that channel (where the oscillations originate), and also for the oscillations everywhere else because of the coupling; and introduction of a pressure drop at the inlet to the lower heat input channel would cause the mass flow rate through the other channel to increase, and the system will be stabilized.

The steady-state experimental data have been plotted as curves of the surge tank to system exit pressure drop versus total mass

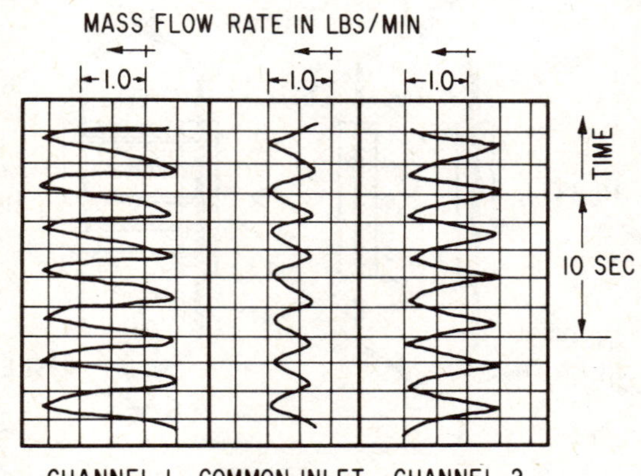

Fig. 11 Recordings of pure density-wave type oscillations [26].

flow rate for various heat input combinations. Figures 12 and 13 show the steady state characteristics and stable-unstable regions of the system for different heat inputs. A study of the two figures with regard to the density-wave oscillations, indicates that: (1) increase in total flow rate increases stability; (2) increase in pressure drop increases stability; and (3) increase in difference between the heat inputs into individual channels decreases stability.

In the experiments with higher heat inputs the pressure-drop type flow and pressure oscillations were encountered, in addition to the density-wave type oscillations. In general, the pressure-drop type oscillations commenced soon after the start of boiling in one of the channels where the pressure drop versus total mass flow rate curve had a large negative slope.

When the pressure-drop type oscillations were established in a channel, the other channel could be stable with respect to the pressure-drop type oscillations or it could have weak (lower amplitude) pressure-drop type oscillations in phase with the oscillations of the former. In many instances, to stabilize the system with respect to the pressure-drop type oscillations, it would be necessary to introduce pressure drops at the inlets of both heaters because the introduction of a pressure drop in the channel originating the oscillations could push the oscillations in the other channel towards instability while stabilizing its own channel.

Whenever the two types of flow oscillations were present in the system, as the pressure drop at inlet to one of the heaters was increased (by closing down the respective inlet valve), first the type of instability which commenced first (as the flow rate was reduced) disappeared, and then the other type of instability disappeared as the valve was closed further.

The region to the left of the dotted curve in figure 13 represents the region where the pressure-drop type oscillations were encountered. The region where the instability regions of the density-wave type and the pressure-drop type instabilities overlap represents the region where both the density-wave type and the pressure-drop type oscillations occur simultaneously. A study of the figure, with regard to the pressure-drop type oscillations, indicates that (1) increase in total flow rate increases stability; (2) increase in system pressure drop decreases stability; and (3) increase in total heat input decreases stability.

4.2 Cross-Connected Parallel Channel Upflow System

The parallel channel systems such as the one discussed in the last section cannot represent all the boiling water reactors from the point of view of fluid dynamics. In many reactor heat removal systems, there are cross-connections between the parallel vertical

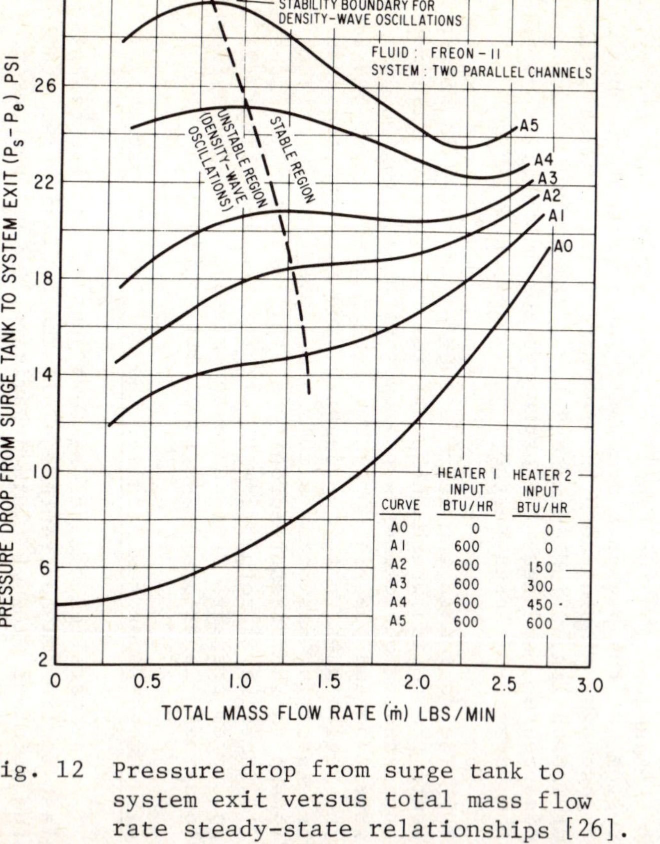

Fig. 12 Pressure drop from surge tank to system exit versus total mass flow rate steady-state relationships [26].

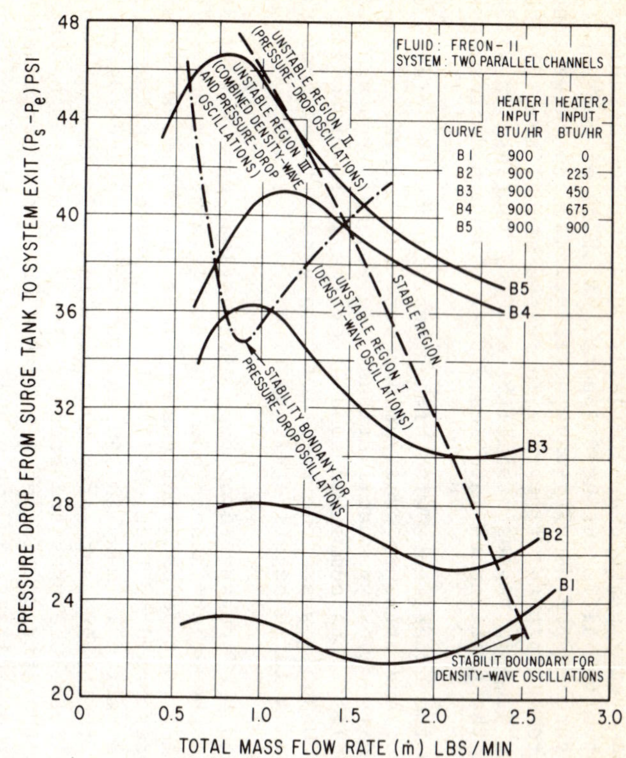

Fig. 13 Pressure drop from surge tank to system exit versus total mass flow rate steady-state relationships [26].

coolant channels.

Some notable differences were observed between the oscillations of the present system and those of the parallel channel and single channel systems with regard to the phase relationships, the superposition of the instability modes and the promotion of stability. The present system was found to be more stable than the others.

Experimental apparatus and procedures were similar to the one used in the experiments discussed in the previous sections except that there were five cross-sections between the two parallel channels.

In each of the tests, the inlet temperature of the test liquid (Freon-11) and the power inputs into one or both of the heaters were fixed. The flow rate through the system was then varied over the range allowed by the system characteristics. In general, the system would be stable within some portions of the flow range and unstable in other portions.

In the unstable regions two fundamental modes of oscillation were observed. They were the same as those observed in single channel and parallel channel (with no cross-connections) upward boiling flows; viz., the density-wave type instabilities and the pressure-drop type oscillations.

Density-wave type oscillations occurred in the positive slope regions of the boiling flow pressure drop versus flow rate curves. These oscillations did not produce any noticeable temperature changes in the heater wall temperatures. Whenever density-wave type (flow and pressure) oscillations started, they would be present in all the channels of the system. The oscillations in the two parallel vertical channels were in phase; the oscillations in the common inlet channel represented the algebraic sum of the oscillations in the parallel channels, and as such their amplitudes were relatively large as compared to the density-wave oscillation amplitudes in the parallel channels.

The density-wave oscillations could be stopped by introducing either pressure drops at the inlets of the two heaters or a pressure drop between the inlet plenum and the surge tank. The latter way of stabilizing was found to be more effective. Introduction of a pressure drop at the inlet of one of the heaters could not stabilize the system; however, it would have a damping effect by reducing the amplitudes of the oscillations in that channel.

The experimental data have been plotted as curves of the surge tank to system exit pressure drop versus total mass flow rate for various heat input combinations. Figure 14 shows the family of curves for equal heat inputs into heaters 1 and 2, and figure 15

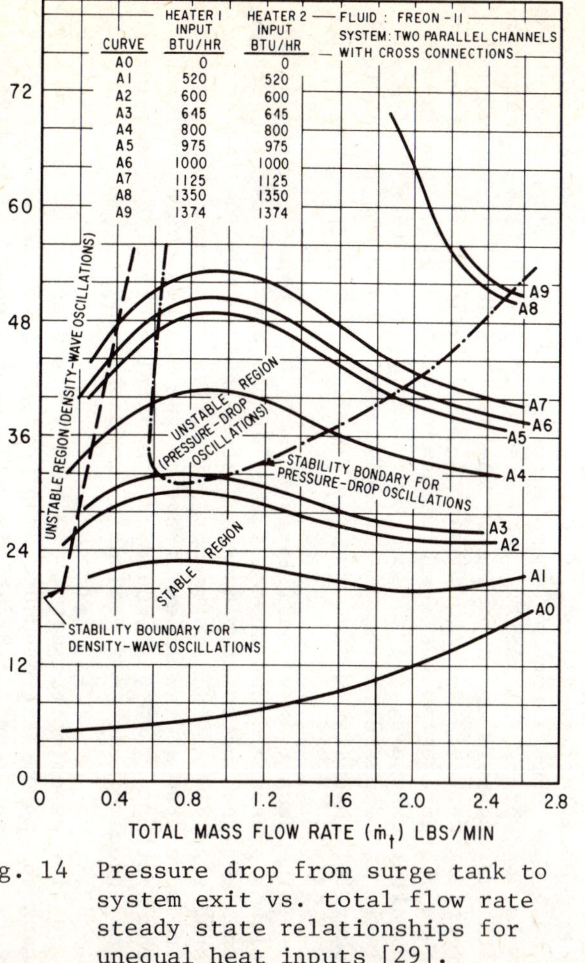

Fig. 14 Pressure drop from surge tank to system exit vs. total flow rate steady state relationships for unequal heat inputs [29].

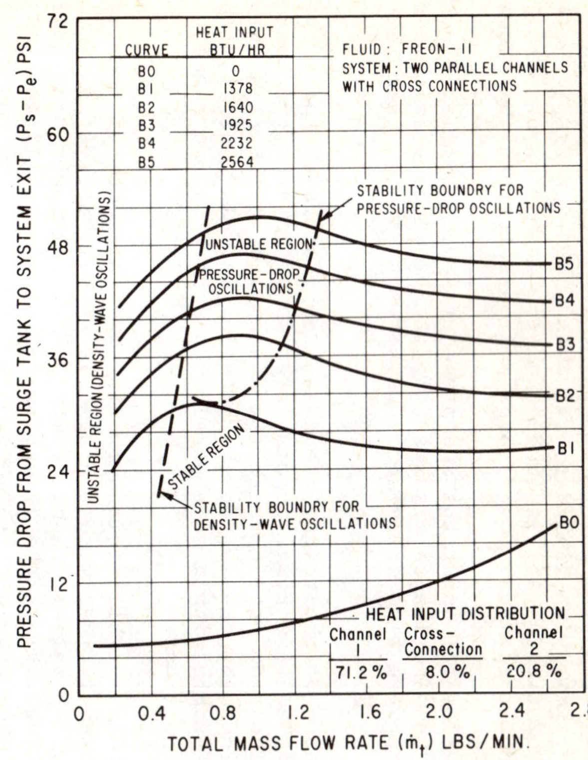

Fig. 15 Pressure drop from surge tank to system exit vs. total flow rate steady state relationships for equal inputs [29].

shows that family of curves for unequal heat inputs. The region to the left of the broken-line curve in each figure represents the region where the density-wave type flow and pressure oscillations were encountered. A study of the two figures, with regard to the density-wave type oscillations, indicated that (a) increase in total flow rate increases stability; (b) increase in system pressure drop decreases stability; (c) increase in heat input decreases stability; and (d) equal heat inputs make the system more stable.

In general, the pressure-drop type oscillations commenced soon after the start of boiling. Whenever the pressure-drop type (flow and pressure) oscillations started, they would be present in all channels of the system. The oscillations in the two parallel vertical channels were in phase; and the oscillations in the common inlet channel represented the algebraic sum of the oscillations in the parallel channels, and as such their amplitudes were relatively large as compared to the pressure-drop oscillation amplitudes in the parallel channels.

The pressure-drop oscillations could be stopped by introducing either pressure drops at the inlets of the two heaters or a pressure drop between the inlet plenum and the surge tank; the latter was found to be more effective. Introduction of a pressure drop at the inlet of one of the heaters was not sufficient to stabilize the system; however, it would have a damping effect by reducing the amplitudes of the oscillations.

4.3 Transient Boiling Flow Instabilities in a Multi-Channel Upflow System

In this study, the transient boiling flow instabilities in a four parallel channel upflow system with and without cross-connections were experimentally investigated using Freon-11 as the test fluid. Several series of transient experiments were run using a step increase in power input with a constant flow control valve setting and a step decrease in flow rate with a constant heat input. Two major modes of oscillations, namely density-wave type and pressure-drop type oscillations, were observed. Total system pressure drop was plotted as a function of mass flow rate for various heat inputs and flow control valve settings, and the stable and unstable regions for transient density-wave type and pressure-drop type oscillations were determined and the boundaries of sustained instabilities were also indicated. Experiments were conducted with equal and unequal heat inputs and the results were compared with each other. A comparison of the transient instabilities observed in four parallel channel system with and without cross connections was also made [168].

The test section consisting of surge tank, inlet valve, inlet plenum, heater, exit plenum and exit restriction is shown in figure 16

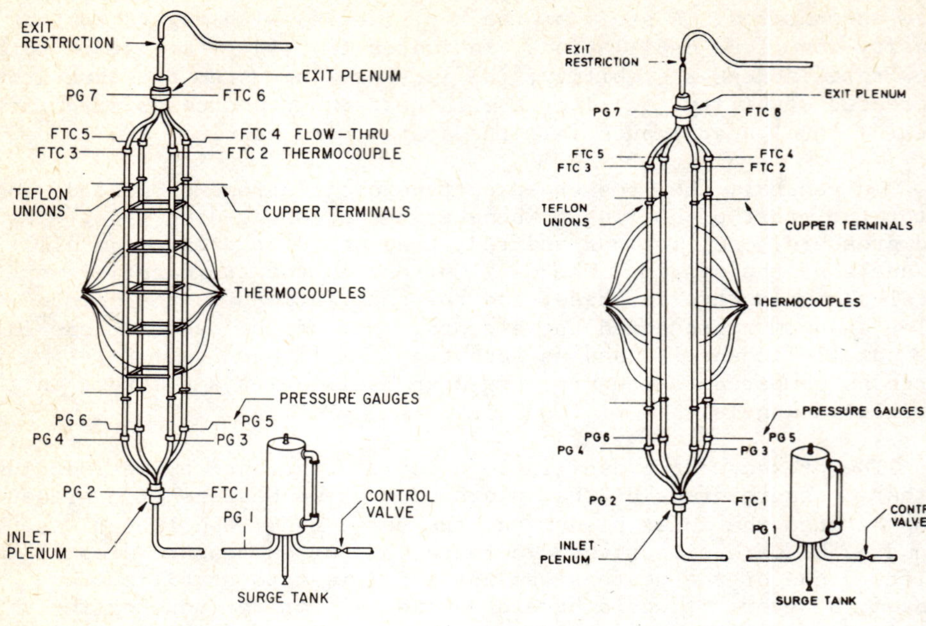

Fig. 16 Schematic diagram of the test section with and without cross-connections.

without and with cross-connections.

Experiments were conducted with Freon-11 as the test fluid since it has the advantages of low boiling point and low burn-out heat flux. These permit operations up into the film boiling regime without causing excessive heater wall temperatures. Also the slope of the temperature-pressure curve for Freon-11 is approximately equal to that of water in the ranges of pressure under consideration.

The experiments performed for investigating the transient two-phase flow instabilities in four-parallel channel upflow system can be separated mainly into three groups: (1) The experiments on cross-connected four parallel channel system with equal heating of channels; (2) the experiments on four parallel channel system without cross-connections with equal heating of channels; (3) the experiments on four parallel channel system without cross-connections with unequal heating of channels. In each of these groups, two series of experiments were performed. In the first series, the heat input to the channels was held constant while the flow rate was suddenly reduced by a certain amount by partially closing the control valve and this was repeated for different heat inputs. In the second series of experiments, the setting of the flow control valve was not changed while the heat input to the channels was incrementally increased.

Typical experimental data are summarized in Tables 1 and 2 for cross-connected four parallel channel for step flow rate decrease and for step heat input increase respectively.

Figure 17 is a typical plot of the test section pressure drop versus total mass flow rate for various constant heat inputs and incrementally decreasing flow rates, for the cross-connected system. On the figure, the curves TPD and TDW are the approximate boundaries

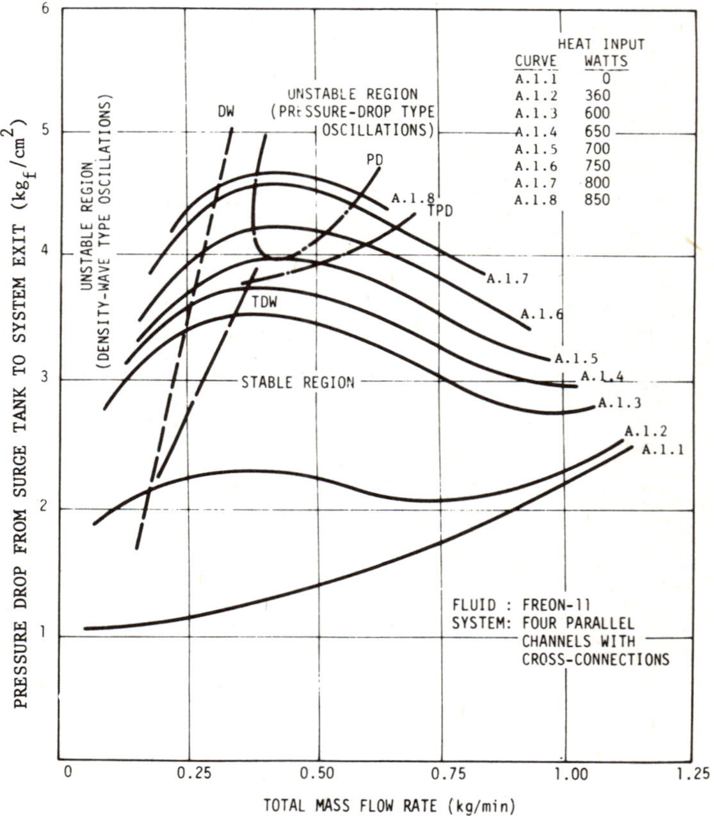

Fig. 17 Pressure drop from surge tank to system exit vs. total mass flow rate for step decreases in flow rates at constant heat inputs [168].

Table 1. Summary of transient instability data for step flow rate decrease for cross-connected four parallel channel system.

Before Step Change			Step Change	After Step Change			Remarks About Instability			
Total Heat Input	Total Mass Flow Rate	Heater Wall Temp.	Flow Rate Decrease	Equilib. Pressure Change	Equilib. Flow Rate Decrease	Heater Wall Temp. Increase	Before Step Change		After Step Change	
W	kg/min	°C	kg/min	kg_f/cm^2	kg/min	°C	τ_{pd}(s)	τ_{dw}(s)	τ_{pd}(s)	τ_{dw}(s)
600	0.340	76	0.136	-0.45	0.136	-4	-	-	-	2(TR)
600	0.204	72	0.068	-0.90	0.068	+90	-	-	-	2
650	0.318	79	0.159	-1.10	0.159	+75	-	-	-	2
580	0.454	83	0.136	-0.25	1.136	-1	-	-	20(TR)	-
680	0.318	82	0.095	-0.40	0.091	-	-	-	-	2(TR)
680	0.227	82	0.091	-1.15	0.091	+81	-	-	-	2
750	0.500	87	0.069	-1.15	0.069	-1	20	-	20	-
750	0.431	86	0.091	-0.20	0.091	-2	20	-	18	-
750	0.340	84	0.090	-0.30	0.090	-1	18	-	22(TR)	2(TR)
750	0.250	83	0.091	-1.25	0.091	+84	-	-	-	2
800	0.431	89	0.091	-0.10	0.091	-1	23	-	20	-
800	0.340	82	0.090	-0.35	0.090	-2	20	-	22(TR)	-
800	0.182	166	0.046	-0.90	0.046	+52	-	-	-	1.7

Table 2. Summary of transient instability data for step heat input increase for cross-connected four parallel channel system.

Before Step Change			Step Change	After Step Change			Remarks About Instability			
Initial Flow Rate	Total Heat Input	Total Mass Flow Rate	Heat Input Increase	Equilib. Pressure Change	Equilib. Flow Rate Decrease	Heater Wall Temp. Increase	Before Step Change		After Step Change	
kg/min	W	kg/min	W	kg_f/cm^2	kg/min	°C	τ_{pd}(s)	τ_{dw}(s)	τ_{pd}(s)	τ_{dw}(s)
0.227	0	0.227	317	+1.30	0.009	+34	–	–	–	3
0.227	317	0.218	71	+0.45	0.005	+5	–	–	–	2.6
0.363	309	0.340	61	+0.80	0.031	+7	–	–	–	3(TR)
0.363	370	0.309	110	+1.00	0.079	+9	–	–	12(TR)	2.5
0.454	390	0.362	105	+0.65	0.039	+6	–	–	–	2.2(TR)
0.454	495	0.323	75	+0.35	0.033	+3	–	–	20(TR)	2
0.567	490	0.400	70	+0.50	0.046	+5	–	–	30(TR)	2(TR)
0.681	490	0.476	65	+0.50	0.055	+4	–	–	25(TR)	–
0.681	555	0.421	85	+0.50	0.067	+3	–	–	27(TR)	0.4(TR)
0.817	555	0.476	85	+0.65	0.063	+5	–	–	40	–
0.953	615	0.554	40	+0.35	0.064	+3	–	–	45(TR)	–
1.090	590	0.431	40	+0.20	0.114	+2	–	–	20(TR)	–
1.090	630	0.317	70	+0.05	0.055	+1	–	–	–	2

for transient pressure-drop type and transient density-wave type oscillations respectively in response to a step decrease in mass flow rate. PD and DW represent the boundaries of the sustained pressure-drop and density-wave type oscillations.

It was found that the cross-connected four-parallel-channel system is more stable than the four-parallel-channel system without cross-connections, with regard to transient pressure-drop and density-wave type oscillations; also the minimum heat input at which this system can operate without any transient pressure-drop type oscillations is higher in the cross-connected system than that of the system without cross-connections.

The periods of the transient pressure-drop type oscillations in the system without cross-connections are relatively larger than those of the system with cross-connections. The periods of the transient density-wave type oscillations are not much different in these two systems; this is due to the fact that, rather than the heater geometry, the exit restriction is the determining factor. In both of the systems, the transient density-wave and pressure-drop type oscillations are in phase in all of the channels.

With regard to the transient and sustained pressure-drop type oscillations, the equally heated system is again more stable than the unequally heated system. But this time, the difference between the two systems is not large, even they show similar transient pressure-drop type stability behavior, if the heat input per channel rather than total heat input is taken into consideration. The periods of the transient pressure-drop oscillations observed in equal heating experiments are larger than those observed in unequal heating experiments.

4.4 Sustained Instabilities in a Cross-Connected Four Parallel Channel Upflow System

In this experimental investigation, the instabilities induced by forced convection upflow boiling in a four parallel channel system with cross-flow mixing have been studied using Freon-11 as the test fluid. Two major modes of oscillation, i.e., density-wave and pressure-drop types, similar to those encountered in single channel and parallel channel boiling flows have been identified. In most instances, some temperature oscillations were observed. Stable and unstable regions for density-wave type and pressure-drop type oscillations have been determined in system pressure drop versus mass flow rate planes for various heat inputs. A comparison of the instabilities of the present system with those of other simpler systems has been made with regard to the phase relationships, the periods of oscillation, and the stability. The present system was found to be more stable than the others [169].

The experimental apparatus was the same as it is shown in figure 3. The test section is shown in figure 16.

It has been observed that whenever the density-wave type and pressure-drop type oscillations occurred, they were present in all channels of the system. The oscillations in the four parallel channels were in phase.

Some typical experimental data are presented in Table 3. It lists the total heat input, total mass flow rate $\dot{m}$, inlet pressure drop fraction y (the pressure drop from surge tank to inlet plenum divided by the pressure drop from surge tank to system exit), heat fraction expended to remove subcooling c, exit quality x, and the inlet plenum to exit plenum density ratio ρ_i/ρ_e under stable conditions. The table also lists the periods of the density-wave type and pressure-drop type oscillations whenever such instabilities were present. From a study of Table 3 it can be seen that (a) density-wave type oscillations occur after the start of boiling for a wide range of high inlet to exit density ratios (25 through 89); (b) pressure-drop type oscillations occur for lower density ratios (7 to 14) and at higher heat inputs (750 Watts or more); (c) the periods of the density-wave type oscillations are relatively low (0.8 to 3 seconds) and the periods of the pressure-drop type oscillations are relatively high (15 to 27 seconds); (d) in general the periods of pressure-drop type oscillations increase with increasing heat input and decreasing flow rate, and the periods of density-wave type oscillations increase with increasing flow rates.

Figure 18 shows typical total pressure drop between the surge tank and the exit as a function of the total mass flow rate for various heat inputs.

4.5 Comparison of Various Channel Systems

In order to compare the results of the various channel system geometries discussed, their stability-instability boundaries have been plotted in a total heat input versus total mass flow rate plane as shown in figure 19 [170]. In the figure stability boundaries shown are those which occur first as the flow rate is reduced. The full-line curves correspond to the boundaries for the density-wave type instabilities, and the broken-line curves correspond to the boundaries for the pressure-drop type instabilities. The region to the left of the curves is unstable region, and the region to the right is stable. A study of the figure shows that for equal heat input the three systems, namely, single channel, two parallel channel and cross-connected two parallel channel (curves A, B, and E) behave in almost the same manner, especially in the pressure-drop instability boundary regions. With regard to density-wave type of oscillations, the most stable system is the cross-connected four parallel channel system with equal heat inputs into the parallel

Table 3. Sustained instability data for cross-connected four parallel channel system.

Total Heat Input Watts	Mass Flow Rate Kg/min $\dot{m}$	Inlet Pressure Drop Fraction y	Heat Fraction To Remove Subcooling c	Exit Quality x	Inlet To Exit Density Ratio ρ_i/ρ_e	Density-Wave Oscillations Period Secs τ_{wd}	Pressure-Drop Oscillations Period Secs τ_{pd}
360	0.127	0.0390	0.203	0.813	72	3	--
360	0.091	0.0421	0.140	Superheat	89	1	--
600	0.204	0.0087	0.335	0.785	43	2	--
600	0.136	0.0254	0.052	Superheat	60	1.5	--
700	0.318	0.0077	0.425	0.492	25	3	--
700	0.227	0.0083	0.299	0.823	42	1	--
700	0.136	0.0178	0.173	Superheat	59	0.8	--
750	0.556	0.0195	0.775	0.115	6	--	--
750	0.424	0.0190	0.590	0.227	13	--	15
750	0.250	0.0200	0.326	0.781	37	1.5	--
750	0.159	0.0215	0.177	Superheat	65	1	--
800	0.704	0.0190	0.917	0.037	3	--	--
800	0.566	0.0180	0.754	0.136	7	--	20
800	0.454	0.0069	0.582	0.294	13	--	27
800	0.250	0.0069	0.324	0.843	37	2.5	--
800	0.182	0.0200	0.224	Superheat	46	1-1.5	--
850	0.510	0.0068	0.530	0.307	14	--	20
850	0.499	0.0066	0.520	0.409	17	--	--
850	0.300	0.0069	0.350	0.744	33	1.5	--
850	0.240	0.0143	0.184	Superheat	60	0.8	--

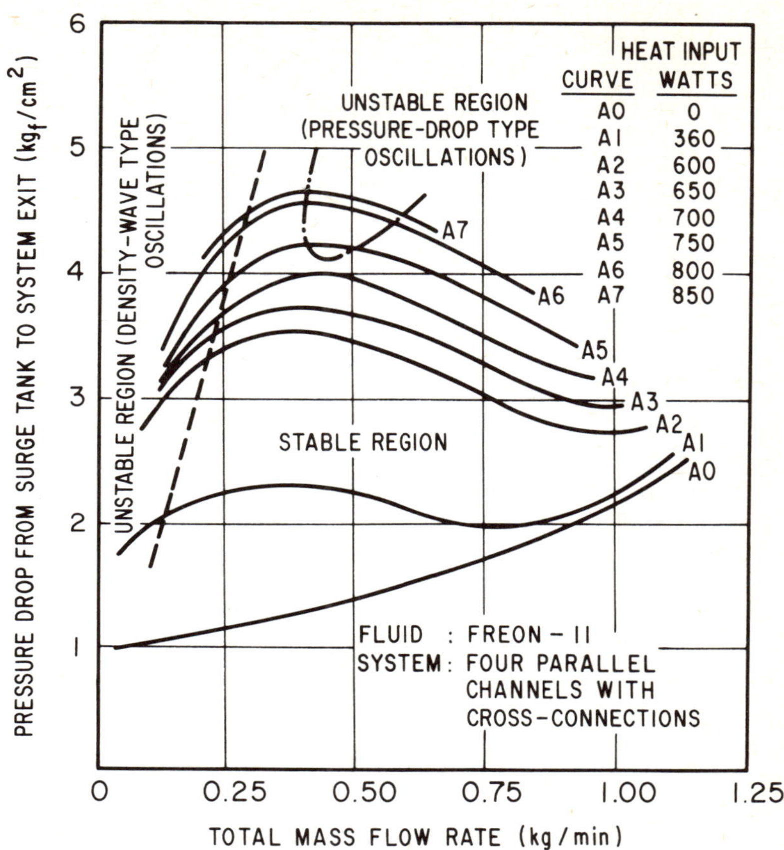

Fig. 18 Pressure drop from surge tank to system exit vs. total mass flow rate steady state relationships [169].

channels (curve G), followed by the single channel system (curve A), the cross-connected parallel channel system with unequal heat inputs (curve F), the parallel channel system with equal heat inputs (curve B), the parallel channel system with a constant heat input of 600 BTU/hr. to channel 1 (curve C) and the parallel channel system with a constant heat input of 900 BTU/hr. to channel 1 (curve D), in that order. The last two systems look very unstable in the figure as compared with the other systems. This is caused by the fact that, while the flow rate in channel 1 in each case was low enough for the density-wave oscillations to occur, the flow rate in channel 2 was high. Consequently, the total flow rate in each case was biased towards higher flow rates. With regard to the pressure-drop type of oscillations, the most stable system is the cross-connected four

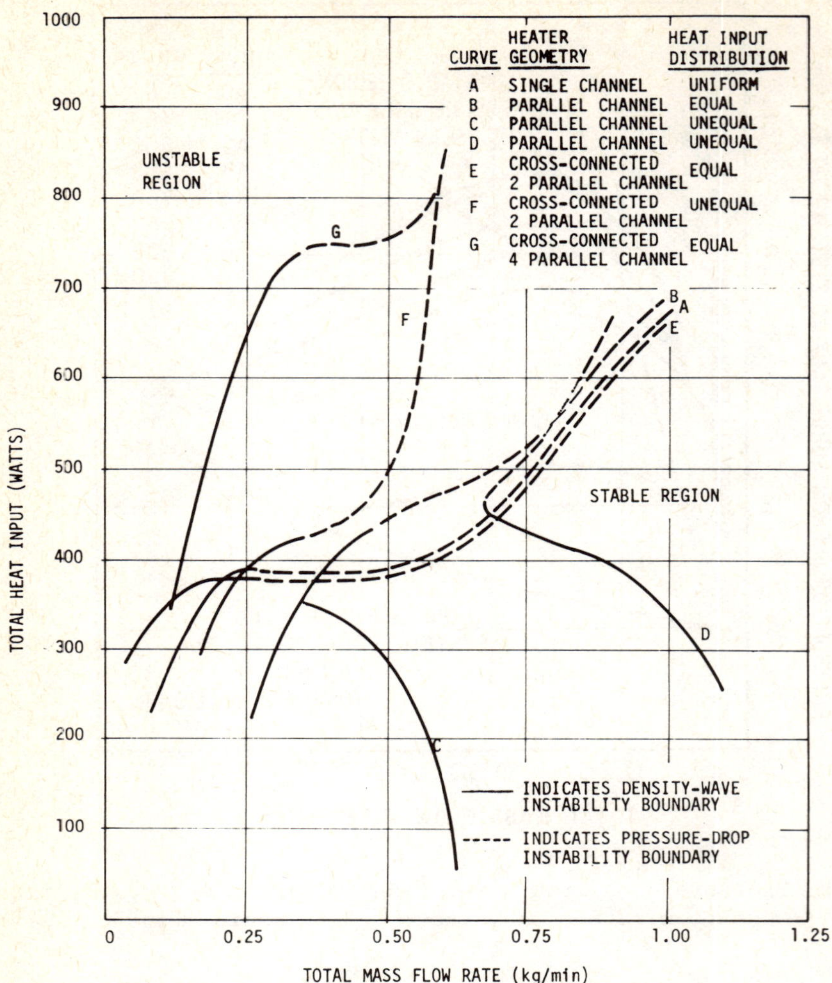

Fig. 19 Stability-instability boundary comparison of various channel systems [170].

parallel channel system with equal heat inputs (curve G) followed by the cross-connected two parallel channel with unequal heat inputs (curve F), the parallel channel system with a constant heat input of 900 BTU/hr. to channel 1 (curve D), the parallel channel system (curve A) and the cross-connected parallel channel system with equal heat inputs (curve E), in that order.

4.6 Effect of Heat Transfer Augmentation on Two-Phase Flow Instabilities

Freon-11 was used as the test fluid and six different heater tubes with various inside surface configurations were tested at five different heat inputs. In addition to temperature and pressure recordings, high speed motion pictures of the two-phase flow were taken for some of the experiments to study the two-phase flow behavior at different operating points [171, 172].

In recent years, there has been increased emphasis on techniques to enhance two-phase flow heat transfer; an excellent survey of the subject is presented by Bergles [173].

A schematic diagram and the basic dimensions of the experimental setup are shown in figure 3. For this investigation, all the tubing except the recovery section was made of 7.493 mm ID Nichrome tube. Six different tubes, which are shown in figure 20, were prepared and used during the experiments. The description of the tubes is in Table 4.

The tubes are classified according to their effective diameters, which is defined as $D_e = (4V/\pi L)^{\frac{1}{2}}$ where V is the net inside volume and L is the length of the heater tube.

Two basic types of oscillations, namely the pressure-drop type and the density-wave type, and various forms of superimposition of these oscillations were observed during the experiments. Experimental results are presented in graphical forms. Figure 21 shows steady-state characteristics of the tube E, with system pressure drop and mass flow rate as coordinates. Oscillation boundaries are also indicated on this figure.

Boundaries for the pure pressure-drop and the superimposed oscillations for different tubes were obtained (figures 22 and 23).

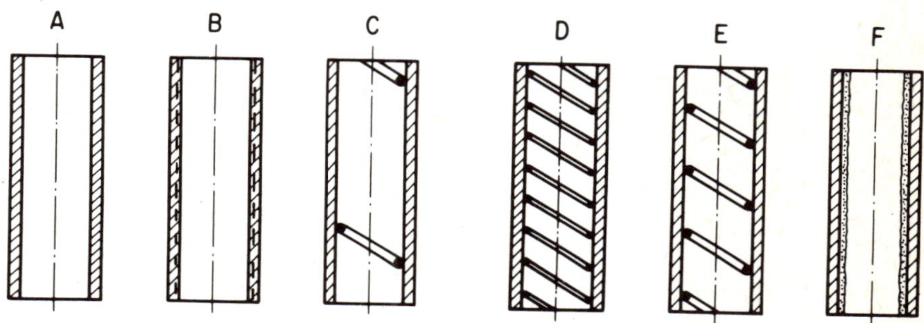

Fig. 20 Heater surface configurations.

Table 4. Description of the Heater Tubes.

TUBE	DESCRIPTION OF THE TUBE	D_e mm
A	Bare	7.493
B	Threaded, 7.938 mm--16 threads per 25.4 mm	7.619
C	With internal spring of 0.794 mm wire diameter and 19.05 mm pitch	7.446
D	With internal spring of 0.432 mm wire diameter and 3.175 mm pitch	7.401
E	With internal spring of 1.191 mm wire diameter and 6.350 mm pitch	7.192
F	Coated with: Union Carbide Linde High Flux Coating	7.073

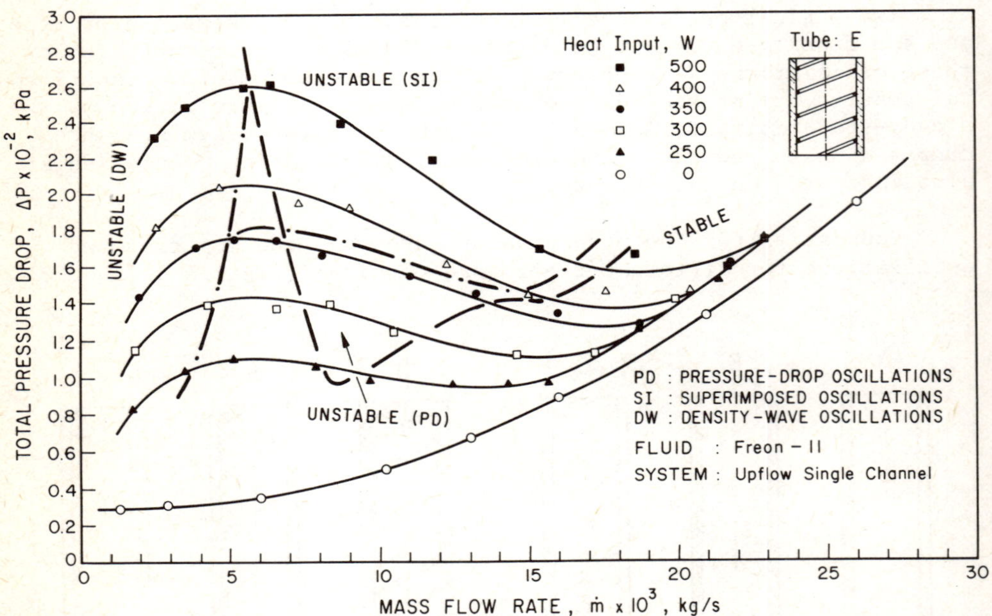

Fig. 21 Steady-state characteristics and stability boundaries. Inlet temperature: 23°C [171, 172].

Table 5. The Surface Condition Coefficient, C_s, for the Tubes Tested.

TUBE	A	B	C	D	E	F
C_s	1.0	1.202	1.012	1.083	1.052	1.938

It was observed that the curves for different heater tubes exhibit a similar pattern, with internally springed tubes having narrower unstable region.

A similar order is also observed at the plots of inlet throttling necessary to stabilize the system against the mass flow as shown on figure 24.

During our studies, the effect of surface configurations on heat transfer coefficient has also been examined. The results have been correlated for use in predicting the oscillating-state boiling heat transfer coefficient as [174]

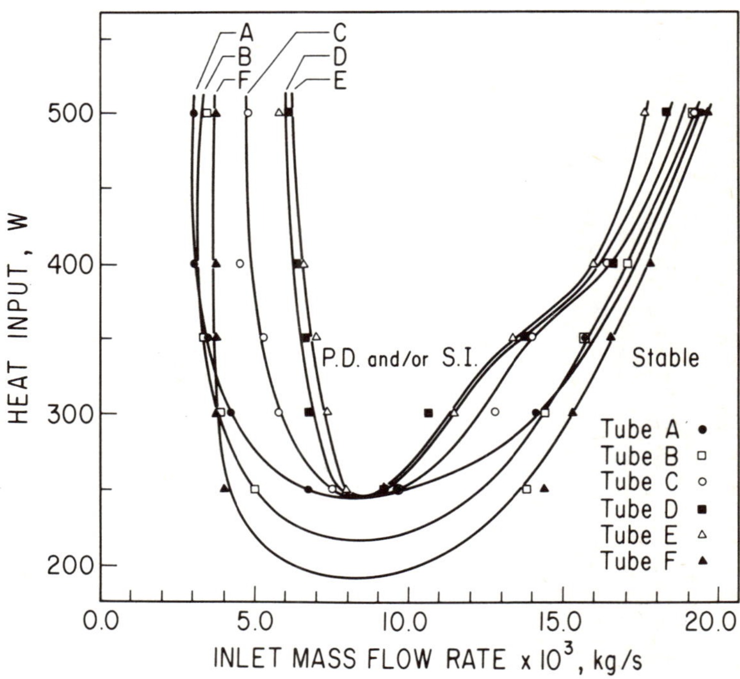

Fig. 22 Boundaries of the pressure-drop type oscillations [171, 172].

$$Nu = 190\ C_s \left(\frac{P}{P_c}\right)^{0.25} \left(\frac{q\,D_e}{\Delta h_v \mu_\ell}\right)^{0.7} e^{-0.125x} \tag{3}$$

where $Nu = \alpha D_e/\lambda_\ell$, α is the time average local heat transfer coefficient (W/K m^2), D_e is the effective diameter of the tube (m), λ_ℓ is the saturated liquid thermal conductivity [W/(mK)], C_s is the surface condition coefficient (Table 5), q is the flux density (W/m^2), x is the vapor quality, P is the working pressure (bar), P_c is the critical pressure of the fluid (bar), μ_ℓ is the saturated liquid viscosity (Kg/ms), Δh_v is the latent heat of vaporization (J/kg).

Calculations show that equation (3) agrees with the current experimental data with an accuracy of $\leq \pm$ 15%.

Comparing the bare tube with the other tubes it has been found out [170] that the time average local heat transfer coefficients during oscillations increased 1 to 8% for different spring configurations, about 20% for the internally threaded tube, and up to 94% for the coated tube.

From the investigation of figures 22, 23 and 24, one can deduce that among the tested tubes, tube with Linde High Flux Coating is the most unstable one.

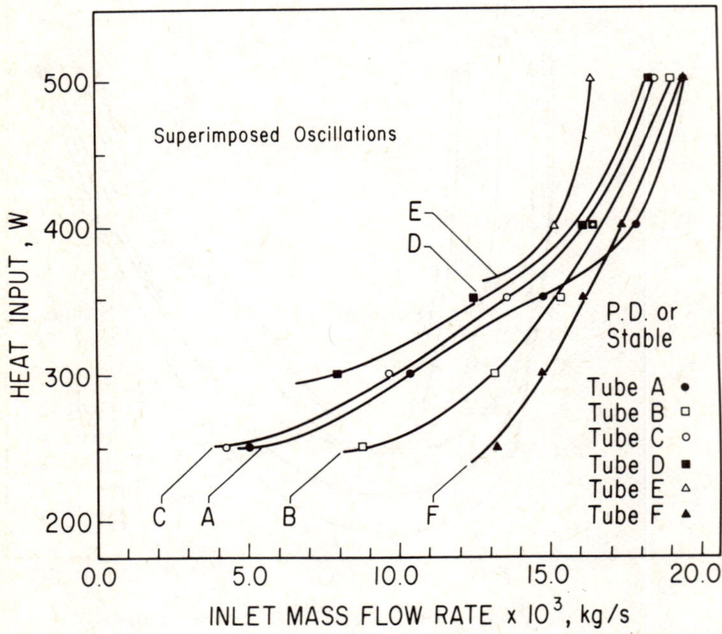

Fig. 23 Boundaries of the pressure-drop type oscillations with superimposed density-wave type oscillations [171, 172].

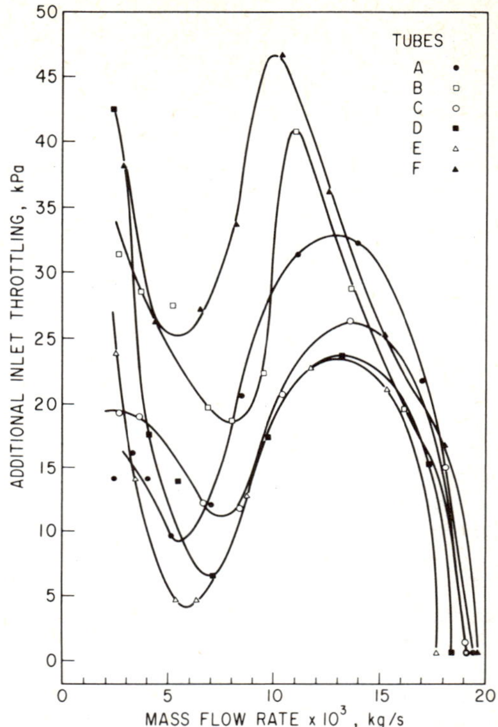

Fig. 24 Inlet throttling necessary to stabilize the system (single channel, upflow boiling; fluid: Freon-11; heat input: 500 W) [171, 172].

5. MATHEMATICAL MODELING OF INSTABILITIES

Modeling of the two-phase flows, in general, starts from the formulation of the conservation equations. Appropriate constitutive relations are necessary to supplement these equations. Boundary conditions are to be written for the particular problem under consideration. At each step of the formulation and the solution physical assumptions and approximations are essential, the accuracy of which can be tested by experiment.

Different models are used to describe the two-phase flow behavior in channels. In the literature, the following models are identifiable:

In the most general formulation of the two-phase flow problem, the conservation equations are written separately for each of the phases. This is called separated-flow model. The resulting equations form the basis of the separated-flow model, or so-called unequal temperature (UVUT) model. In most of the practical problems, the one-dimensional time dependent equations are used; and the problem of the correct statistical and cross-sectional averaging in flow passages arises. The various forms of the conservation equations have been given [158, 175] and [176]. This model can be written using either time averages [142] or space averages[147]. If it is written using space averages, the two-fluid model can deal only with transient flow problems with a single space variable. To solve the six phase equations written for each phase, seven constitutive laws, that is four at the wall (friction and heat transfer), three at the interface (mass, momentum and energy transfer), must be known. Knowledge of the constitutive laws permits the calculation of the flow parameters, such as the phase velocities and phase enthalpies. As a result, slip velocity and thermodynamic nonequilibria for each phase can be computed. Although the two-fluid model is the most satisfactory in theory, it has been thought difficult to use because seven constitutive laws are required.

To reduce the degree of complexity, several authors [55, 161] have suggested a drift velocity, a void fraction, or a slip ratio correlation. In the solution of two-phase flow problems by the use of one of these correlations, the six phase equations are written separately and then combined for the mixture [175].

In our studies of two-phase flow instabilities the homogeneous equilibrium model and the drift-flux model are used. The former assumes a single fluid which is equivalent to a compressible single-phase fluid; the temperature is assumed to be the saturation temperature and the phase velocities are equal. These assumptions are valid for rapid interfacial rates of heat and momentum transfer.

The homogeneous model is the simplest one which provides valuable insights into the physics of the problem; but the latter provides more accurate results.

The homogeneous model's applicability mostly depends on the flow regimes experienced; for those flow regimes such as bubbly, churn or mist flow where the two phases are expected to be well-mixed, the model is most applicable. For annular or slug flow where the phases flow in two separate streams with different velocities, the model does not give accurate results as compared with experimental results.

Drift-flux model is an unequal velocity (UV) model which takes the relative velocity between the phases into account; the phase velocities are related to each other and other flow variables by a drift-flux formulation [55].

In the prediction of the density-wave type and pressure-drop type oscillations and the thresholds of the oscillations are obtained by the use of two models which will be summarized below [164, 172, 177, and 183-185].

5.1 Homogeneous Model

The conservation equations are the same as a single phase flow of a compressible fluid. The conservation laws for mass, momentum and energy as applied to one dimensional, unsteady, homogeneous viscous flows can be written, respectively, as

$$\frac{\partial \rho}{\partial t} = - u \frac{\partial \rho}{\partial z} - \rho \frac{\partial u}{\partial z} \tag{4}$$

$$\rho \frac{\partial u}{\partial t} = -\rho u \frac{\partial u}{\partial z} - \frac{\partial p}{\partial z} - 2 \frac{f}{d} \rho u^2 - \rho g \tag{5}$$

$$\frac{\partial}{\partial t} (\rho h) + \frac{\partial}{\partial z} (Gh) = \frac{4}{d} \alpha(T_w - T_\ell) \tag{6}$$

Although the equations are similar to single phase flow, however the constitutive relations are quite different. These relations, i.e., two-phase friction and heat transfer correlations, account for the two-phase phenomena.

The two-phase flow oscillations have periods in the order of seconds, and the changes in flow properties take place relatively slow so that thermodynamic equilibrium is pertained easily. Experiments conducted showed that the void fraction becomes one very shortly after the boiling point.

The friction factor between the pipe wall and the liquid-vapor mixture is defined for two-phase flow as

$$f = C \left(\frac{\rho u d}{\mu_{tp}} \right)^{-0.25} \tag{7}$$

where μ_{tp} is the effective two-phase viscosity and C is a constant that depends on the roughness of the pipe. Dukler's definition for μ_{tp} is found to be in better agreement with the experimental data which is given by the following expression [178]

$$\mu_{tp} = (1 - x) \frac{\rho}{\rho_\ell} \mu_\ell + x \frac{\rho}{\rho_v} \mu_v \tag{8}$$

The value of C in equation (7) for two-phase flow was almost twofold compared with the value for a smooth tube.

The power input into the fluid is given by

$$q = \alpha A_h (T_w - T_f) \tag{9}$$

It is assumed that the heat transfer coefficient α is linearly dependent on the mass flow rate of the two-phase mixture [33]

$$\alpha = \alpha_o \left[b + (1 - b) \frac{G}{G_o} \right] \tag{10}$$

where α_o is the steady state heat transfer coefficient, G_o is the steady-state mass flux, and b is the fraction of heat transfer independent of the flow rate. b may range from zero to unity and can be evaluated experimentally for a given system or predicted from available correlations of boiling heat transfer.

Initial conditions for both types of oscillations are determined by the steady-state solution of the conservation equations.

The system under study is shown in figure 25. The part of the system from surge tank to the inlet of the heater, including the inlet plenum, is called region 1. The rest of the system is divided into regions according to the regime of flow

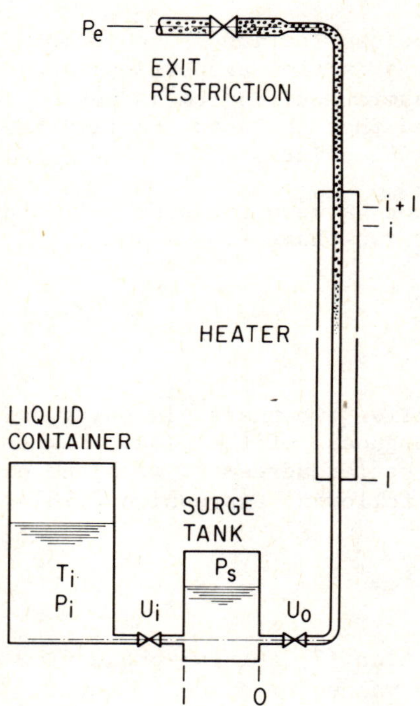

Fig. 25 Schematic of upward boiling flow system.

and boundary conditions.

An empirical relation is used to estimate the pressure-drop across the exit restriction [33]:

$$\Delta p_e = 142\left(1 + x\,\frac{v_{\ell v}}{v_\ell}\right)\frac{G^2 v_\ell}{2} \tag{11}$$

Equations (4), (5) and (6) together with the constitutive relations are solved by the finite difference technique for steady-state characteristics of the two-phase flow system for constant exit pressure condition. The properties are assumed to be known functions of pressure.

The results of the numerical solution together with the experimental findings are shown in figure 26 for various values of inlet temperature.

Pressure-drop type oscillations. The period of the pressure-drop type oscillations is much larger than the residence time of a

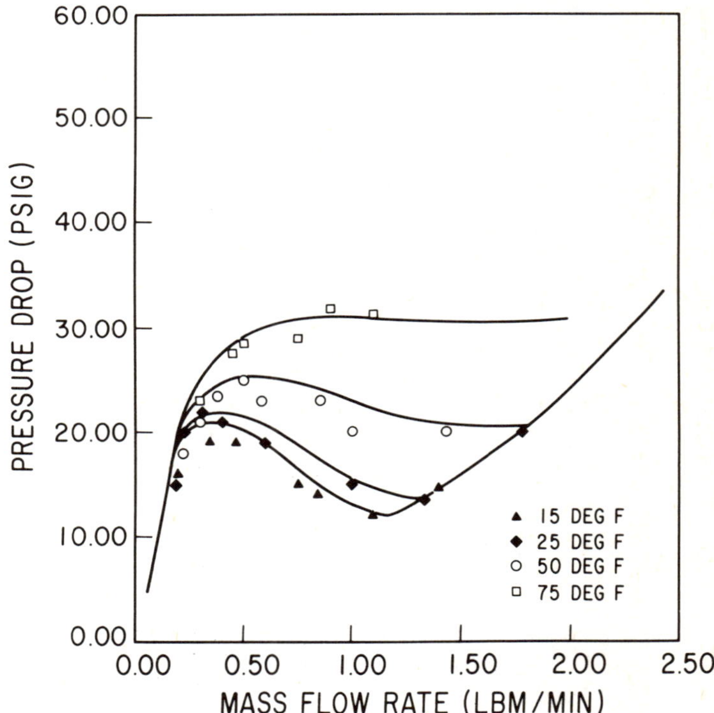

Fig. 26 Comparison of theoretical and experimental steady-state. Heat input: 1023 Btu/h [164, 177].

fluid particle in the heater; therefore, it is reasonable to assume that quasi-steady-state flow conditions prevail in the heater, and that each point in the oscillation corresponds to a steady-state operating point on the set of pressure-drop curves.

For the formulation of the governing equations, an additional equation which describes the surge tank dynamics is included.

For the surge tank, the continuity equation may be written as

$$\frac{dV_\ell}{dt} = A_p(u_i - u_o) - \frac{dV_{ev}}{dt} \tag{12}$$

where V_{ev} is the loss of liquid volume due to evaporation in the surge tank, V_ℓ is the volume of liquid in the surge tank, and A_p is the cross sectional area of the pipe. Since the sum of the liquid volume V_ℓ and gas volume V_g in the surge tank is constant, we write

$$\frac{dV_\ell}{dt} = - \frac{dV_g}{dt} \tag{13}$$

The rate of loss of liquid volume due to evaporation in the surge tank is given by,

$$\frac{dV_{ev}}{dt} = \frac{\rho_v}{\rho_\ell} \frac{dV_g}{dt} \tag{14}$$

Substituting Eqs. (13) and (14) into (12), and rearranging, we obtain,

$$u_i - u_o = - \frac{1}{A_p} (1 - \rho_v/\rho_\ell) \frac{dV_g}{dt} \tag{15}$$

The gas in the surge tank is a mixture of air and saturated vapor. Assuming a perfect gas, and if the liquid temperature is constant and thermal equilibrium prevails, one can get

$$\frac{dP_s}{dt} = P_{sa}^2 \frac{A_p(u_i-u_o)}{P_{sao}V_{go}(1-\rho_v/\rho_\ell)} \tag{16}$$

The rate of heat transfer into fluid is time dependent, the electrical power input into the heater tube $\dot{q}_o$ can be assumed to be time independent. Therefore

$$\dot{q}_o - \dot{q} = (mc_v)_w \frac{dT_w}{dt} \tag{17}$$

where w stands for the heater wall.

Under the quasi-steady-state flow conditions, the mass flow rate into the heater is always equal to the mass flow rate out. Therefore, inlet velocity can be found from

$$\frac{du_i}{dt} = \frac{1}{\rho_i L_i} (P_i - P_s) - K_1 \left(2 \frac{f}{d} u_i^2\right) \tag{18}$$

where K_1 is a constant indirectly proportional to the opening through the inlet valve; and the velocity of fluid out of the surge tank is given by

$$\frac{du_o}{dt} = \frac{1}{\rho_\ell L_t} [(P_s - P_e) - (P_o - P_e)] \tag{19}$$

The $(P_o - P_e)$ term is the steady-state pressure drop across the system. It is a function of u_o and is found from the steady-state solution.

During the oscillations P_i, P_e and T_i are constants and the system is perturbed by increasing the value of K_1 in Equation (17) which means decreasing the inlet velocity by closing the inlet valve.

The equations (16), (17), (18) and (19) are solved for the above boundary conditions by the use of finite difference technique.

A typical result of the theoretical analysis together with the experimental findings are given [33, 177] in figure 27.

Density-wave type oscillations. The previous theoretical analysis [33, 177] has shown that the compressibility of the two-phase region has an important role in generating and sustaining the density-wave oscillations, therefore must be included in the formulation. However, there is no reason why the liquid at the upstream side of the boiling interface cannot be assumed incompressible. A formulation based on this assumption requires different solution methods for each region. This brings up the problem of matching the two solutions at the boiling interface. A technique which solves this problem has been developed [164, 177].

The geometry of the analysis is given in figure 25. The surge tank was excluded in the analysis. The reason is that the liquid level inside the surge tank remains constant during the oscillations, which has been observed in every density-wave instability experiment.

The system is divided into two domains. The point where the vapor quality is around ten percent is defined as the boundary

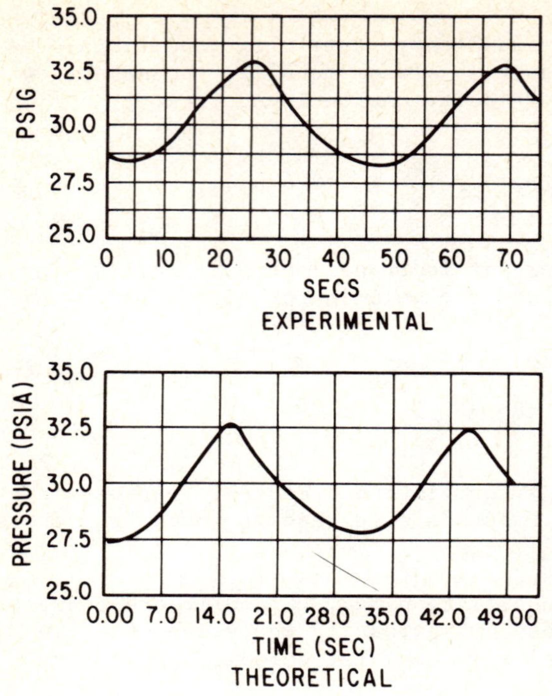

Fig. 27 Numerical prediction of pressure-drop type oscillations. Inlet pressure (heat input: 1023 Btu/h, inlet temp: 25°F) [177].

between the two domains. The upstream of this boundary is assumed incompressible. The downstream side of the interface is a vapor-liquid mixture with an increasing vapor quality and is assumed to be compressible. The governing equations for the compressible region are equations (4)-(6), and the exit pressure boundary condition is constant. The boundary conditions at the interface are open. The equations for the incompressible region are:

$$\frac{\partial(G)}{\partial z} = 0 \tag{20}$$

and the momentum and the energy equations are same as given in equations (5) and (6), respectively. G, the product of ρ and u, is the mass flux at the interface and is taken from the compressible domain solution.

The pressure at the inlet, P_o, and the temperature T_o were assumed to be constant. The initial conditions are determined from the steady-state solution.

The governing equations for the compressible domain are non-linear and hyperbolic. These are rewritten in conservative form as suggested by Ritchmyer and Morton [179]:

$$\frac{\partial A}{\partial t} + \frac{\partial B}{\partial z} + C = 0 \tag{21}$$

where

$$A = \begin{bmatrix} \rho \\ \rho u \\ \rho h \end{bmatrix}, \quad B = \begin{bmatrix} \rho u \\ \rho u^2 + P \\ \rho u h \end{bmatrix}, \quad C = \begin{bmatrix} 0 \\ F \\ -\Phi \end{bmatrix} \tag{22}$$

F is the two-phase friction loss and Φ is the heat input. The Alternating Grid Method is adapted to solve these equations [180].

The pressure is found from the state equation: $P = f(\rho,h)$. This scheme requires that the computational time step is found from Courant-Friedrich stability restriction [179]:

$$\Delta t \leq \frac{\Delta z}{a+|u|} \tag{23}$$

where a is the acoustic speed in the mixture and u is the maximum fluid velocity.

The governing equations for the incompressible flow domain are solved by the Integral Momentum Method and which is described in detail elsewhere [11]. The main assumption of this method is that all the fluid properties are evaluated at some reference pressure, usually the steady-state pressure. This is a good assumption if the flow is incompressible. The method is modified here as follows: The energy equation is written in finite difference form for each segment. The time derivatives are approximated by forward differences. The enthalpy distribution is determined by solving the resulting equations explicitly. The densities are then determined from the equation of state for the reference pressure and the calculated enthalpy. The velocities are determined by using equation (20) where G is the previously determined mass flow rate. Finally, the pressure at the node just before the interface is found by considering the total resistance to the fluid flow.

A computer program has been developed which adopts the integral momentum method for the whole system. The inlet velocity oscillations obtained by this program are shown in figure 28 together with the experimentally observed oscillations.

Lumped parameter analysis of instabilities has also been made to predict both the steady state and

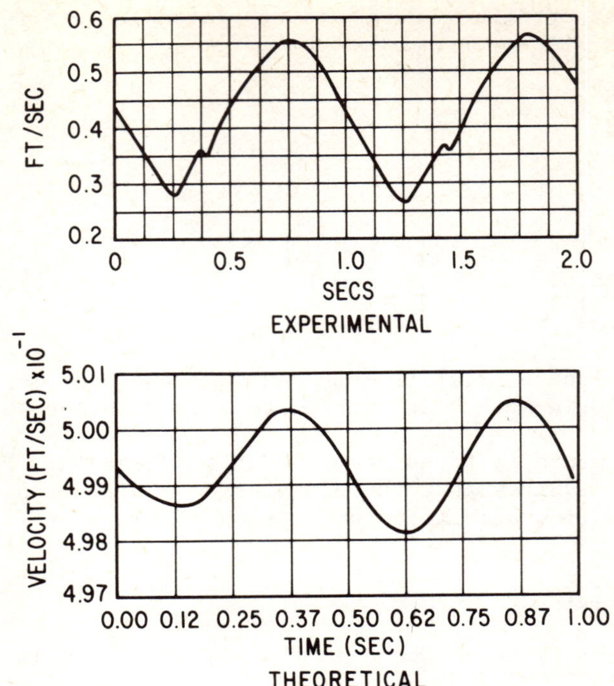

Fig. 28 Numerical prediction of density-wave oscillations--inlet velocity (heat input: 1023 Btu/h, inlet temp: 25°F) [177].

transient behavior of forced convection boiling two-phase flow in a single channel. The model is based on the assumption of homogeneous two-phase flow and thermodynamic equilibrium of the two phases. Compressibility effects in the two-phase region, motion of the bulk boiling interface and the thermal capacity of the heater wall have been included in the analysis. The model is used to study the effects of heat input, inlet subcooling and flow rate on the system behavior. For comparison purposes, an experimental investigation was conducted using a single channel, electrically heated, forced-convection upflow system. Steady state operating characteristics and stable and unstable regions are determined as a function of heat flux, inlet subcooling and mass flow rate. Different modes of oscillations and their characteristics have been investigated [165, 181].

Conservation laws for mass, momentum and energy are given by equations (4), (5) and (6).

The total wall friction force per unit length can be expressed as

$$\tau_w C = \tau_{w\ell} C_\ell + \tau_{wv} C_v \tag{24}$$

where C_ℓ and C_v are the liquid and vapor wetted perimeters given by

$$C_\ell = (1 - \psi)C , \quad C_v = \psi C \tag{25}$$

It was assumed that, a section of two-phase region can be imagined to be composed of alternate slices of liquid and vapor.

Assuming that $\tau_{w\ell}$ and τ_{wv} wall shear stresses for the liquid and vapor are given by the Fanning's relation for the corresponding phases, then equation (24) can be put into a form similar to the single-phase relationship of

$$\tau_w = \frac{1}{2} f\rho u^2 \tag{26}$$

where

$$f = (1 - \psi) \frac{\rho_\ell}{\rho} f_\ell + \psi \frac{\rho_v}{\rho} f_v \tag{27}$$

with

$$f_p = 0.079 \left(\frac{\rho_p ud}{\mu_p} \right)^{-0.25} , \quad p = \ell, v \tag{28}$$

The heat input into the fluid is given by equation (9). The effective heat transfer coefficient α was assumed to be composed of a microconvective term, α_b, and macroconvective term α_c based on Chen's correlation [182]. Thus $\alpha = \alpha_b + \alpha_c$.

The macroconvective heat transfer coefficient was assumed to be contributed proportionately from the liquid and vapor regions:

$$\alpha_c = (1 - \psi) \alpha_\ell + \psi\alpha_v \tag{29}$$

where α_ℓ and α_v are calculated from the Dittus-Boelte equation.

The heater wall temperature, T_w, in equation (9) is obtained from an energy balance of the heater itself,

$$(\rho c A)_w \frac{dT_w}{dt} = \dot{q}A_w - \alpha C(T_w - T_f) \tag{30}$$

In applying the above set of equations, distinction must be made between the two-phase and liquid regions. This is done by defining a bulk boiling boundary where the bulk liquid temperature reaches the local saturation temperature. Assuming a linear variation of both h and h_ℓ in the vicinity of boiling boundary, the expression for the change in the subcooled length, Δb, can be obtained as [165, 181]. The additional two equations are the pressure-drop through the exit restriction and the surge tank dynamic which are given by equations (11) and (16) respectively.

Equations (4), (5), (6), (11), (16), (24) through (30) are solved using a lumped-parameter approach. The basic idea of the method is to view the flow system to be composed of some finite number of segments. The flow variables then are assumed to be represented by appropriate average values over each segment.

In this study, the test section of figure 3 is represented by five lumps. If boiling occurs then that segment is further divided into two segments to separate the liquid region from the two-phase region. The resulting lumped parameter representation of the system is shown in figure 29. The differential equations are replaced with their finite difference forms and steady- and unsteady-state solutions are obtained. The steady-state solution gives the initial and boundary conditions of the problem.

By giving a perturbation to the inlet flow rate the transient response of the system is obtained by solving the algorithm developed.

Typical results of this analysis are shown in figures 30 through

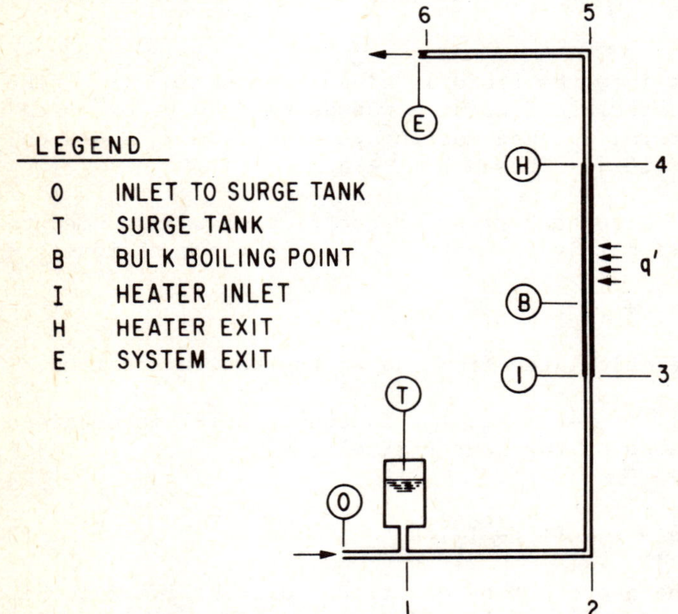

Fig. 29 Lumped-parameter equivalence of the actual flow system.

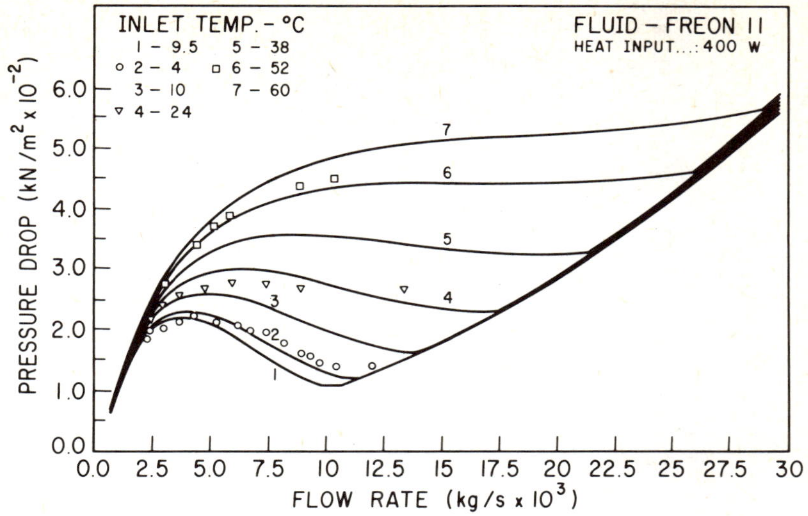

Fig. 30 Steady-state pressure-drop characteristics [33, 181].

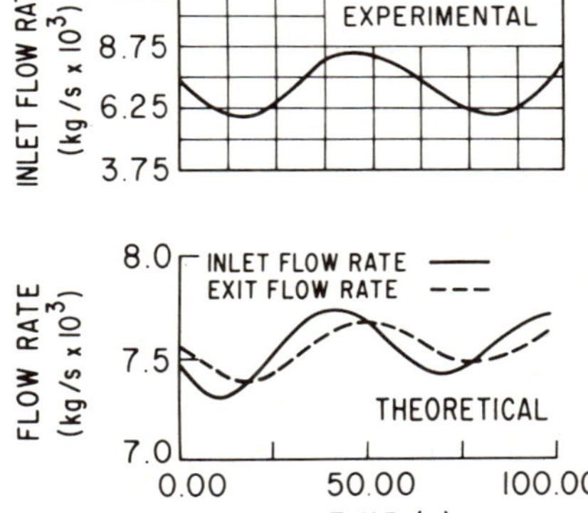

Fig. 31 Experimental and predicted pressure-drop type oscillations (400W, 10°C, 7.6 x 10^{-3}kg/s) [33, 181].

32. Figure 30 shows the steady-state characteristics of the system. Both experimental and theoretical results indicate a region of negative slope which is the characteristic of boiling flow system.

Figure 31 illustrates experimental and predicted pressure-drop type oscillations in mass flow rate in the negative slope region of the pressure-drop curve.

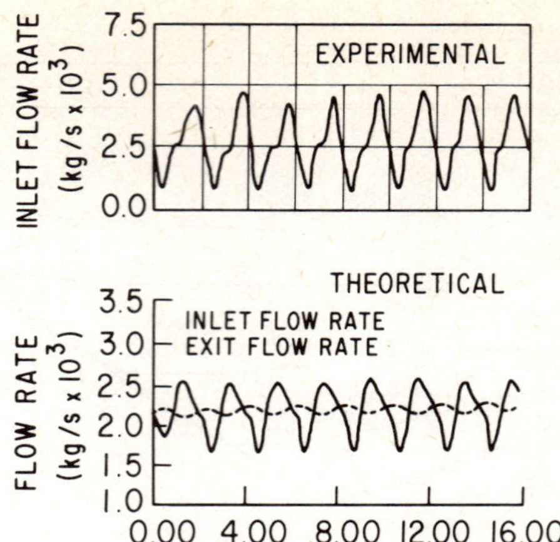

Fig. 32 Experimental and predicted density-wave type oscillations (400W, 24°C, 2.12 x 10^{-3} kg/s) [33, 181].

At higher qualities, as the saturated vapor conditions are approached, the slope of the pressure-drop curve changes sign becoming positive again. This region is also stable and is predicted so by the model. However, as the flow rate is further decreased, new oscillations are observed. Figure 32 is a plot of such oscillations.

The results of these theoretical and experimental analyses verify the following facts:

(i) The variations of heat input into the fluid play an important role in generating and sustaining the pressure-drop oscillations.

(ii) The compressible air-vapor mixture located at the upstream side of the system is equally important in sustaining the pressure-drop type oscillations.

(iii) The total pressure drop in the system is constant and the variations in heat transfer into the fluid are relatively small in the case of density-wave type oscillations.

(iv) The compressibility of the two-phase mixture must be included in the mathematical formulations of the density-wave type oscillations.

(v) The agreement between the experimental results and the prediction of the theory is satisfactory. The predicted frequencies are the same as the experimentally observed frequencies. However, the model predicts lower amplitudes than experiments.

5.2 Instability Thresholds in a Boiling Channel

Two different two-phase flow models, namely a constant-property homogeneous flow model and a variable-property drift-flux model, have been employed to study the instability thresholds in a single upflow boiling system operating between constant pressures with upstream compressibility introduced through a surge tank [172] and [183-185].

5.2.1. Homogeneous model. The three conservation equations given by equations (4), (5) and (6) can be written as follows.

Continuity: $$\frac{\partial\rho}{\partial t} + \frac{\partial(\rho u)}{\partial z} = 0 \qquad (31)$$

Momentum: $$\rho\frac{\partial u}{\partial t} = -\rho u\frac{\partial u}{\partial z} - \frac{\partial p}{\partial z} - 2\frac{f}{d}\rho u^2 - \rho g \qquad (32)$$

Energy: $$\frac{\partial h}{\partial t} = -u\frac{\partial h}{\partial z} + \frac{q'}{\rho A} \qquad (33)$$

These add up to three equations with four dependent variables, ρ, u, p and h. The additional equation is the equation of the state. In the actual state of affairs, the state equation is of the form

$$\rho = \rho(p,h) \qquad (34)$$

However, by neglecting the pressure dependence in this equation, the homogeneous flow model can be simplified to a large extent. In this case, the state equation is written as

$$\rho = \rho(h) \qquad (35)$$

Under conditions of thermodynamic equilibrium and constant-property flow, the specific form of this equation becomes

$$\rho = \rho_\ell, \quad \text{when } h \le h_\ell$$
$$\frac{1}{\rho} = \frac{1}{\rho_\ell} + \frac{v_{\ell v}}{h_{\ell v}}(h - h_\ell), \quad \text{when } h_\ell < h \le h_v \qquad (36)$$

The continuity equation can be written as

$$\frac{\partial\rho}{\partial h}\left[\frac{\partial h}{\partial t} + u\frac{\partial h}{\partial z}\right] + \rho\frac{\partial u}{\partial z} = 0 \qquad (37)$$

Substituting Eq. (32) into Eq. (37), one gets the following equation:

$$\frac{\partial u}{\partial z} = \frac{q'}{A} \frac{d(1/\rho)}{dh} \tag{38}$$

A time constant Θ is defined such that

$$\frac{\partial u}{\partial z} = \frac{1}{\Theta} \tag{39}$$

where

$$\begin{aligned} &\Theta \to \infty && \text{when } h \leq h_\ell \\ &\Theta = Ah_{\ell v}/v_{\ell v}q' && \text{when } h_\ell < h \leq h_v \end{aligned} \tag{40}$$

By combining Eqs. (31) and (39), one obtains

$$\frac{1}{\rho}\left[\frac{\partial \rho}{\partial t} + u \frac{\partial p}{\partial z}\right] = -\frac{1}{\Theta} \tag{41}$$

Then the field equations can be written as

$$\frac{\partial u}{\partial z} = \frac{1}{\Theta} \tag{42}$$

$$\frac{1}{\rho}\left[\frac{\partial \rho}{\partial t} + u \frac{\partial \rho}{\partial z}\right] = -\frac{1}{\Theta} \tag{43}$$

These two equations define a system with $\rho(z,t)$ and $u(z,t)$ as the dependent variables. The boundary condition on this system is provided by the momentum equation, Eq. (32).

The equations (32), (42) and (43) are made dimensionless by defining new variables as follows:

$$\text{Length : } \bar{z} = z/L_h \tag{44}$$

$$\text{Time : } \bar{t} = t/\Theta_s \tag{45}$$

$$\text{Density: } \bar{\rho} = \rho/\rho_o \tag{46}$$

The other terms are non-dimensionalized accordingly, to give

$$\frac{\partial \bar{u}}{\partial \bar{z}} = \frac{1}{\bar{\Theta}} \tag{47}$$

$$\frac{1}{\bar{\rho}}\left[\frac{\partial \bar{\rho}}{\partial \bar{t}} + \bar{u} \frac{\partial \bar{\rho}}{\partial \bar{z}}\right] = -\frac{1}{\bar{\Theta}} \tag{48}$$

and as the boundary condition

$$\frac{\partial \bar{p}}{\partial \bar{z}} = - \bar{\rho} \frac{\partial \bar{u}}{\partial \bar{t}} - \bar{\rho}\bar{u} \frac{\partial \bar{u}}{\partial \bar{z}} - 2 \frac{f}{\bar{d}} \bar{\rho}\bar{u}^2 - \bar{\rho}\bar{g} \tag{49}$$

Since the heat capacity of the wall is neglected, the heat input does not vary with time so that $\bar{\Theta} = \Theta/\Theta_s = 1$ in the above equations.

5.2.2. Drift-flux model. For a number of two-phase flow regimes such as annular flow or slug flow, the assumption of equal phase velocities is not very reasonable. One may take into account the relative velocity between the phases by using a drift-flux formulation, such as the one proposed by Zuber and Findlay. The primary assumptions of the drift-flux model are stated as follows:

(a) The flow is one-dimensional.

(b) The phases are in thermodynamic equilibrium.

(c) The pressure and other fluid properties are uniform across each section taken along the channel. The effect of the radial distribution of the void fraction is represented by a distribution parameter, C_o.

(d) The heat input may be non-uniformly distributed along the channel and the effect of the heat capacity of the wall is taken into account.

(e) The energy equation is limited to a thermal balance.

(f) In the boiling region, the fluid is treated as a mixture of saturated liquid and vapor phases traveling at different velocities.

(g) The axial conduction in the wall is neglected.

(h) The flow is not allowed to reverse.

In addition to these basic assumptions, the liquid phase is assumed to be incompressible so as to simplify the ensuing algebra. The dimensionless conservation equations are written as follows.

Dimensionless continuity equation:

$$\frac{\partial}{\partial \bar{t}} [\bar{\rho}_\ell (1-\psi) + \bar{\rho}_v \psi] + \frac{\partial}{\partial \bar{z}} [\bar{\rho}_\ell \bar{u}_\ell (1-\psi) + \bar{\rho}_v \bar{u}_v \psi] = 0 \tag{50}$$

Dimensionless energy equation:

$$q' = \frac{\partial}{\partial \bar{t}} [\bar{\rho}_\ell \bar{h}_\ell (1-\psi) + \bar{\rho}_v \bar{h}_v \psi] + \frac{\partial}{\partial \bar{z}} [\bar{\rho}_\ell \bar{u}_\ell \bar{h}_\ell (1-\psi) + \bar{\rho}_v \bar{u}_v \bar{h}_v \psi] \qquad (51)$$

Dimensionless momentum equation:

$$- \frac{\partial \bar{p}}{\partial \bar{z}} = \frac{\partial \bar{G}}{\partial \bar{t}} + \left[\frac{\partial \bar{p}}{\partial \bar{z}} \right]_{tp,fric} + \frac{\partial}{\partial \bar{z}} \left\{ \bar{G}^2 \left[\frac{(1-x)^2}{\bar{\rho}_\ell (1-\psi)} + \frac{x^2}{\bar{\rho}_v \psi} \right] \right\} +$$

$$\bar{g}[\bar{\rho}_\ell (1-\psi) + \bar{\rho}_v \psi] = 0 \qquad (52)$$

where

$$\bar{G} = \bar{\rho}_\ell \bar{u}_\ell (1-\psi) + \bar{\rho}_v \bar{u}_v \psi \qquad (53)$$

The dimensionless variables in these equations are related to the actual variables through the following relations:

Time	:	$\bar{t} = t/\tau$ where $\tau = \rho_o h_{\ell vo} A/q'_s$	(54)
Length	:	$\bar{z} = z/L_h$	(55)
Density	:	$\bar{\rho} = \rho/\rho_o$	(56)
Enthalpy	:	$\bar{h} = h/h_{\ell vo}$	(57)
Velocity	:	$\bar{u} = u\tau/L_h$	(58)
Pressure	:	$\bar{p} = p\tau^2/\rho_o L_h^2$	(59)
Gravitational acceleration	:	$\bar{g} = g\tau^2/L_h$	(60)
Heat input	:	$\bar{q}' = q'/q'_s$	(61)
Temperature	:	$\bar{T} = T\, c_{po}/h_{\ell vo}$	(62)

Since the effect of the wall heat capacity is included in the model, a separate energy equation should be written for the wall. Neglecting the axial heat conduction term, an energy balance for the tube wall leads to the following dimensionless equation:

$$\lambda_w \frac{d\bar{T}_w}{d\bar{t}} = 1 - \bar{q}' \qquad (63)$$

where the dimensionless constant, λ_w, is defined as

$$\lambda_w = \frac{\rho_w c_w A_w}{\rho_o c_{po} A} \tag{64}$$

Thus, Eqs. (50), (51), (52) and (63) constitute the governing equations for the flow. The quality, x, and the void fraction, ψ, in these equations are defined in the following way:

$$x = \frac{\rho_v u_v A_v}{\rho_\ell u_\ell A_\ell + \rho_v u_v A_v} \tag{65}$$

and

$$\psi = A_v / A \tag{66}$$

where A_ℓ and A_v are the flow areas for the liquid and vapor, respectively, and A is the total cross-sectional area.

The vapor velocity, u_g, is related to the volumetric flux, j, by the following expression of Zuber and Findlay:

$$u_g = C_o + u_{gj} j \tag{67}$$

where C_o is a distribution parameter and u_{gj} is the drift velocity [55, 186].

The two-phase frictional pressure gradient in the momentum equation is related to the single-phase pressure gradient through the use of a two-phase friction multiplier which is related to the quality, x, by the following relation:

$$F_M = c_1 x^2 + c_2 x + 1 \tag{68}$$

where c_1 and c_2 are constants.

The system stability was studied both in the frequency domain and the time domain. A schematic diagram of the system is provided in figure 25. In the frequency-domain analysis, the governing equations for both models were first linearized for small perturbations and the stability of the linearized equation set was investigated. Using the homogeneous-flow model, simple algebraic expressions have been developed for the steady-state system pressure drop and the pressure-drop instability boundaries. Computer solutions have been generated for the drift-flux model to the same purposes. A reasonable agreement with experimental results is indicated in both cases. Steady-state predictions of the homogeneous and the drift-flux models are given in figures 33 and 34. The pressure-drop stability predictions of the two models are presented in figures 35 and 36.

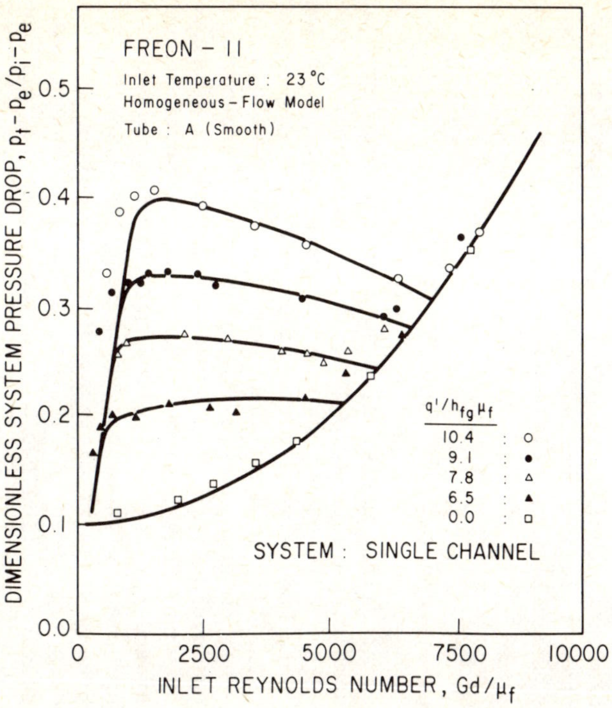

Fig. 33 Dimensionless system pressure drop as a function of inlet Reynolds number at different heat inputs as predicted by homogeneous-flow model; comparison with experiments (inlet temperature = 23°C) [172, 185].

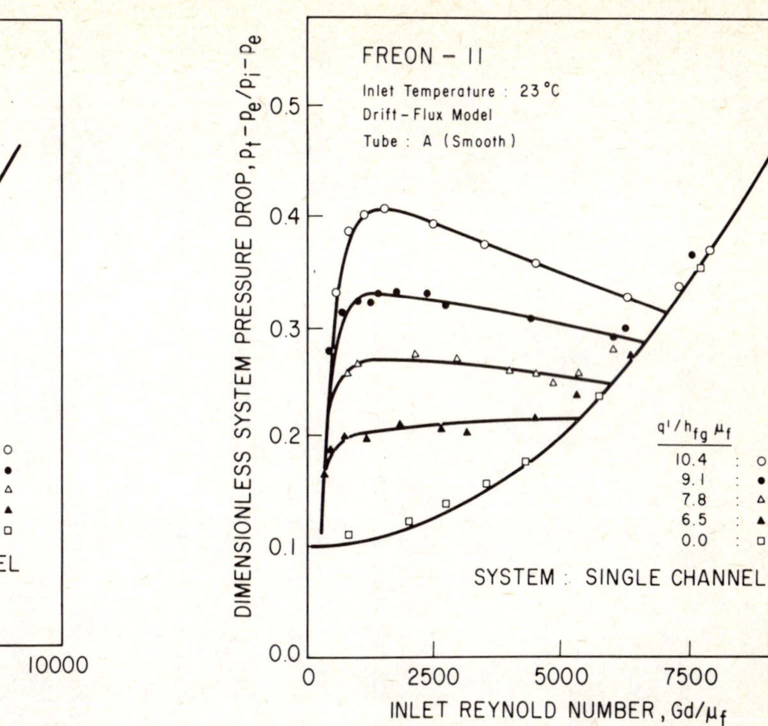

Fig. 34 Dimensionless system pressure drop as a function of inlet Reynolds number at different heat inputs as predicted by drift-flux model; comparison with experimental results (inlet temperature = 23°C) [172, 185].

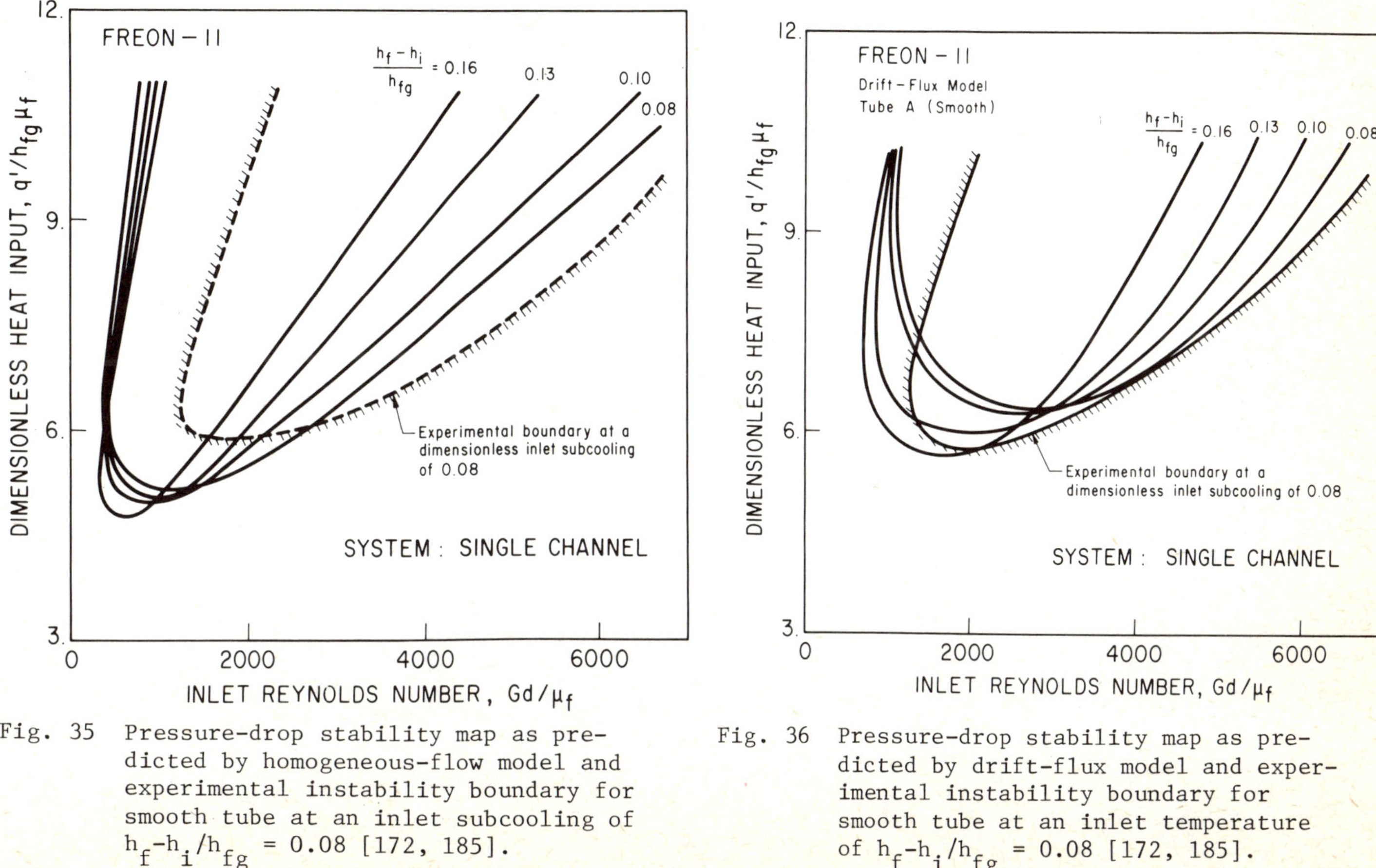

Fig. 35 Pressure-drop stability map as predicted by homogeneous-flow model and experimental instability boundary for smooth tube at an inlet subcooling of $h_f - h_i / h_{fg} = 0.08$ [172, 185].

Fig. 36 Pressure-drop stability map as predicted by drift-flux model and experimental instability boundary for smooth tube at an inlet temperature of $h_f - h_i / h_{fg} = 0.08$ [172, 185].

A unified criterion with respect to both the pressure-drop and the density-wave oscillations was developed. As a measure of the relative instability of the system under different conditions, the amounts of inlet throttling necessary to stabilize the system at particular operating points were calculated. Obviously, the bigger the magnitude of the stabilizing inlet throttling coefficient at a particular operating point, the less stable is the system at that point. The computations were performed for both models. In the homogeneous-flow model, the effects of the wall heat storage and the variation of the fluid properties with pressure were neglected and the system pressure drop after the surge tank was assumed to be concentrated at the two restrictions placed before and after the heater. Due to these simplifications, the quantitative predictions of the model were not accurate. This model was mainly used for a qualitative description and understanding of the problem. A sufficiently complex and realistic representation of the system was realized through using a drift-flux model. In the drift-flux model, the heat flux changes in time and along the channel; the distributed

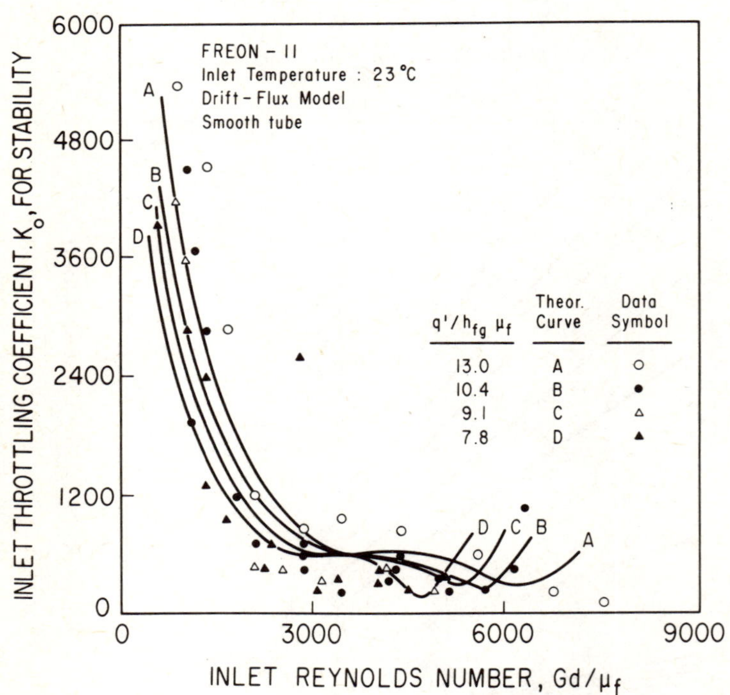

Fig. 37 Inlet throttling required to stabilize system as a function of inlet Reynolds number and dimensionless heat input; comparison between predictions of drift-flux model and experimental results (inlet temperature = 23°C) [172, 185].

pressure drop terms due to two-phase gravitational, accelerational and inertial pressure drops were included in the momentum equation and all the fluid properties were evaluated at local pressures. The result of this analysis is shown in figure 37.

Although analytical solutions under the reasonable conditions and complex codes are now available for predicting two-phase flow instabilities, the analytical modeling of two-phase flow instability phenomena still remains limited.

The presently developed models, i.e., the homogeneous model and the drift-flux model, can be extended to parallel channels with augmented surfaces. The conservation equations for one-dimensional two-phase flows in parallel channel systems can be solved. The effects of the radial distribution of velocities, enthalpies and void fractions could be taken into account through the use of correlation coefficients. New alternative forms of the conservation equations that might be more useful for parallel channel flow systems could be formulated and solved.

ACKNOWLEDGEMENTS

The authors gratefully acknowledge the financial support of the National Science Foundation and the Scientific Affairs Division of NATO. The authors wish to thank Messrs. A. Menteş, H. Gürgenci, and O. T. Yıldırım for their work and assistance in the preparation of the manuscript.

NOMENCLATURE

A area, cross section, m^2

a speed of sound, m/s

C perimeter, m

c specific heat capacity, kJ/kgK

c_p specific heat capacity at constant pressure, kJ/kgK

D_e equivalent diameter, m

d diameter, m

f friction factor, dimensionless

G mass flux, kg/m^2s

g –	gravitational acceleration, m/s^2
h	fluid enthalpy, kJ/kg
L	tubing length, m
$\dot{m}$	mass flow rate, kg/s
p	pressure, kN/m^2
p_{sa}	unsteady-state pressure of the air, kN/m^2
p_{sao}	steady-state pressure of the air
q	rate of heat transfer, kJ/s
q'	rate of heat transfer per unit length, W/m
$\dot{q}$	rate of heat generation per unit volume, W/m^3
T	temperature, K
t	time, s
u	velocity, m/s
V	volume, m^3
V_{go}	steady-state pressure of the air, m^3
x	flow mixture quality, dimensionless
z	axial distance, m
α	heat transfer coefficient, W/m^2k
μ	dynamic viscosity, kg/ms
λ	thermal conductivity, W/mk
ν	kinematic viscosity, m^2/s
ρ	density, kg/m^3
ρ_ℓ	density of liquid, kg/m^3
ρ_v	density of vapor, kg/m^3
σ	surface tension, N/m

τ shear stress, N/m^2

ψ void fraction, dimensionless

Θ reference time, $Ah_{\ell v}/v_{\ell v}q'_s$, dimensionless

Nu Nusselt number, $\alpha d/\lambda$, dimensionless

Pr Prandtl number, $c_p\mu/\lambda$, dimensionless

Re Reynolds number, ud/ν, dimensionless

Subscripts

a air

b boiling

c convective

e exit

f liquid

g gas

ℓ saturated liquid

i inlet, spatial index

j temporal index

p phase, pipe

s steady-state, surge tank

t total

tp two-phase

v saturated vapor

w wall

REFERENCES

1. Lorentz, H., "Die arbeitsweise und berechnung der druckluft-flüssigkeitsheber," Zeitschr. des ver. Deutscher Ingenieure (1909).

2. Ledinegg, M., "Instability of flow during natural and forced circulation," Die Wärme, 61, 891 (1938).

3. Profos, P., "Die stabilität der wasserverteilung in zwanglauf-heizflächen," Sulzer Technical Review, 1, 1 (1947).

4. Chilton, H., "A theoretical study of stability in water flow through heated passages," J. Nuclear Energy, 5, 271 (1957).

5. Fraas, A. P., "Flow stability in heat transfer matrices under boiling conditions," ORNL Report 59-11-1 (1961).

6. Zuber, N., "Flow excursions and oscillations in boiling two-phase flow systems with heat additions," Proc. Symp. Two-Phase Flow Dynamics, Eindhoven, v. 1, 1091 (1967).

7. Wissler, E. H., Isbin, H. S., and Amundson, N. R., "Oscillatory behavior of a two-phase natural circulation loop," AIChE J., 2 (2), 157-162 (1956).

8. Wallis, G. B. and Heasley, J. H., "Oscillations in two-phase flow systems," J. Heat Transfer, Trans. ASME, Series C, 83, 363 (1961).

9. Quandt, E. R., "Analysis and measurements of flow oscillations," Chem. Eng. Prog., Monograph and Symposium Series, 32 (57), 111-126 (1961).

10. Bouré, J., "Hydrodynamic instabilities limiting the power of boiling water reactors," EURAEC 1464 (1965).

11. Meyer, J. and Rose, R., "Application of a momentum integral model to the study of parallel channel boiling flow oscillations," J. of Heat Transfer, Trans. ASME, Series C, 85, 1 (1963).

12. Gouse, S. W. and Andrysiak, C. D., "Flow oscillations in a closed loop with transparent, parallel, vertical heated channels," M.I.T., NSF Grants G11355 and G19771, Report No. 8973-2 (1963).

13. Stenning, A. H. and Veziroglu, T. N., "Flow oscillation modes in forced convection boiling," Proc. 1965, Heat Transfer and Fluid Mechanics Institute, Stanford University Press, 301 (1965).

14. Zuber, N., "An analysis of thermally induced flow oscillations in the near-critical and super-critical thermodynamic region," NAS 8-11422, Final Report (1966).

15. Villeneuve, J., "Study on flow oscillation in a boiling channel," NASA n65-26838 (1964).

16. Bouré, J., "The oscillatory behavior of heated channels--An analysis of density effects," CEAR 3049, Centre d'Etudes Nucleaires de Grenoble, France (1966).

17. Bouré, J. and Mikaila, A., "The oscillatory behavior of heated channels," Proc. Symp. Two-Phase Flow Dynamics, Eindhoven, Netherlands (1967).

18. Bouré, J. A., Bergles, A. E., and Tong, L. S., "Review of two-phase flow instability," Nuclear Engineering and Design, 25, 165 (1973).

19. Ishii, M., "Study on flow instabilities in two-phase mixtures," ANL-76-23 (1976).

20. Bergles, A. E., "Review of instabilities in two-phase systems," Eds. S. Kakac, F. Mayinger, T. N. Veziroglu, Two-Phase Flows and Heat Transfer, Hemisphere Publishing Corp., USA (1977).

21. Yadigaroglu, G., "Two-phase flow instabilities and propagation phenomena," Thermohydraulics of Two-Phase Systems for Industrial Design and Nuclear Engineering, Eds. J. M. Delhaye, M. Giot, M. L. Riethmüller, Hemisphere Publishing Corporation (1981).

22. Bergles, A. E., Goldberg, P., and Maulbetsch, J. S., "Acoustic oscillations in high pressure single channel boiling system," Symp. on Two-Phase Flow Dynamics, Eindhoven, Netherlands (1967).

23. Stenning, A. H., Veziroglu, T. N., and Callahan, G. M., "Pressure-drop oscillations in forced convection flow with boiling," EURATOM Symp. on Dynamics of Two-Phase Flows, September 1967.

24. Veziroglu, T. N., "Oscillations in boiling upward flow," AEC Oak Ridge National Laboratory-Union Carbide Subcontract No. 2785, Final Report (1967).

25. Veziroglu, T. N. and Lee, S. S., "Instabilities in boiling upward flow," International Symp. on Cocurrent Gas-Liquid Flow, September 1968.

26. Veziroglu, T. N. and Lee, S. S., "Boiling upward flow instabilities," AEC-Oak Ridge National Laboratory Subcontract No. 2975,

Final Report (1968).

27. Veziroglu, T. N. and Lee, S. S., "Boiling flow instabilities in a two parallel channel upflow system," AEC-Oak Ridge National Laboratory Subcontract No. 2975, Final Report (1969).

28. Veziroglu, T. N. and Lee, S. S., "Boiling flow instabilities in a cross-connected parallel channel upflow system," AEC-Oak Ridge National Laboratory Subcontract No. 2975, Final Report (1970).

29. Veziroglu, T. N. and Lee, S. S., "Sustained and transient boiling flow instabilities in a cross-connected parallel channel upflow system," AEC-ORNL-Union Carbide Subcontract No. 2975, Final Report (1971).

30. Kakaç, S., Veziroglu, T. N., Akyüzlü, K. and Berkol, O., "Sustained and transient boiling flow instabilities in a cross-connected four parallel channel upflow system," 5th Int. Heat Transfer Conference, Paper No. B5.11, Tokyo, Japan (1974).

31. Veziroglu, T. N., Lee, S. S., and Kakac, S., "Fundamentals of two-phase oscillations and experiments in single-channel systems," Eds. S. Kakac, F. Mayinger, T. N. Veziroglu, Two-Phase Flows and Heat Transfer, Hemisphere Publishing Corp., USA (1977).

32. Veziroglu, T. N., Kakaç, S., Williams, A., and Lee, S. S., "Two-phase flow instabilities and effect of inlet subcooling," prepared for National Science Foundation, Final Report (1978).

33. Veziroglu, T. N. and Kakaç, S., "Two-phase flow instabilities and effect of inlet subcooling," NSF Project Eng 75-16618, Final Report (1980).

34. Veziroglu, T. N. and Kakaç, S., "Two-phase flow instabilities," NSF Project CME 79-20018, First Annual Report (1981).

35. Maulbetsch, J. S. and Griffith, P., "System-induced instabilities in forced convection flow with subcooled boiling," Proc. 3rd Int. Heat Transfer Conference, Chicago, Illinois, v. 4, 247 (1966).

36. Hayama, S., "A study of hydrodynamic instability in boiling channels," Bull. JSME, vol. 6, no. 23, pp. 549-56 (1963); vol. 7, no. 25, pp. 129-135 (1964); vol. 10, no. 38, pp. 308-319, 320-327 (1967).

37. Akagawa, K., Sakaguschi, T., Kono, M., and Nishimura, M., "Study on distribution of flow rates and flow stabilities in

parallel long evaporators," Bull. JSME, vol. 14, p. 837 (1971).

38. Marris, A. W., "Hydraulic stability in the simple surge tank," Canadian J. of Tech., 34, 182-198 (1956).

39. Jeglic, F. A. and Grace, T. M., "Onset of flow oscillations in forced flow subcooled boiling," NASA TN-D-2821 (1965).

40. Berenson, P. J., "An experimental investigation of flow stability in multitube forced convection vaporizers," ASME Paper 65-HT-61 (1965).

41. Schuster, J. R. and Berenson, P. J., "Flow stability of a five-tube forced-convection boiler," ASME Paper 67-WA/HT-20 (1967).

42. D'Arcy, D. F., "An experimental investigation of boiling channel flow instability," Proc. Symp. Two-Phase Flow Dynamics, Eindhoven, Netherlands (1967).

43. Davies, A. L., and Potter, R., "Hydraulic stability, an analysis of the causes of unstable flow in parallel channels," Proc. Symp. Two-Phase Flow Dynamics, Eindhoven, Netherlands (1967).

44. Friedly, J. C., Manganaro, J. L., and Kroeger, P. G., "Stability investigation of thermally induced flow oscillations in cryogenic heat exchangers," NASA Contract 21045, Final Report (1967).

45. Zuber, N. and Staub, F. W., "The propagation and the wave form of the vapor volumetric concentration in boiling, forced convection system under oscillatory conditions," Int. J. Heat Mass Transfer, vol. 9, pp. 871-895 (1966).

46. Carver, M. B., D'Arcy, D. F., Wikhammer, G. A., Casterline, J. E., and Matzner, B., "Instabilities in flow through multirod bundles and their effect on dryout," Int. Symp. on Research in Cocurrent Gas-Liquid Flow, Waterloo, Canada (1968).

47. Crowley, J. D., Deane, C. and Gouse, S. W., Jr., "Two-phase oscillations in vertical, parallel, heated channels," Proc. Symp. on Two-Phase Flow Dynamics, Commission of the European Communities, Brussels, 1131-1172 (1969).

48. Crowley, J. D. and Bergles, A. E., "Fluid-to-fluid modeling of the hydrodynamic stability of flow in boiling channels," ASME Paper, No. 70-HT-28 (1970).

49. Collins, D. B., Gacesa, M. and Parsons, C. B., "Study of the onset of premature heat-transfer crisis during hydrodynamic

instability in a full-scale reactor channel," ASME Paper, 71-HT-11 (1971).

50. Matsui, G., "Flow instability in boiling channel systems," Int. Chem. Eng., 11 (3), 554-560 (1971).

51. Matsui, G., "Experimental study of flow instability in boiling channel systems," Heat Transfer--Japanese Research, 1 (1), 46-52 (1972).

52. Yadigaroglu, G. and Bergles, A. E., "Fundamental and higher mode density-wave oscillations in two-phase flows," Trans. ASME, J. Heat Transfer, 189-195 (May 1972).

53. Friedly, J. C., "RS164: An experimental study of oscillations in a liquid nitrogen evaporator," UKAEA Report, AERE-R8487 (1976).

54. Ishii, M. and Zuber, N., "Thermally induced flow instabilities in two-phase mixtures," 4th International Heat Transfer Conference, Paris, Paper No. B5-11 (1970).

55. Zuber, N. and Findlay, J., "Average volumetric concentration in two-phase flow systems," Trans. ASME, Series C, J. of Heat Transfer, 87, 453 (1965).

56. Saha, P., Ishii, M. and Zuber, N., "An experimental investigation of the thermally induced flow oscillations in two-phase systems," Trans. ASME, J. Heat Transfer, 616-622 (1976).

57. Lahey, R. T., Jr., and Moody, F. J., "The thermal-hydraulics of boiling water reactor," American Nuclear Society, Hinsdale, Ill. (1977).

58. Blumenkrantz, A. and Taborek, J., "Application of stability analysis for design of natural circulation boiling systems and comparison with experimental data," AIChE Symp. Ser., vol. 68, no. 118, pp. 136-146 (1972).

59. Fukuda, K. and Kobori, T., "Two-phase flow instability in parallel channels," 6th Int. Heat Transfer Conference, Toronto, Ontario, Canada (1978).

60. Fukuda, K. and Hasegawa, S., "Analysis on two-phase flow instability in parallel multichannels," J. Nuc. Sci. Technol., 16 (3), 190-199 (1979).

61. Takitani, K., "Density-wave instability in once-through boiling flow system--I," J. Nuc. Sci. Technol., 15 (5), 355 (1978).

62. Takitani, K., "Density-wave instability in once-through boiling flow system--II," J. Nuc. Sci. Technol., 15 (6), 389 (1978).

63. Nakanishi, S., Ishigai, S., Ozawa, M. and Mizuta, Y., "Analytical investigation of density wave oscillation," Technology Reports of the Osaka University, v. 28, No. 1421 (1978).

64. Ozawa, M., Akagawa, K., Sakaguchi, T., Tsukahara, T. and Fujii, T., "Oscillatory flow instabilities in air-water two-phase flow system--1. Pressure drop oscillation," Bull. JSME, 22 (174), 1763-1770 (1979).

65. Ünal, H. C., "Density-wave oscillations in sodium-heated once-through steam generator tubes," Trans. ASME, J. Heat Transfer, 103, 485-491 (1981).

66. Romberg, T. M. and Rees, N. W., "Multivariate hydrodynamic analysis of boiling channel flow stability from inherent noise measurements," Int. J. Multiphase Flow, 6, 523-551 (1980).

67. Romberg, T. M. and Rees, N. W., "On the modelling of boiling channel dynamics using spectral methods," Int. J. Control, 34 (2), 259-284 (1981).

68. Bankoff, S. G., "Some condensation studies pertinent to LWR safety," Int. J. Multiphase Flow, 6 (1-2), 51-67 (1980).

69. Bhatt, B. L. and Wedekind, G. L., "A self-sustained oscillatory flow phenomenon in two-phase condensing flow systems," Trans. ASME, J. Heat Transfer, 102, 694-700 (1980).

70. Wedekind, G. L., Bhatt, B. L. and Beck, B. T., "A system mean void fraction model for predicting various transient phenomena associated with two-phase evaporating and condensing flows," Eds. S. Kakac, F. Mayinger and T. N. Veziroglu, Two-Phase Flows and Heat Transfer, Hemisphere Publishing Corporation, New York, pp. 1201-1226 (1977).

71. Pitts, J. H., "Steam chugging in a boiling water reactor pressure-suppression system," Int. J. Multiphase Flow, 6 (4), 329-344 (1980).

72. Takeda, Y. and Toda, S., "Pressure oscillation in subcooled decompression under temperature gradient," J. Nuc. Sci. Technol., 16 (7), 484-495 (1979).

73. Henry, R. E., Grolmes, M. A. and Fauske, H. K., "Propagation velocity of pressure waves in gas-liquid mixtures," Int. Symp. on Research in Cocurrent Gas-Liquid Flow, Waterloo, Canada (1968).

74. Henry, R. E., "Pressure wave propagation in two-phase mixtures," AIChE-ASME, 11th National Heat Transfer Conference, Minneapolis (1969).

75. Mori, Y., Hijikata, K. and Komine, A., "Propagation of pressure waves in two-phase flow," Int. J. Multiphase Flow, 2, 139-152 (1975).

76. Zenz, F. A., "Conveyability of materials of mixed particle size," I and EC Fund., 3, 65-75 (1964).

77. Canada, G. S., McLaughlin, M. H. and Staub, F. W., "Flow regimes and void fraction distribution in gas fluidization of large particles in beds without tube banks," AIChE Symp. Ser. v. 74, no. 176 (1978).

78. Tomita, Y., Jotaki, T. and Hayashi, H., "Wavelike motion of particulate slugs in a horizontal pneumatic pipeline," Int. J. Multiphase Flow, 7 (2), 151-166 (1981).

79. Cumo, M., Palazzi, G. and Rinaldi, L., "An experimental study on two-phase flow instability in parallel channels with different heat flux profiles," Camitato Nazionale Energia Nucleare, Report CNEN-RT/ING (1981).

80. Nakanishi, S., Ozawa, M. and Ishigai, S., "The modes of flow oscillations in multi-channel two-phase flow systems," in Advances in Two-Phase Flows and Heat Transfers, Eds. S. Kakaç and M. Ishii, Sijhoff & Noordhoff International Publisher BV (1983).

81. Akagawa, K., "Recent research on two-phase flow in universities and colleges in Japan," in Two-Phase Flow Dynamics, Eds. A. E. Bergles and S. Ishigai, Hemisphere Publishing Corp., New York (1981).

82. Nakanishi, S., "Recent Japanese researchers on two-phase flow instabilities," in Two-Phase Flow Dynamics, Eds. A. E. Bergles and S. Ishigai, Hemisphere Publishing Corp., New York (1981).

83. Bergles, A. E. and Ishigai, S. (Eds.), Two-Phase Flow Dynamics, Hemisphere Publishing Corp., New York (1981).

84. Davidov, A. A., "The study of flow pulsation in tubes of the boiling part of once through boilers," Hydrodynamics and Boiling Heat Transfer in High Pressure Boilers, U.S.S.R. (1955).

85. Davidov, A. A., "Elimination of pulsation in once through boilers," Elektricheskie Stantzu, No. 3, U.S.S.R. (1956).

86. Davidov, A. A., and Shenin, B. I., "Hydrodynamics of once through boilers," Teploenergetika, No. 11, pp. 53-57, U.S.S.R. (1959).

87. Semenovkel, I. I., "On the onset of pulsation in the boiling tubes of high pressure boilers," Hydrodynamics and Boiling Tubes of High Pressure Boilers, U.S.S.R. (1955).

88. Armand, A. A., "The flow mechanism of a two-phase mixture in a vertical tube," Hydrodynamics and Heat Transfer During Boiling in High Pressure Boilers, ed. M. A. Styrikovich, trans. from Russian (1955).

89. Petrov, B. A., "Hydrodynamics of once through boilers," U.S.S.R. (1960).

90. Morozov, I. I., "The stability of the steam generation process in a turn of a once through boiler," Inzhener Fiz. Zhur. Akad. Nauk., U.S.S.R., No. 8, pp. 49-53 (1961).

91. Klaziakova, L. V., and Gloskel, B. N., "On flow stability in U type tube element of once through boilers," Teploenergetika, No. 11, pp. 41-46, U.S.S.R. (1963).

92. Treshev, G. G., "Water-steam flow oscillation in heated tubes," Energo-Machinostroenie, No. 3, pp. 4-7, U.S.S.R. (1964).

93. Celov, I. B., Smilnov, O. K., and Zerkov, L. A., "An experimental study on the flow stability boundary of steam-generating tubes in parallel with uneven heat inputs," Teploenergetika, No. 10, U.S.S.R. (1964).

94. Dolgov, V. V. and Sudnitsyn, O. A., "On hydrodynamic instability in boiling water reactors," Teploenergetika, No. 10, pp. 51-55, U.S.S.R. (1955).

95. The Standard of Hydraulic Calculations of Steam Boilers, U.S.S.R. (1961).

96. Baldina, O. M. and Kalinin, B. I., "The oscillation phenomenon in boiling tube elements," Energo-Machinostroenie, No. 7, pp. 10-13, U.S.S.R. (1966).

97. Baldina, O. M., Kalinin, R. K., and Baitina, I. Z. M., "Flow oscillations in the horizontal elements of steam generator," Teploenergetika, No. 8, pp. 35-39, U.S.S.R. (1968).

98. Habenski, V. B., and Baldina, O. M., "On analysis of discharge pulsation in the system of steam-generating tubes in parallel," Inzhner Zhur. Akad. Nauk., No. 5, pp. 820-828, U.S.S.R. (1969).

99. Habenski, V. B., and Betilson, D. V., "Determine the flow oscillation boundary in system of parallel horizontal steam generating tubes," Teploenergetika, No. 7, pp. 78-80, U.S.S.R. (1970).

100. Habenski, V. B., Baldina, O. M., and Kalinin, B. I., "Fluid oscillation in parallel vertical heated tubes," Energo-Machinostroenie, No. 3, pp. 17-18, U.S.S.R. (1971).

101. Habenski, V. B., Baldina, O. M., and Kalinin, B. I., "The mechanism of pulsation and the influence of constructive and operating parameters on fluid stability boundary," Achievements in the Studying Region of Two-Phase Flow Hydrodynamics and Heat Transfer in Elements of Power Devices, U.S.S.R. (1976).

102. Koshelev, I. I., Sulnov, A. V., and Nikitina, L. V., "Inception of pulsations using a model of water-wall tubes," Heat Transfer Soviet Research, No. 2 (1970).

103. Leleev, N. S., "The pulsation of steam-water mixture in steam-generating tubes," Izvestiia Vysohikh Uchebnylah Zavedenii, No. 11, pp. 71-76, U.S.S.R. (1973).

104. Shevalz, A. L., Kozmin, V. V., and Glagolev, A. S., "The model study of the stability of once through tube elements at supercritical pressure," Teploenergetika, No. 3, pp. 75-77, U.S.S.R. (1971).

105. Nikmatolin, B. E. and Evannof, A. E., "A study of hydrodynamic crisis phenomena in two-phase flow," High Temperature Thermal-Physics, Vol. 15, No. 1, U.S.S.R. (1977).

106. Drizius, M. R., Kairys, O., Skema, R., "Fluctuation of wall temperature of heat educing tubes in subcooled boiling," Academy of Science of the Lithuanian S.S.R. Vilniss, U.S.S.R., Series B, Vol. 6 (115), pp. 79-85 (1979).

107. Drizius, M. R., Skema, R. and Slanciauskas, A., "Pressure fluctuation in heat educing channels at surface boiling," Institute of Physical and Technical Problems of Energetics of the Academy of Science of the Lithuanian S.S.R., Series B, Vol. 1 (98) (1977).

108. Kalinin, A. V., "On stability of trans-sonic two-phase flow," Journal of Applied Mechanics and Mechanical Physics, No. 1 (1976).

109. Miloslafski, A. E., "On two-phase flow stability in pipelines,' Energetics (Izv. vyzov. Energetics) No. 9 (1980).

110. Glikman, B. V. and Ralitnev, V. E., "Compelled oscillations of two-phase gas-liquid flow in channel. Part I. Theory," Information of Academy of Science of U.S.S.R., Energetics and Transport, No. 4 (1975).

111. Glikman, B. V., Ralitnev, V. E., Earokova, E. E., "Compelled oscillations of two-phase gas-liquid flow in channel. Part II. Experiment," Information of Academy of Science of U.S.S.R., Energetics and Transport, No. 5 (1975).

112. Nesis, I. E., Dolofeev, B. M., "On high frequency pressure oscillations in the case of surface boiling," High Temperature Thermal-Physics, Vol. 14, No. 1 (1976).

113. Gerliga, V. A. and Prohorov, U. F., "The mechanism of occurring thermal-acoustic oscillations in the case of below critical pressure," Information of the Academy of Science of U.S.S.R., Energetics and Transport, No. 6 (1974).

114. Frankel, Y. E., "Cavitation on helical surface," Inzhner Zhur. Akad. Nauk., No. 2 (1969).

115. Nesis, I. E., "Volumetric oscillation of steam bubbles in liquid," Izvestiia vyzov. physics, No. 1 (1965).

116. Nesis, I. E., "On causes of noise occurring in boiling clean liquids and mixtures," Inzhener Fiz. Zhur. Akad. Nauk., U.S.S.R., No. 9 (1964).

117. Fedotkin, E. M., et al., "On dynamics of boiling steam phase," Thermalphysics and Thermaltechnics, Vol. 26 (1974).

118. Gerliga, V. A., "On high frequency oscillation in heated channels," TBT, Vol. 6, No. 4 (1968).

119. Gerliga, V. A., et al., "On acoustic oscillations in boiling channels," Symposium Hydrodynamics. Solid Object, U.S.S.R. (1971).

120. Akepman, "Heat transfer of pseudo-boiling water in subcritical region when fluid flowing along smooth and ribbed tubes," Heat Transfer, U.S.S.R., Vol. 92, No. 4 (1970).

121. Sirota, A. M., "On transition at subcritical pressure in one component system," Teploenergetika, No. 8 (1972).

122. Chen, X. J., and Chen, L. X., "The study of flow characteristics in a boiling tube when two-phase medium flows upwards and downwards," Scientific and Technical Report of Xian Jiao-Tong University, China (1972).

123. Bi, K. S., "The experimental study of the hydrodynamics instability of a once-through evaporator," Paper presented at the Second National Fluid Mechanics Conference, China (1979).

124. Fan, K. S., "The experimental study of two-phase flow instability in a low pressure natural circulating boiling loop," Scientific and Technical Report of Qing-Hua University, China (1979).

125. Wang, Z. M. and Wang, B. H., "The multivalue problem of the hydrodynamics characteristics of La-Mont boiler," Paper presented at the Third National Power Engineering Conference, China (1979).

126. Shanghai Central Experimental Institute of Electric Power Industry, "Experiments and improvements of a once-through boiler with a normal discharge of 240 tons/hr," Internal Report (1973).

127. Beijing Central Experimental Institute of Electric Power Industry, "The oscillation problem of once-through boiler no. 4," Internal Report (1975).

128. Lee, R. S., and Chi, K. S., "The hydrodynamic characteristics and film boiling of a once-through boiler with a normal discharge of 1000 tons/hr," Paper presented at the First National Thermal Engineering Physics Conference (1980).

129. Lin, Z., "Measurements of vapor quality and flow rate in two-phase flow with sharp edged circular orifices," Selected Papers of Scientific Research, Xian Jiaotong University, Xian, P.R.C. (1981).

130. Chen, X., "Analysis of the principal characteristics of circulation and internal process for the large capacity boilers used in electric power stations," Selected Papers of Scientific Research, Xian Jiaotong University, Xian, P.R.C. (1981).

131. Wu, P., "The criterial equation of heat transfer for the condensation of Freon 12 and 113 on a horizontal finned tube," Selected Papers of Scientific Research, Xian Jiaotong University, Xian, P.R.C. (1981).

132. Barenblatt, G. I., "On motion of suspended particles in turbulent flow," Prikl. Mat. Mekh., 17, 261 (1953).

133. Serov, E. P., "The operation of once-through boilers in variable regimes," Tr. Mosk. Energ. Inst., 11 (1953).

134. Jones, A. B., "Hydrodynamic stability of a boiling channel,"

KAPL-2170 (1961) and KAPL-3070 (1964).

135. Stenning, A. H., "Instabilities in the flow of a boiling liquid," J. Basic Engineering, Trans. ASME, Series D, 86, 213 (1964).

136. Stenning, A. H. and Veziroglu, T. N., "A parametric study of boiling instability," ASME Paper 64-WA/FE-28, presented at the Annual Meeting, December 1964.

137. Serov, E. P., "Analytical investigation of the boundary conditions for the formation of pulsation in steaming pipes during forced circulation," High Temperature, 3, 545 (1965).

138. Serov, E. P., "Transient processes in steam generators," Teploenergetika, 13 (9), 50 (1966).

139. Thurston, R. S., USAEC Report LA-3616, Los Alamos Scientific Laboratory, Los Alamos, New Mexico (1967).

140. Soo, S. L., Fluid Dynamics of Multiphase Systems, pp. 250-253, Blaisdell, Waltham (1967).

141. Wallis, G. B., One-Dimensional Two-Phase Flow, McGraw-Hill Book Co., Inc., N.Y. (1969).

142. Kocamustafaoğulları, G., "Thermo-fluid dynamics of separated two-phase flow," Ph.D. Thesis, School of Mechanical Engineering, Georgia Institute of Technology, Atlanta, Georgia (1971).

143. Friedly, J. C. and Krishnan, V. S., "Prediction of nonlinear flow oscillations in boiling channels," AIChE Symp. Ser., 68 (118), 127-135 (1972).

144. Delhaye, J. M., "Jump conditions and entropy sources in two-phase systems: Local instant formulation," Int. J. Multiphase Flow, 1, 395 (1973).

145. Wallis, G. B., "The terminal speed of single drops or bubbles in an infinite medium," Int. J. Multiphase Flow, 1, 491-511 (1974).

146. Harlow, F. H., and Amsden, A. A., "Numerical calculation of multiphase flow," J. of Comp. Phys., 17, 19 (1975).

147. Ishii, M., Thermo-Fluid Dynamic Theory of Two-Phase Flow, Eyrolles, Paris (Scientific and Medical Publications of France, N.Y.) (1975).

148. Solbrig, C. W., and Hughes, E. D., "Governing equations for a

seriated continuum: An unequal velocity model for two-phase flow," ANCR-1193 (1975).

149. Bouré, J. A., and Latrobe, A., "On well-posedness of two-phase flow problems," Paper No. 76-CSME/CSChE-9, 16th National Heat Transfer Conference, St. Louis, Missouri (1976).

150. Soo, S. L., "On one-dimensional motion of a single component in two phases," Int. J. Multiphase Flow, 3, 79 (1976).

151. Gidaspow, D., "A thermodynamic theory of two-phase flow with unequal phase velocities," Eds. S. Kakac, F. Mayinger, T. N. Veziroglu, Two-Phase Flows and Heat Transfer, Hemisphere Publishing Corp., USA (1977).

152. Gidaspow, D., "Hyperbolic compressible two-phase flow equations based on stationary principles and the Fick's law," Eds. T. N. Veziroglu, S. Kakac, Two-Phase Transport and Reactor Safety, Hemisphere Publishing Corp., USA (1977).

153. Sha, W. T., and Soo, S. L., "Multidomain multiphase fluid mechanics," ANL-CT-77-3 (1977).

154. Solbrig, C. W., and Hughes, E. D., "Governing equations for a seriated continuum: An unequal velocity model for two-phase flow," Eds. T. N. Veziroglu, S. Kakac, Two-Phase Transport and Reactor Safety, Hemisphere Publishing Corp., USA (1977).

155. Soo, S. L., "Multiphase mechanics and distinctions from continuum mechanics," Eds. T. N. Veziroglu, S. Kakac, Two-Phase Transport and Reactor Safety, Hemisphere Publishing Corp., USA (1977).

156. Weisman, J., and Choe, W. G., "Methods for calculation of pressure drop in cocurrent gas-liquid flow," Eds. T. N. Veziroglu, S. Kakac, Two-Phase Transport and Reactor Safety, Hemisphere Publishing Corp., USA (1977).

157. Lyczkowski, R. W., "Numerical techniques for the computation of transient unequal phase velocity, unequal phase temperature, two-phase flow and heat transfer," Eds. T. N. Veziroglu, S. Kakac, Two-Phase Transport and Reactor Safety, Hemisphere Publishing Corp., USA (1977).

158. Ishii, M., "Drift-flux model and derivation of kinematic constitutive laws," Eds. S. Kakac, F. Mayinger, T. N. Veziroglu, Two-Phase Flows and Heat Transfer, Hemisphere Publishing Corp., USA (1977).

159. Chao, B. T., Sha, W. T., and Soo, S. L., "On inertial coupling

in dynamic equation of components in a mixture," Int. J. Multiphase Flow, 4, 219 (1978).

160. Achard, J. L., Drew, D. A. and Lahey, R. T., Jr., "The analysis of linear and nonlinear instability phenomena in heated channels," NUREG/CR-1718 (1980).

161. Martinelli, R. C. and Nelson, D. B., and Schenectady, N.Y., "Prediction of pressure drop during forced circulation boiling of water," Trans. ASME, 70, 695 (1948).

162. Yeşin, O., "Pressure-drop type oscillations in a horizontal boiling channel," Habilitation thesis, Middle East Technical University, Ankara (1975).

163. Yüncü, H., "Dynamic instabilities in two-phase flow systems," Habilitation thesis, Middle East Technical University, Ankara (1982).

164. Akyüzlü, K., "Mathematical modeling of two-phase flow oscillations," Ph.D. thesis, University of Miami, Florida, USA (1979).

165. Doğan, T., "Lumped-parameter analysis of two-phase flow oscillations," Ph.D. thesis, University of Miami, Florida, USA (1979).

166. Park, G. C., Podowski, M., Becker, M. and Lahey, R. T., Jr., "The modeling of density-wave oscillations in boiling water nuclear reactors," in Advances in Two-Phase Flows and Heat Transfer, Sijthoff & Noordhoff International Publishers BV (1983).

167. Yadigaroglu, G., "Two-phase flow instabilities and propagation phenomena," in Thermodynamics of Two-Phase Systems for Industrial Design and Nuclear Engineering, Hemisphere Publishing Corp., New York (1981).

168. Kakaç, S., Veziroglu, T. N., Özboya, N. and Lee, S. S., "Transient boiling flow instabilities in a multi-channel upflow system," Wärme-und Stoffübertragung 10, pp. 175-188 (1977).

169. Kakaç, S. and Akyüzlü, K., "Sustained boiling flow instabilities in a cross-connected four parallel channel upflow system," METU Journal of Pure and Applied Sciences, Vol. 10, No. 2, pp. 157-178, Ankara (1977).

170. Lee, S. S., Veziroglu, T. N., and Kakaç, S., "Sustained and transient boiling flow instabilities in two parallel channel systems," in Two-Phase Flows and Heat Transfer, Eds. S. Kakac, F. Mayinger and T. N. Veziroglu, Hemisphere Publishing Corp.,

USA (1977).

171. Menteş, A., Gürgenci, H., Yıldırım, O. T., Kakaç, S., and Veziroglu, T. N., "Effect of heater surface configurations on two-phase flow instabilities in a vertical boiling channel," Proceedings of the 16th Southeastern Seminar on Thermal Sciences, Hemisphere Publishing Corporation, 19-21 April 1982, Miami, Florida.

172. Veziroglu, T. N., and Kakaç, S., "Two-phase flow instabilities," Final Report, NSF Project CME 79-20018, 1982.

173. Bergles, A. E., "Principles of heat transfer augmentation--II, Two-phase heat transfer," in Heat Exchangers: Thermal-Hydraulic Fundamentals and Design, Eds. S. Kakac, A. E. Bergles and F. Mayinger, Hemisphere Publishing Corp. pp. 857-914 (1981).

174. Lin, Z. H., Veziroglu, T. N., Kakaç, S., Gürgenci, H., and Mentes, A., "Heat transfer in oscillating two-phase flows and effect of heater surface conditions," 7th Int. Heat Transfer Conference, München, F.R.G. (1982).

175. Yadigaroglu, G., and Lahey, R. T., Jr., "On the various forms of the conservation equations in two-phase flow," Int. J. of Multiphase Flow, 2, pp. 477-484 (1976).

176. Bouré, J. A., "Mathematical modeling and the two-phase constitutive equations," European Two-Phase Flow Group Meeting, Haifa, Israel (1975).

177. Akyüzlü, K., Veziroglu, T. N., Kakaç, S. and Doğan, T., "Finite difference analysis of two-phase flow pressure-drop and density-wave type oscillations," Wärme-und Stoffübertragung 14, 253-267 (1980).

178. Dukler, A. E. et al., "Pressure drop and hold-up in two-phase flow," AIChE, Vol. 10, No. 1, pp. 38-51 (1961).

179. Ritchmyer, R. D., and Morton, K. W., Difference Methods for Initial Valve Problems, New York, J. Wiley and Sons (1967).

180. Lyczkowski, R. W., et al., "Pipe blowdown analysis using explicit numerical schemes," 15th International Heat Transfer Conference, San Francisco (1975).

181. Doğan, T., Kakaç, S. and Veziroglu, T. N., "Lumped parameter analysis of two-phase flow instabilities," 7th Int. Heat Transfer Conference, München, F.R. Germany (1982).

182. Chen, J. C., "A correlation for boiling heat transfer to

saturated fluids in convective flow," 6th National Heat Transfer Conference, ASME Paper No. 63-HT-34 (1963).

183. Gürgenci, H., Veziroglu, T. N., and Kakaç, S., "Pressure-drop and density-wave instability thresholds in boiling channels," Proceedings of the 16th Southeastern Seminar on Thermal Sciences, Hemisphere Publishing Corporation, 19-21 April 1982, Miami, Florida.

184. Gürgenci, H., Veziroglu, T. N. and Kakaç, S., "Simplified nonlinear descriptions of two-phase flow instabilities in a vertical boiling channel," Proceedings of the 16th Southeastern Seminar on Thermal Sciences, Hemisphere Publishing Corporation, 19-21 April 1982, Miami, Florida.

185. Gürgenci, H., "Two-phase flow instabilities," Ph.D. Thesis, University of Miami, Coral Gables, Florida (1982).

186. Delhaye, J. M., "Two-phase flow modeling," in Thermodynamics of Two-Phase Systems for Industrial Design and Nuclear Engineering, Hemisphere Publishing Corp. (1981).

THE MODELING OF DENSITY-WAVE OSCILLATIONS IN BOILING WATER NUCLEAR REACTORS

G.C. Park, M. Podowski, M. Becker, R.T. Lahey, Jr.

Rensselaer Polytechnic Institute
Department of Nuclear Engineering
Troy, New York - USA

ABSTRACT

Density-wave oscillations constitute a potential instability model in Boiling Water Nuclear Reactors (BWR's). Simplified models using conservative assumptions may lead to unnecessary operating restrictions. This paper presents a state-of-the-art-one-dimensional thermal-hydraulic model which can be used for the linear analysis of BWR stability. The model accounts for phasic slip, heated wall dynamics, distributed spacers, and a detailed model of the loop. The axial heat flux profile is replaced by a stair-case function such that the linearized, Laplace-transformed equations, can be integrated analytically over each node. In its final form, the coupled neutron-kinetics/thermal-hydraulics model constitutes a multi-input/multi-output linear system of a BWR. The model was parametrically evaluated with a digital computer code.

INTRODUCTION

Density-wave oscillations constitute a potential instability mode in boiling water nuclear reactors (BWR's). Accordingly, these oscillations have been the object of considerable study. Due to the complexity of the phenomena, many analyses have been done with simplified models, having conservative assumptions, which may lead to unnecessary operating restrictions.

In this paper, a state-of-the-art one-dimensional model is presented. The new model accounts for phasic slip, distributed spacers, space/time-dependent power distributions, and heated wall dynamics. It should be noted that the time constants characterizing both the reactor kinetics and heated wall thermal inertia are comparable to the time constants characterizing the system thermal-hydraulic phenomena. Thus considerable care has also gone into the modelling of these effects.

In order to allow for an arbitrary axial power distribution, the actual heat flux curve was replaced by a stair-case (i.e., piece-wise constant) function such that the linearized, Laplace transformed, conservation equations could be integrated analytically over the length of each node. Analytical models were also developed for heated wall dynamics for both plate type, cylinderical and BWR fuel rod geometries. In addition to a parallel channel instability analysis, the model can be used for BWR loop analysis, for both natural circulation and forced circulation modes of operation.

In its final form, the coupled neutron-kinetics/thermal-hydraulics model constitutes a multi-input/multi-output linear system. The stability characteristics of a BWR/4 has been investigated with this model using the Nyquist criterion (1).

The present paper is focused on two points. One deals with the derivation of the model's thermal-hydraulics. A formulation of the characteristic equation of the system is presented, together with a discussion of the assumptions made. The second point is concerned with application of the coupled neutron-kinetics/thermal-hydraulics model to the analysis of BWR instability modes. The computer implementation of the model, NUFREQ-N, was used for the parametric study of a typical BWR/4. The analysis revealed several new and important factors influencing BWR stability margins.

PARALLEL CHANNEL STABILITY - MODEL DEVELOPMENT

In order to treat arbitrary axial power distributions we can divide the heated channel into N axial nodes, as shown in Figure 1. The internal generation rate (q_k''') is assumed to be uniform in each axial node. Let us now consider heated wall dynamics.

Heated wall dynamics is important in the stability analysis of a BWR. In the single-phase region, heated wall dynamics affects the boiling boundary dynamics and thus introduces a lag between the inlet flow perturbation and the void fraction perturbations in the two-phase region. In the two-phase region it also plays an important role because of the time lag between the local power perturbation ($\hat{\delta q}_k'''$) and the heat flux perturbation ($\hat{\delta q}_k''$).

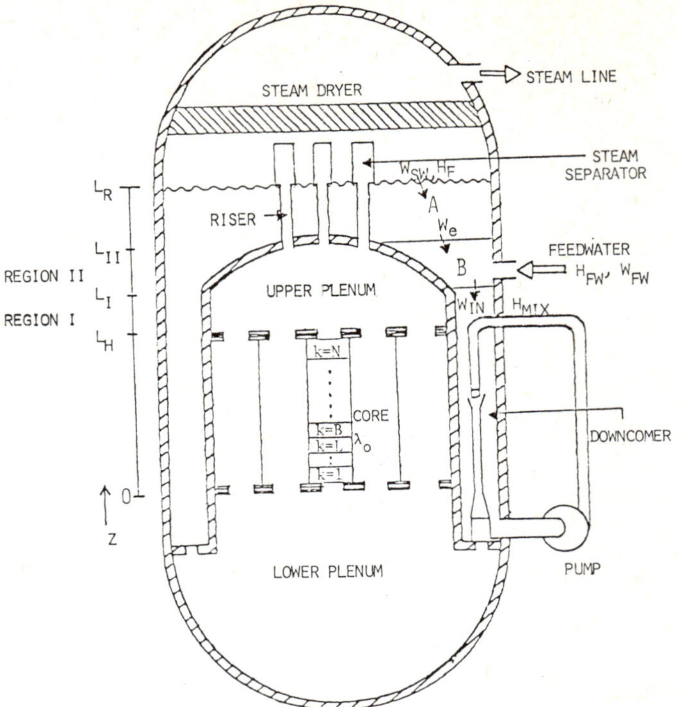

Fig. 1 BWR reactor configuration.

Previous investigators (3,4) have considered heated wall dynamics for special cases or with a simple lumped parameter model. In this paper we developed distributed parameter models for three typical heater geometries (i.e., a slab, a cylindrical tube and a BWR fuel rod).

For a parallel plate heater, we assume an insulated outer surface and constant thermal conductivity and heat capacity. Solving the perturbed and Laplace transformed time-dependent heat conduction for the j-th node,

$$\frac{d^2 \delta \hat{T}_j}{dx^2} - \frac{s}{\alpha} \delta \hat{T}_j + \frac{1}{k} \delta \hat{q}_j''' = 0 \tag{1}$$

with the perturbed boundary conditions,

$$- k \left. \frac{d\delta\hat{T}_j}{dx} \right| = 0 \tag{2}$$

$$- k \left. \frac{d\delta\hat{T}_j}{dx} \right|_{x=b} = \delta q_j'' \tag{3}$$

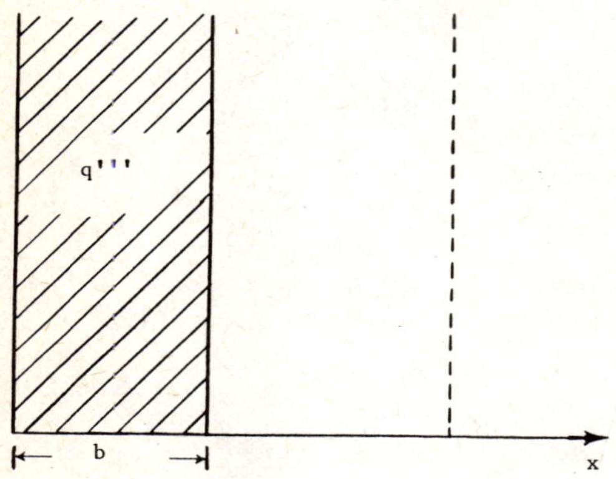

we obtain,

$$\delta\hat{T}_j(x_j,s) = - \frac{\delta\hat{q}_j''}{\varepsilon k \mathrm{Sinh}(\varepsilon b)} \mathrm{Cosh}\ (\varepsilon x_j) + \frac{\delta\hat{q}_j'''}{\rho c_p s} \tag{4}$$

where,

$$\varepsilon \overset{\Delta}{=} \sqrt{s/\alpha}$$

From Eq. (4), the heated wall temperature of the j-th node can be expressed as,

$$\delta\hat{T}_{w_j} = Z_{1p}\ \delta\hat{q}_j'' + Z_{2p}\ \delta\hat{q}_j''' \tag{5}$$

where,

$$Z_{1_p} \stackrel{\Delta}{=} - \frac{\text{Coth}\ (\varepsilon b)}{\varepsilon k}$$

$$Z_{2_p} = \frac{1}{\rho c_p s}$$

For a cylindrical tube, the perturbed and Laplace transformed conduction equation forms an inhomogeneous modified Bessel equation,

$$\frac{1}{r}\frac{d}{dr}\left(r\frac{d\delta\hat{T}}{dr}\right) - \frac{s}{\alpha}\,\delta\hat{T} + \frac{1}{k}\,\delta\hat{q}''' = 0 \tag{6}$$

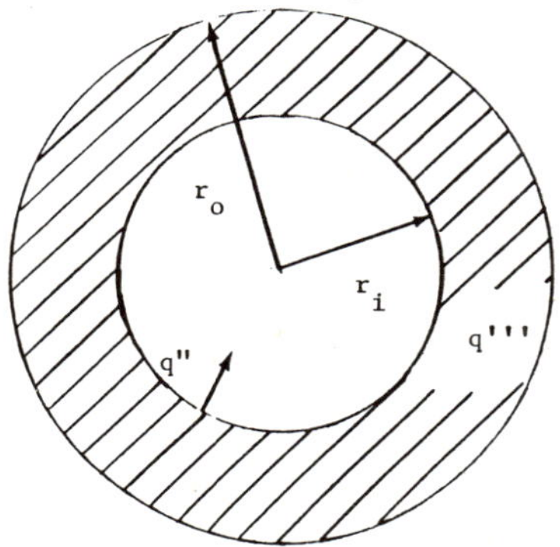

Integrating Eq. (6) with the boundary conditions,

$$- k \left.\frac{d\delta\hat{T}}{dr}\right|_{r=r_i} = \delta\hat{q}'' \tag{7}$$

$$k \left.\frac{d\delta\hat{T}}{dr}\right|_{r=r_o} = 0 \tag{8}$$

the heater wall temperature perturbation of the j-th node is,

$$\delta\hat{T}_{w_j} = Z_{1c}(s)\ \delta\hat{q}_j'' = Z_{2c}\ \delta\hat{q}_j''' \tag{9}$$

where,

$$Z_{1c} \overset{\Delta}{=} - \frac{[K_1(\varepsilon r_o)I_o(\varepsilon r_i)+I_1(\varepsilon r_o)K_c(\varepsilon r_i)]}{[I_1(\varepsilon r_i)K_1(\varepsilon r_o)-K_1(\varepsilon r_i)I_1(\varepsilon r_o)]} \frac{1}{\varepsilon k}$$

$$Z_{2c} \overset{\Delta}{=} \frac{1}{\rho_k c_p s}$$

Let us next consider a BWR fuel rod, in which the gap conductance is assumed to be constant. The perturbed heat conduction equations for the fuel and the cladding are, respectively,

<u>Fuel</u>

$$\frac{1}{r}\frac{d}{dr}\left(r\frac{d\delta\hat{T}_f}{dr}\right) - \frac{s}{\alpha_f}\delta\hat{T}_f + \frac{1}{k_f}\delta\hat{q}''' = 0 \tag{10}$$

<u>Clad</u>

$$\frac{1}{r}\frac{d}{dr}\left(r\frac{d\hat{T}_c}{dr}\right) - \frac{s}{\alpha_c}\delta\hat{T}_c = 0 \tag{11}$$

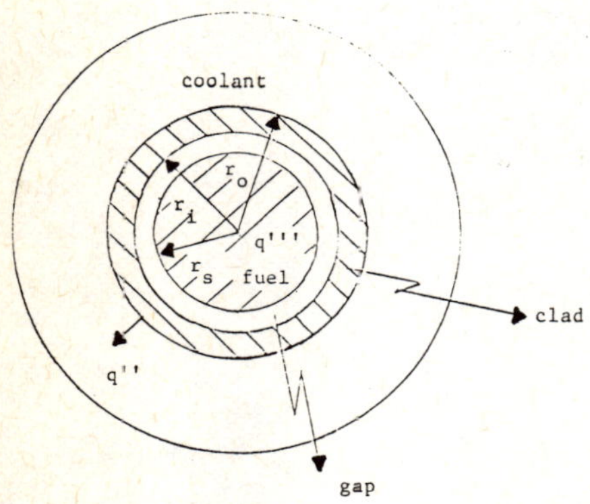

The perturbed boundary conditions are,

$$- k_f \left. \frac{d\hat{\delta T}_f}{dr} \right|_{r=r_s} = \left(\frac{r_i}{r_s}\right) \hat{\delta q}''_s \tag{12}$$

$$- k_c \left. \frac{d\hat{\delta T}_c}{dr} \right|_{r=r_i} = \hat{\delta q}''_s = H_g(\hat{\delta T}_s - \hat{\delta T}_i) \tag{13}$$

$$- k_c \left. \frac{d\hat{\delta T}_c}{dr} \right|_{r=r_o} = \hat{\delta q}'' \tag{14}$$

$$k_f \left. \frac{d\hat{\delta T}_c}{dr} \right|_{r=o} = 0 \tag{15}$$

These equations give the outer cladding surface temperature perturbation for the j-th node as,

$$\hat{\delta T}_{w,j} = Z_{1f}(s)\ \hat{\delta q}''_j + Z_{2f}(s)\ \hat{\delta q}'''_j \tag{16}$$

where,

$$Z_{1,f} \stackrel{\Delta}{=} \frac{1}{A} \left[\frac{a_1 K_o(\varepsilon_c r_o) - a_2 I_o(\varepsilon_c r_o)}{\varepsilon_c k_c}\right]$$

$$Z_{2,f} \stackrel{\Delta}{=} \frac{1}{A} \left[\frac{I_1(\varepsilon_c r_o) K_o(\varepsilon_c r_o) + K_1(\varepsilon_c r_o) I_o(\varepsilon_c r_o)}{\rho_f c_{p_f} s}\right]$$

$$a_1 = I_o(\varepsilon_c r_i) - W_s I_1(\varepsilon_c r_i) \tag{17a}$$

$$a_2 = K_o(\varepsilon_c r_i) + W_s K_1(\varepsilon_c r_i) \tag{17b}$$

$$A = a_1 K_1(\varepsilon_c r_o) + a_2 I_1(\varepsilon_c r_o) \tag{17c}$$

$$W_s = \varepsilon_c k_c \left\{\frac{I_o(\varepsilon_f r_s)}{k_f \varepsilon_f I_1(\varepsilon_f r_s)} \left(\frac{r_i}{r_s}\right) + \frac{1}{H_g}\right\} \tag{17d}$$

In order to express $\delta \hat{q}_j''$ in terms of $\delta \hat{q}_j'''$, we must eliminate $\delta \hat{T}_w$ from Eqs. (5), (9) and (16) by using a convective heat transfer equation. In particular, the perturbed and Laplace-transformed Dittus-Boelter equation can be applied for the single-phase region, and the Jens-Lottes equation for the two-phase region. Both can be expressed as,

$$\delta \hat{q}'' = \chi \, \delta \hat{T}_w \tag{18}$$

where χ is $H_{1\phi}$ for the single-phase region and is equal to $4q_o''/(T_{w_o} - T_{sat})$ for the two-phase region. Using Eqs. (5), (9) or (16) with Eq. (18), the final form of $\delta \hat{q}_j''$ is,

$$\delta \hat{q}_j'' = Z_{3,j} \, \delta \hat{q}_j''' \tag{19}$$

where,

$$Z_{3,j} = \frac{\chi_j Z_2}{1 - \chi_j Z_1} \tag{20}$$

The appropriate momentum equations can be used to calculate the pressure drop perturbation $(\delta \hat{\Delta p}_k)$ for each axial node. The total pressure drop perturbation, $\delta \hat{\Delta p}_{total}$, (including all local losses), is obtained by summing the nodal $\delta (\hat{\Delta p})_k$.

The enthalpy perturbation in the single-phase region has been previously derived (1) as,

$$\frac{d(\delta \hat{h})}{dz} + \phi(s) \, \delta \hat{h}(s,z) = \theta(s,z) \frac{\delta \hat{j}_i}{j_{i_o}} + \beta(s,z) \left(\frac{\delta \hat{q}'''}{q_o'''}\right) \tag{21}$$

where,

$$\phi(s) \stackrel{\Delta}{=} \frac{s}{j_{i_o}} + \frac{P_h}{\rho_f j_{i_o} c_{pf} A_{x-s}} \left(\frac{H_{1\phi o}}{1-H_{1\phi o} Z_1(s)}\right)$$

$$\theta(s,z) \stackrel{\Delta}{=} \frac{P_h V_h q_o'''(z)}{\rho_f A_{x-s} j_{i_o} A_h} \left[\frac{a}{(1-H_{1\phi o} Z_1(s)} -1\right]$$

$$\beta(s,z) \stackrel{\Delta}{=} \frac{P_h q_o'''(z)}{\rho_f j_{i_o} A_{x-s}} \left(\frac{H_{1\phi o}}{1-H_{1\phi o} Z_1(s)}\right) Z_2(s)$$

The parameters Z_1 and Z_2, are the transfer functions of $\delta \hat{T}w_k/\delta \hat{q}_k''$ and $\delta T_{wk}/\delta q'''$ respectively. Since $q'''(z)$ is discretized and assumed axially uniform within each node, Eq. (21) can be integrated subject to the boundary condition, $\delta \hat{h}(z=o)=\delta h_i$. We can obtain $\delta \hat{\lambda}$, an expression for the boiling boundary dynamics, from the enthalpy perturbation at the boiling boundary, $\delta h(s,\lambda_o)$, as (1),

$$\delta \hat{\lambda} = \frac{-\delta \hat{h}(s,\lambda_o)}{\left[\frac{dh(\lambda_o)}{dz}\right]} = \Lambda_1(s)\delta \hat{j}_i + \Lambda_2(s) \sum_{\ell=1}^{L} H_\ell' \delta \hat{q}_{t,\ell}'''$$

$$+ \Lambda_3(s)\delta \hat{h}_i \tag{22}$$

where,

$$\Lambda_k(s) = -\frac{\rho_f A_{x-s} j_{i_o}}{q_{B_o}'' P_H} H_k(s) \qquad \text{(for } k=1,2,3\text{)}, \tag{23a}$$

$$H_1(s) = \frac{\theta'(s)}{\phi(s)} \frac{e^{-\phi(s)\lambda_o}}{j_{i_o}} \sum_{k=1}^{L} \left(e^{\phi(s)z_k} - e^{\phi(s)z_{k-1}}\right) q_{o,k}''' \tag{23b}$$

$$H_2(s) = \frac{\beta'(s)}{\phi(s)} e^{-\phi(s)\lambda_o} \tag{23c}$$

$$H_3(s) = e^{-\phi(s)\lambda_o} \tag{23d}$$

$$H'_\ell(s) = (e^{\phi(s)z_k} - e^{\phi(s)z_{k-1}}) \tag{23e}$$

$$\theta'(s) = \theta(s,z)/q_o'''(z) \tag{23g}$$

$$\beta'(s) = \beta(s,z)/q_o'''(z) \tag{23h}$$

The total static pressure drop perturbation in the single-phase region, $\delta(\Delta\hat{p}_{1\phi})$, can be obtained by perturbing and Laplace transforming the single-phase momentum equation (1),

$$\delta(\Delta\hat{p}_{1\phi}) = \int_0^{\lambda_o} [\frac{\rho_f}{g_c} s\, \delta\hat{j}_i + \frac{f\rho_f j_{i_o}}{g_c D_H} \delta\hat{j}_i]dz$$

$$+ (\sum_{m=1}^{M_s} K_{spacer}) \frac{\rho_f j_{i_o}}{g_c} \delta\hat{j}_i + (\frac{f\rho_f j_{i_o}^2}{2g_c D_H} + \frac{g}{g_c}\rho_f)\delta\hat{\lambda} \tag{24a}$$

Combining Eqs. (22) and (24);

$$\delta(\Delta\hat{p}_{1\phi}) = \Gamma_1(s)\delta\hat{j}_i + \Gamma_2(s) \sum_{k=1}^{L} H'_k\, \delta\hat{q}_{tk}''' + \Gamma_3(s)\delta\hat{h}_i \tag{24b}$$

where,

$$\Gamma_1(s) \overset{\Delta}{=} (\frac{\rho_f}{g_c} s + \frac{f\rho_f j_{i_o}}{g_c D_H})\lambda_o + (\sum_{m=1}^{M_s} K_{spacer}) \frac{G_o}{g_c} + C_2\Lambda_1(s)$$

$$\Gamma_k(s) \stackrel{\Delta}{=} C_2\Lambda_k(s) \quad ; \quad k=2,3$$

$$C_2 \stackrel{\Delta}{=} [\frac{fG_o^2}{2g_c D_H \rho_f} + \frac{g}{g_c}\rho_f]$$

In the two-phase region we have used a drift-flux model and allowed for the heated wall dynamics (i.e., $\delta\hat{q}''$) in that region. In order to obtain the two-phase pressure drop perturbation. $\delta(\Delta p_{2\phi})$, the mixture momentum equation was investigated, perturbed and Laplace transformed as:

$$\delta(\Delta\hat{p}_{2\phi}) = \int_{\lambda_o}^{L_H} [\frac{1}{g_c}(s\delta\hat{G} + \frac{\partial\delta[G^2/<\rho>]}{\partial z}) + \frac{fG_o\phi^2_{\ell o}}{g_c\rho_f D_H}\delta\hat{G}$$

$$+ \frac{fG_o^2}{2g_c D_H \rho_f}\delta\hat{\phi}^2_{io} + \frac{g}{g_c}\delta\hat{<\rho>} + \frac{d}{dz}\hat{\delta}\{(\frac{\rho_f - \overline{<\rho>}}{\overline{<\rho>} - \rho_g})\frac{\rho_f\rho_g}{\overline{<\rho>}}(V'_{gj})\}]dz$$

$$- [\frac{1}{g_c}(\frac{\partial\delta[G^2/<\rho>}{\partial z})_o + \frac{fG_o^2}{2g_c\rho_f D_H} + \frac{g}{g_c}\rho_f$$

$$+ \frac{1}{g_c}\frac{d}{dz}\{(\frac{\rho_f - \overline{<\rho>}}{\overline{<\rho>} - \rho_g})\frac{\rho_f\rho_g}{\overline{<\rho>}}(V'_{gj})^2\}\Big|_{z=\lambda_o}]\delta\hat{\lambda}$$

$$+ \sum_{m_s=1}^{M} K_{spacer_m}\{\frac{G_o\Phi_o}{g_c\rho_f}\Big|_{z=z_m}\delta\hat{G}(s,z_m) + \frac{G_o^2}{2g_c\rho_f}\Big|_{z=zm}\delta\hat{\Phi}(s,z_m)\}$$

$$+ K_{exit}\{\frac{G_o\Phi_o}{g_c\rho_f}\Big|_{z=L_H}\delta\hat{\Phi}(s,L_H\} \qquad (25)$$

We analytically performed the integration node-by-node with the boundary conditions at the boiling boundary given by,

$$\delta\langle\hat{\alpha}\rangle_{\lambda_o} = -\frac{\Omega_{o,B}}{\rho_g v_{fg}(C_{k,o})_B^{in}}\,\delta\hat{\lambda} \tag{26}$$

and,

$$\delta\langle\hat{j}\rangle_{\lambda_o} = \delta\hat{j}_i - \Omega_{o,B}\,\delta\hat{\lambda} \tag{27}$$

In Appendix I, we show that the total two-phase pressure drop summed over all nodes in the two-phase region, can be expressed as,

$$\delta(\Delta\hat{p}_{2\phi}) = \Pi_1(s)\delta\hat{j}_i + \sum_{k=1}^{N} \Pi_2^k(s)\delta\hat{q}'''_{t,k} + \Pi_3(s)\delta\hat{h}_i \tag{28}$$

In addition to the calculation of the pressure drop perturbation, the node-averaged void fraction perturbation $\delta\langle\hat{\alpha}\rangle_k$ (for use in the power-to-void transfer function), can be calculated with the void propagation Eq., (2). The resultant node-averaged void, derived in Appendix-II is,

$$\delta\langle\hat{\alpha}\rangle_k = T_{1,k}(s)\delta\hat{j}_i + \sum_{i=B}^{k} (T_{2,i}^k(s)\delta\hat{q}'''_{t,i}) + T_{3,k}(s)\delta\hat{h}_i \tag{29}$$

BWR CORE STABILITY ANALYSIS

Equations (24), (28) and (29) and the associated boundary condition, $\delta(\Delta\hat{p}_{1\phi}) + \delta(\Delta\hat{p}_{2\phi}) = 0$, form the basis of a frequency domain parallel channel stability model corresponding to input perturbations, $\delta\hat{j}_i$, $\delta\hat{h}_i$ and $\delta\hat{q}'''_k$. When combined with a suitable model for neutron dynamics these same transfer functions can be used for the linear stability analysis of a BWR core.

A parametric study of BWR/4 core stability predictions based on this model will be presented. Figure-2 shows the effect of power for various power shapes. It can be seen, as expected, that channels having the highest power and longest boiling length are the most unstable.

Figure-3 shows the effect of slip on BWR core stability. Since slip decreases the void fraction, and thus the pressure drop in the two-phase region, slip increases core stability.

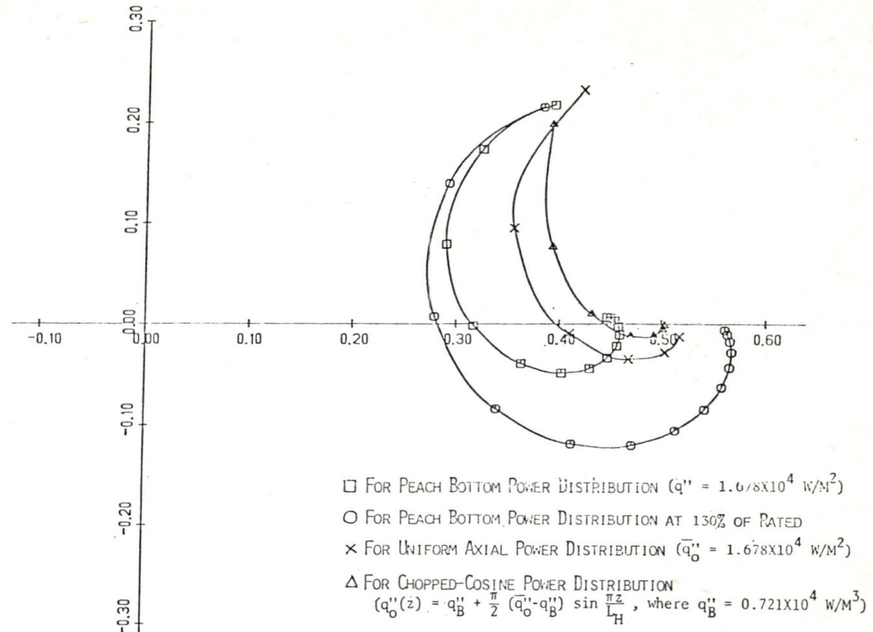

Fig. 2 The effect of various shapes of power and increasing power.

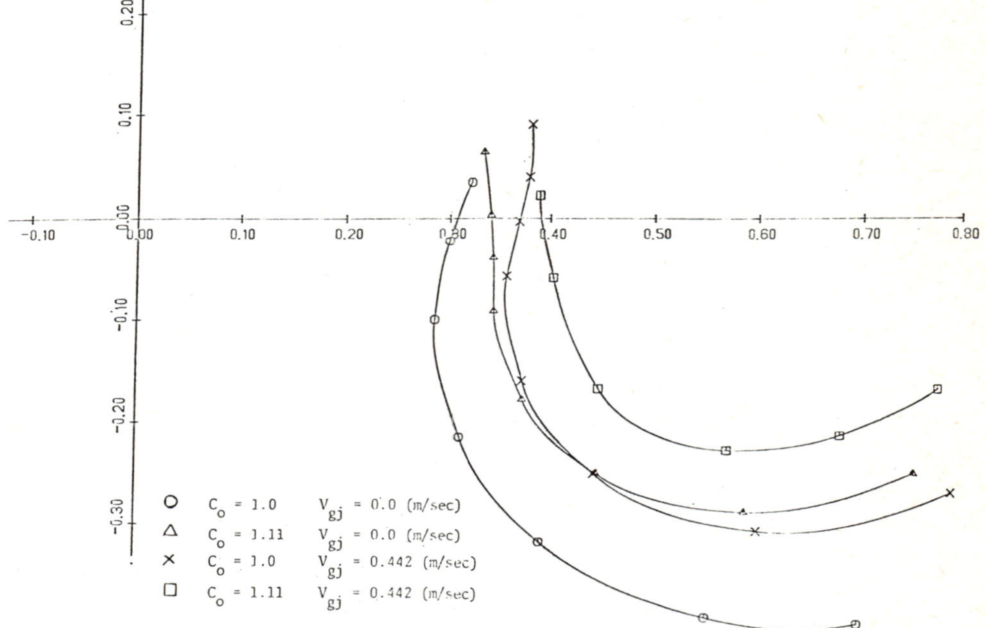

Fig. 3 The effect of slip for $\delta\hat{j}_i$ with nuclear coupling.

Figure-4 shows a comparison between a distributed parameter BWR fuel rod model and a lumped parameter model previously used (1) for BWR stability analysis. As shown in the figure, the lumped parameter model has a significant time-delay and is more conservative than the distributed model. This comparison indicates that an accurate treatment of fuel rod heat transfer dynamics is necessary.

BWR LOOP MODEL DEVELOPMENT

In order to investigate BWR loop stability, the pressure drop relationships developed for the core must be supplemented by other single and two-phase pressure drop perturbations. The recirculating loop, shown schematically in Figure-1, can be divided into four regions, the core, the upper plenum, the adiabatic riser/separator

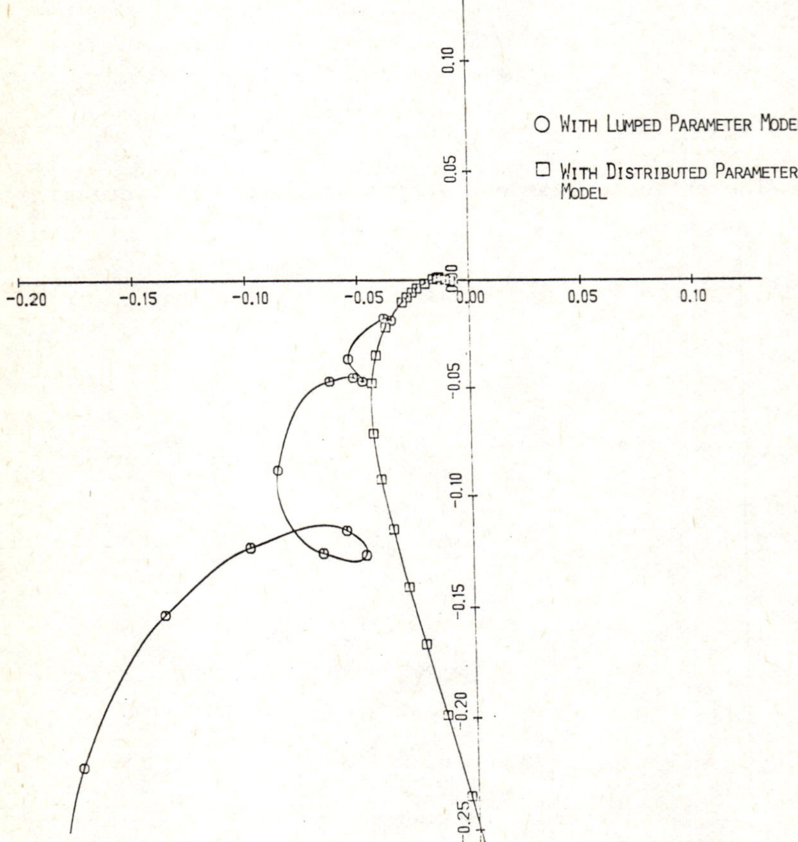

Fig. 4 Comparison between the lumped parameter model and the distributed parameter model for $\delta\hat{q}'''_{ext}$.

and the single-phase portion of the loop (which includes the downcomer, lower plenum and pump). The upper plenum is assumed to be composed of the two regions shown in Figure-1. In Region-I, the flow area is constant, while in Region-II it varies in the axial direction. We can assume a homogeneous flow in the upper plenum because the mixing is quite large.

In the upper plenum, the adiabatic mixture continuity equation and the void propagation equation yield,

$$\delta\langle\hat{j}_I\rangle^{out} = \delta\langle\hat{j}_p\rangle^{in} \tag{30}$$

$$\delta\langle\hat{\alpha}_J\rangle^{out} = \delta\langle\hat{\alpha}_p\rangle^{in} e^{-\frac{s}{\langle j_p\rangle_o^{in}}(L_I - L_{II})} \tag{31}$$

Since there is a sudden expansion in the flow area at the inlet to the upper plenum, the plenum inlet perturbations must be related to the exit perturbations from the fuel rod bundles. This is done subject to the constraint of constant quality and flow rate. Using the homogeneous flow assumption for the upper plenum, the perturbed inlet values are given by,

$$\delta\langle\hat{\alpha}_p\rangle^{in} = F_1\ \delta\langle\hat{j}(L_H)\rangle + F_2\ \delta\langle\hat{\alpha}(L_H)\rangle \tag{32}$$

$$\delta\langle\hat{j}_p\rangle^{in} = J_1\ \delta\langle\hat{j}(L_H)\rangle + J_2\ \delta\langle\hat{\alpha}(L_H)\rangle \tag{33}$$

$$F_1 \stackrel{\Delta}{=} -\frac{\rho_f\rho_g V_{gj}}{G_o^2}\ \frac{1 - \frac{\Delta\rho}{\rho_f}\langle\alpha_o\rangle}{\frac{\rho_g}{\rho_f} + \frac{\Delta\rho}{\rho_f}\langle x_o(L_H)\rangle}\ \langle\alpha_o(L_H)\rangle \tag{34a}$$

$$F_2 \stackrel{\Delta}{=} \frac{C_o\frac{\rho_g}{\rho_f} + \frac{\rho_g V_{gj}}{G_o} + C_o\frac{\Delta\rho}{\rho_f}\langle x_o(L_H)\rangle + \frac{\Delta\rho V_{gj}\rho_g}{G_o^2}\langle\alpha_o(L_H)\rangle\langle C_{k-o}(L_H)\rangle}{1 - C_o\frac{\Delta\rho}{\rho_f}\langle\alpha_o(L_H)\rangle}$$

$$\times\ \frac{1 - \frac{\Delta\rho}{\rho_f}\langle\alpha_o\rangle_p^{in}}{\frac{\rho_g}{\rho_f} + \frac{\Delta\rho}{\rho_f}\langle x_o(L_H)\rangle} \tag{34b}$$

$$J_1 \stackrel{\Delta}{=} \frac{(A_{x-s}/A_p)(\rho_f - \Delta\rho C_o <\alpha_o(L_H)>) + \Delta\rho <j_o>_p^{in} F_1}{\rho_f - \Delta\rho <\alpha_o>_p^{in}} \tag{34c}$$

$$J_2 \stackrel{\Delta}{=} \frac{\Delta\rho <j_o>_p^{in} F_2 - \Delta\rho\{C_{k-o}(L_H)\}(A_{x-s}/A_p)}{\rho_f - \Delta\rho <\alpha_o>_p^{in}} \tag{35d}$$

In Region-II, the plenum flow area is no longer constant, but A(z). The appropriate mass continuity equation and the void propagation equation for homogeneous flow are,

$$\frac{\partial <j>}{\partial z} = - <j> \frac{d(\ell nA)}{dz} \tag{36}$$

$$\frac{\partial <\alpha>}{\partial t} + C_k \frac{\partial <\alpha>}{\partial t} = 0 \tag{37}$$

Perturbing, Laplace transforming and integrating Eqs. (36) and (37),

$$\delta<\hat{j}_{II}>^{out} = \frac{A_P}{A_R} \delta<\hat{j}_{II}>^{in} \tag{38}$$

$$\delta<\hat{\alpha}_{II}>^{out} = [e^{-\frac{s}{A_p <j_p>_o^{in}} \int_{L_I}^{L_{II}} A(z)dz}] \delta<\hat{\alpha}_{II}>^{in} \tag{39}$$

Now in order to find A(z), we can assume that Region-II has the form of a chopped hemisphere, as shown below:

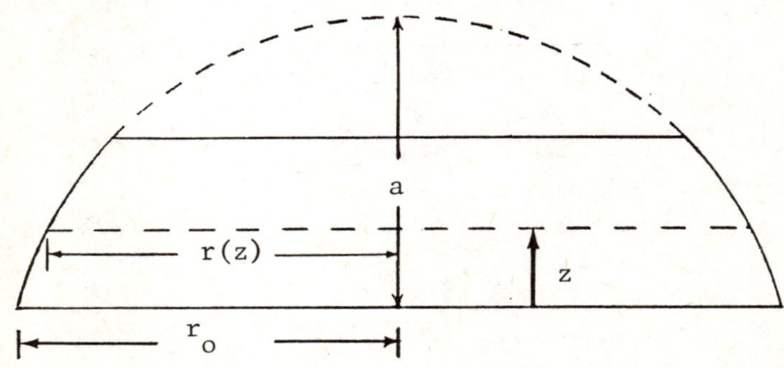

Then Eq. (39) becomes,

$$\delta\langle\alpha_{II}\rangle^{out} = \delta\langle\hat{\alpha}_{II}\rangle \, e^{-\frac{s}{\langle j_p\rangle_o^{in}} \frac{V_{II}}{A_p}} \tag{40}$$

where,

$$V_{II} = A_p(L_{II}-L_I) + (\Pi a - \frac{A_p}{a}) \frac{(L_{II}^2-L_I^2)}{2} - \frac{\Pi}{3}(L_{II}^3-L_I^3)$$

In the adiabatic riser, the homogeneous flow assumption is no longer appropriate. Moreover, we have to convert the exit perturbations from the upper plenum, Eqs. (38) and (39), to appropriate inlet perturbations of the riser. Thus, proceeding as before, the inlet values of the riser are given by,

$$\delta\langle\hat{j}_R\rangle^{in} = E_{1,1} \frac{A_p}{A_R} \delta\langle\hat{j}_p\rangle^{in} + E_{1,2}\, e^{-\frac{s}{\langle j_p\rangle_o^{in}} (L_I-L_{II}+\frac{V_{II}}{A_p})} \delta\langle\hat{\alpha}_p\rangle^{in} \tag{41}$$

$$\delta\langle\hat{\alpha}_R\rangle^{in} = E_{2,1} \frac{A_p}{A_R} \delta\langle\hat{j}_p\rangle^{in} + E_{2,2}\, e^{-\frac{s}{\langle j_p\rangle_o^{in}} (L_I-L_{II}+\frac{V_{II}}{A_p})} \delta\langle\hat{\alpha}_p\rangle^{in} \tag{42}$$

where,

$$E_{1,1} \stackrel{\Delta}{=} \frac{1}{0_D} [\rho_f - \Delta\rho\langle\alpha_{II}\rangle_o^{out}] F_{1,R} \tag{43a}$$

$$E_{1,2} \stackrel{\Delta}{=} \frac{1}{0_D} [\Delta\rho(C_{o,R}\langle j_R\rangle_o^{in} + V_{gj,R}) F_{1,II} - \Delta\rho\langle j_{II}\rangle_o^{out} F_{1,R}] \tag{43b}$$

$$E_{2,1} \stackrel{\Delta}{=} -\frac{1}{0_D} [\rho_f - \Delta\rho\langle\alpha_{II}\rangle_o^{out}] F_{2,R} \tag{43c}$$

$$E_{2,2} \stackrel{\Delta}{=} \frac{1}{0_D} [\Delta\rho\langle j_{II}\rangle_o^{out} F_{2,R} + (\rho_f - \Delta\rho C_{o,R}\langle\alpha_R\rangle_o^{in}) F_{1,II}] \tag{43d}$$

$$O_D \overset{\Delta}{=} \Delta\rho(C_{o,R}\langle j_R\rangle_o^{in} + V_{gj,R})\ F_{2,R} + [\rho_f - \Delta\rho C_{o,R}\langle\alpha_R\rangle_o^{in}]\ F_{1,R} \quad (43e)$$

$$F_{1,II} = \frac{\rho_g + \Delta\rho\langle x_o\rangle_o^{out}}{\rho_f - \Delta\rho\langle\alpha_o\rangle_p^{out}} \quad (43f)$$

$$F_{1,R} = \frac{C_{o,R}\frac{\rho_g}{\rho_f} + \frac{\rho_f V_{gj,R}}{G_{o,R}} + C_{o,R}\frac{\Delta\rho}{\rho_f}\langle x_o\rangle_R^{in} + \frac{\Delta\rho\,\rho_g V_{gj,R}}{\{G_{o,R}\}^2}\langle\alpha_o\rangle_R^{in}(C_{k,o})_R^{in}}{(1 - C_{o,R}\frac{\Delta p}{\rho_f}\langle\alpha_R\rangle_o^{in})} \quad (43g)$$

$$F_{2,R} = -\frac{\rho_f\rho_g V_{gj,R}}{\{G_{o,R}\}^2}\langle\alpha_o\rangle_R^{in} \quad (43h)$$

The mixture momentum equation for the riser is the same as Eq. (25) with the exception of the spacers and the spatial acceleration. For the riser, all parameters can be derived as in Appendix-I. However, the drift-flux model must be evaluated for adiabatic conditions, so that the mixture velocity and void fraction are constant along the riser. By integrating term-by-term, perturbing, and the Laplace-transforming the mixture momentum equation, we obtain the pressure drop perturbation of the riser as,

$$\delta(\Delta\hat{p}_{riser}) = R_1\delta\hat{j}_i + \sum_{k=1}^{K} R_{2,k}\,\delta\hat{q}'''_{t,k} + R_3\delta\hat{h}_i \quad (44)$$

where,

$$R_1(s) \overset{\Delta}{=} T_1^R(1-\Omega_{0,B}\Lambda_1) + T_2^R X_{1,R} \quad (44a)$$

$$R_{2,k}(s) \overset{\Delta}{=} \begin{cases} (-T_1^R\Omega_{0,B}\Lambda_2 + T_2^R X_{2,R}^{1\phi})\ H'_k\ , & k \le L \\ T_1^R \frac{v_{fg}P_H}{A_{x-s}h_{fg}} Z_{3,k}\Delta z_k + T_2^R X_{2,k}^R\ , & B \le k \le k \end{cases} \quad (44b)$$

$$R_3(s) \overset{\Delta}{=} -\Omega_{0,B}\Lambda_3 T_1 + T_2 X_{3,R} \tag{44c}$$

$$T_1^R(s) = \frac{A_p}{A_R} J_1 \, (0_{1,1}R_1' + 0_{2,1}R_2') + (0_{1,2}R_1' + 0_{2,2}R_2') \, F_1 \, e^{-\frac{s}{\langle j_p\rangle_o^{in}}\{(L_I - L_{II}) + \frac{A}{A_p}\}} \tag{44d}$$

$$T_2^R(s) = \frac{A_p}{A_R} J_2 \, (0_{1,1}R_1' + 0_{2,1}R_2') + (0_{1,2}R_1' + 0_{2,2}R_2') \, F_2 \, e^{-\frac{s}{\langle j_p\rangle_o^{in}}\{(L_I - L_{II}) + \frac{A}{A_p}\}} \tag{44e}$$

$$R_1' \overset{\Delta}{=} \left(\frac{s}{g_c} + \frac{f G_{R,o} (\phi_{\ell o}^2)_{R\,o}}{g_c \rho_f D_{H,R}}\right)(\rho_f - \Delta\rho C_{o,R} \langle\alpha_R\rangle_o^{in}) \Delta L_R + C_{R,3} \, \Delta L_R + (K_{inlet,R} + K_{exit,R}) \, C_{R,5} \tag{44f}$$

$$R_2' \overset{\Delta}{=} \left(\frac{s}{g_c} + \frac{f G_{R,o} (\phi_{\ell o}^2)_{R\,o}}{g_c \rho_f D_{H,R}}\right) \frac{\Delta\rho (C_{K,R})_o^2}{s} \left\{e^{-\frac{s}{(C_{K,R})_o}\Delta L_R} - 1\right\} + \frac{1}{g_c}\left\{C_{R,1} + V_{R,2} + \frac{g\Delta\rho (C_{K,R})_o}{s} + g_c C_{R,4}\right\}\left\{e^{-\frac{s}{(C_{K,R})_o}\Delta L_R} - 1\right\} + C_{R,6}\left\{K_{inlet,R} + K_{exit,R} \, e^{-\frac{s}{(C_{K,R})_o}\Delta L_R}\right\} \tag{44g}$$

The single-phase part of the loop (i.e., the downcomer region and the lower plenum) may have various cross-sectional flow areas, A_i, of length segments, L_i. Thus, for convenience, all velocities have been referenced to the flow area of the fuel rod bundle, A_{x-s}. The pressure drop perturbation along the loop can be expressed as,

$$\delta(\Delta\hat{p}_{1\phi,loop}) = \frac{\rho_f}{g_c} A_{x-s} \left(\sum_{i=i}^{J} \frac{L_i}{A_i}\right) s\delta\hat{j}_i$$

$$+ \frac{\rho_f A_{x-s}^2 j_{i0}}{g_c} \left(\sum_{i=i}^{J} \frac{f_i L_i}{D_{H,i} A_i^2}\right)\delta\hat{j}_i - \frac{g}{g_c}\rho_f\ \delta\hat{L}_D - \zeta A_{x-s}\ \delta\hat{j}_i \tag{45}$$

where the individual terms represent the perturbations in loop inertia, friction, hydrostatic head and pump head, respectively. The parameter ζ is the slope of recirculation pump head-flow curves. In the downcomer, the liquid level (L_D) can oscillate due to riser exit flow rate perturbations.

In order to obtain the downcomer length perturbation, we can perturb and Laplace transform the mass continuity equation in control volume-A of Figure-1. By assuming perfect mixing, and a constant flow rate and enthalpy of the feedwater into control volume-B, we obtain,

$$\delta\hat{L}_D = \frac{1}{\rho_f A_D S}\left[\delta\hat{w}_{sw} - \rho_f A_{x-s}\delta\hat{j}_i\right] \tag{46}$$

Since $w_{sw} = (1 - \langle x_R\rangle^{out})\ A_R G_R^{out}$, the perturbed flow rate of the saturated water can be calculated from Eq. (I.2), (41), (42) and (32) as,

$$\delta\hat{w}_{sw} = L_1(s)\ \delta\langle\hat{j}(L_H)\rangle + L_2(s)\ \delta\langle\alpha(L_H)\rangle \tag{47}$$

where,

$$L_i(s) = L_i'\left[E_{i1}\frac{A_p}{A_R}F_1 + E_{i2}\, e^{-\frac{s}{\langle j_o\rangle_p^{in}}\left(L_{II}-L_I+\frac{A}{A_p}\right)} J_1\right]$$

$$- L_2'(s)\left[E_{i1}\frac{A_p}{A_R}F_2 + E_{i2}\, e^{-\frac{s}{\langle j_o\rangle_p^{in}}\left(L_{II}-L_I+\frac{A}{A_p}\right.} J_2\right],$$

$$i=1,2 \tag{47a}$$

$$L' = -\frac{A_R}{v_{fg}} + \frac{A_R \rho_f}{\Delta\rho}(\rho_f - \Delta\rho\, C_{o,R} \langle\alpha_o\rangle_R^{in}) \tag{47b}$$

$$- \frac{s}{(C_{K,0})_R} \Delta L_R$$

$$L'_s(s) = A_R \rho_f (C_{K,0})_R \; e \tag{47c}$$

Equations (46), (47), (I.5) and (I.7) now yield the desired expression for $\delta\hat{L}_D$ in terms of $\delta\hat{j}_i$, $\delta\hat{q}'''_k$ and δh_i.

The pressure drop perturbations which are calculated in each part of the loop can be summed to generate the total loop pressure drop perturbation,

$$\delta(\hat{\Delta p}_{loop,total}) = \overline{W}_1(s)\delta\hat{j}_i + \sum_{k=1}^{N} \overline{W}_{2,k}(s)\delta\hat{q}'''_{t,k}$$

$$+ \overline{W}_3(s)\delta\hat{h}_i \tag{48}$$

where,

$$\overline{W}_1(s) \stackrel{\Delta}{=} R_1(s) + \frac{\rho_f}{g_c} A_{x-s} \left(\sum_{i=1}^{J} \frac{L_i}{A_i}\right) s$$

$$+ \frac{\rho_f A^2_{x-s} j_{i_o}}{g_c} \left(\sum_{i=1}^{J} \frac{f_i L_i}{D_{H,i} A^2_i}\right) - \xi A_{x-s}$$

$$+ (I_{f1}-I_{f2})\{L_1 C1-\Omega_{o,B})\Lambda_1 + L_2 X_{1,R}\} - I_{f2} A_{x-s} \rho_f \tag{49a}$$

$$\overline{W}_{2,k}(s) \stackrel{\Delta}{=} \begin{cases} (I_{f1}-I_{f2})(L_2 X^{1\phi}_{2,R} - L_1 \Omega_o \Lambda_2) H'_k , & \text{for } 1 \leq k < B \\ (I_{f.}-I_{f2}) \dfrac{P_H v_{fg}}{A_{x-s} h_{fg}} (Z_{3,k} \Delta z_k L_1 + X^R_{2,k} L_2), & \text{for } B \leq k \leq N \end{cases} \tag{49b}$$

$$\overline{W}_3(s) \stackrel{\Delta}{=} (I_{f1} - I_{f2})(X_{3,R} L_2 - \Omega_o L_1 \Lambda_3) \tag{49c}$$

$$I_{f1} \overset{\Delta}{=} \frac{w_{sw,o}}{g_c A_p} + \frac{f w_{sw,o}}{g_c D_H A_D^2 \rho_f} \tag{49d}$$

$$I_{f2} \overset{\Delta}{=} \left(\frac{f w_{sw,o}^2}{2 g_c D_H A_D^2 \rho_f} + \frac{g}{g_c} \rho_f\right) \frac{1}{\rho_f A_D S} \tag{49e}$$

In order to close the loop we must eliminate $\delta\hat{h}_i$ in terms of δj_i and $\delta q_k'''$. In order to accomplish this, we must use the time-dependent single-phase energy equation including a lumped parameter pressure vessel wall dynamics,

$$\frac{d(\delta\hat{h})}{dz} + \phi(s)\delta\hat{h} = \theta(s)\, e^{-b_1 z} \frac{\delta\hat{w}}{w_o} \tag{50}$$

where,

$$b_1 \overset{\Delta}{=} \frac{H_{1\phi,o}}{\left(1 + \frac{d_1 H_{1\phi,o}}{k_v}\right)} \frac{P_H}{w\, c_{p_f}}$$

Now if we solve this Bernoulli-type ordinary differential equation, the exit enthalpy perturbation from the downcomer has the form,

$$\delta\hat{h}_{DC}^{out} = D(s)\ \delta\hat{w}_i \tag{51}$$

where,

$$D(s) = \frac{\theta(s)\{e^{-bL_{DC}} - e^{-\phi L_{DC}}\}}{w_o\{\phi(s) - b_1\}} + \frac{\langle x_o\rangle_R^{out}}{w_o} (h_f - h_{fw,o}) e^{-\phi L_{DC}} \tag{51a}$$

In order to take into account the thermal inertia of the lower plenum, one can assume perfect mixing and apply a lumped parameter vessel wall dynamics model. The lower plenum energy equation is thus,

$$M_{LP} c_{pf} \frac{dT_{Lp}}{dt} = w_{LP}^{in} h_{LP}^{in} - w^i h_{LP}^{out} - q_{LP}'' A_{H,LP} \tag{52}$$

where,

$$w_{LP}^{in} = w_{DC}^{out} = w_{LP}^{out} \quad w_i \tag{53a}$$

$$h_{LP}^{in} = h_{DC}^{out} \tag{53b}$$

$$h_{LP}^{out} = h_i \tag{53c}$$

and, in accordance with the perfect mixing assumption,

$$c_{pf} dT_{LP} = dh_i \tag{53d}$$

By perturbing and Laplace transforming, Eq. (52) yields,

$$\delta\hat{h}_i = \frac{1}{(M_{LP}s + w_o)} [w_o \; \delta\hat{h}_{DC}^{out} + (h_{DC,o} - h_{io}) \; \delta\hat{w}_{in}$$

$$- \; \delta\hat{q}''_{LP} \quad A_{H,LP}] \tag{54}$$

Using Eq. (51), and the transfer function derived for the plate heater wall dynamics to obtain $\delta\hat{q}''_{LP}$ we obtain,

$$\delta\hat{h}_i = Y(s) \; \delta\hat{j} \tag{55}$$

$$Y(s) = \left[\frac{w_o D(s) + (h_{DC,o} - h_{io}) - t(s) \; A_{H,LP} \dfrac{aq''_{LP,o}}{w_o}}{M_{LP}s + w_o + t(s) \dfrac{H_{1\phi,o}}{c_{pf}} \; A_{H,LP}} \right] \rho_f A_{x-s}$$

$$t(s) = \frac{k_v \varepsilon_v}{k_v \varepsilon_v + H_{1\phi,o} \tanh(\varepsilon_v d)}$$

Combining Eqs. (48) and (55), we obtain the total loop pressure drop perturbation corresponding to external forcing functions, $\delta\hat{j}_i$ and $\delta\hat{q}'''$ as,

$$\delta(\Delta\hat{p}_{loop}) = \{\overline{W}_1(s) + \overline{W}_3(s)\ Y(s)\}\ \delta\hat{j}_i + \sum_{k=1}^{N} \overline{W}_{2,k}(s)\delta\hat{q}'''_{1,k} \quad (56)$$

BWR LOOP STABILITY ANALYSIS

The results obtained from these transfer functions is shown in Figure-5. It can be seen that slip tends to improve BWR loop stability.

In addition to the previously discussed thermal-hydraulic feedbacks, in a BWR one has void-reactivity. In previous work (4) a point neutron kinetics model was used, in which the only perturbed quantity was core-average void fraction. Actually, the local void fraction perturbations affected many nuclear parameters, e.g., the effective delayed neutron fraction (β_i), neutron generation (Λ),

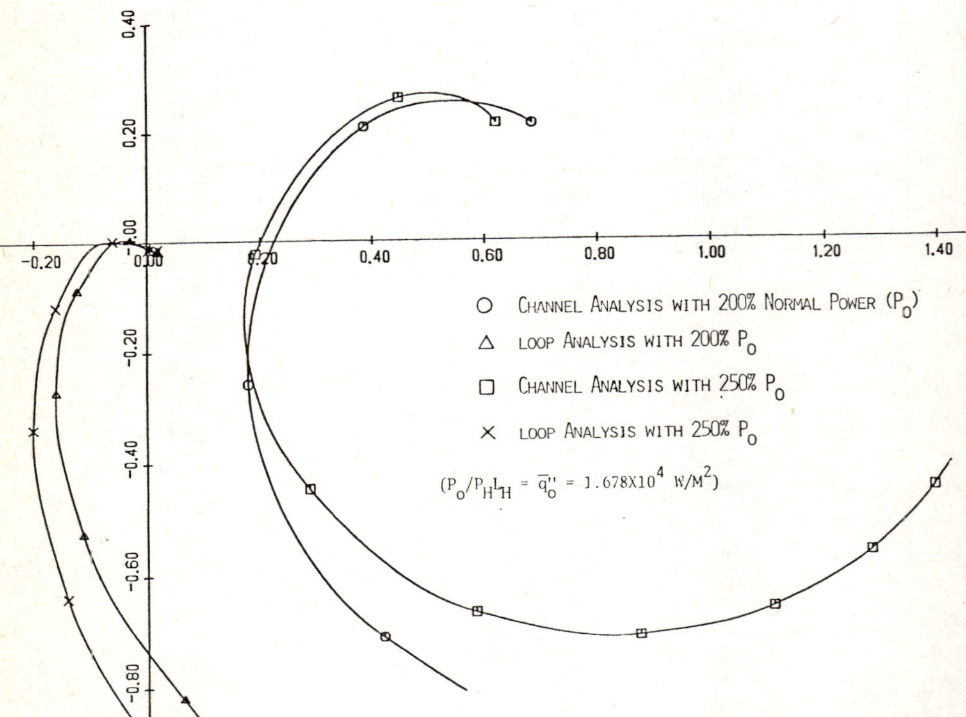

Fig. 5 The comparison between the loop analysis and the channel analysis for various power levels.

fission cross section ($\nu\Sigma_f$), and neutron velocity (v). In this study we employ a more accurate nuclear coupling transfer function derived from the 6-group point kinetics equation in which all relevant nuclear parameters were perturbed,

$$\frac{\delta\hat{n}}{n_o} = [\delta\hat{\rho} - s \sum_{k=1}^{6} \frac{\delta\hat{\beta}_k}{s+\lambda_k} + \frac{\delta\hat{\Lambda}}{\Lambda_o} s \sum_{k=1}^{6} \frac{\beta_{k,0}}{s+\lambda_k}] / s(\Lambda_o - \sum_{k=1}^{6} \frac{\beta_{k,0}}{s+\lambda_k})] \tag{57}$$

The core-average volumetric heat generation rate (q'''_{av}) was related to the neutron density by,

$$q'''_{av} = \frac{\kappa}{\nu} (\nu\Sigma_f) vn \tag{58}$$

where,

$$q'''_{av} = \frac{1}{L_H} \sum_{k=1}^{N} q'''_k \Delta z_k$$

Perturbing and Laplace transforming Eq. (58),

$$\frac{\delta\hat{q}'''_{av}}{q'''_{av,o}} = \frac{\delta\hat{n}}{n_o} + \frac{\delta(\widehat{\nu\Sigma_f})}{\nu_o\Sigma_{fo}} + \frac{\delta\hat{v}}{v_o} \tag{59}$$

In Eqs. (57) and (59), the perturbed parameters, e.g., the reactor reactivity perturbation ($\delta\hat{\rho}$), $\delta\hat{\Lambda}$, $\delta\hat{\beta}$, $\delta\widehat{\nu\Sigma}_f$ and δv, can be expressed in terms of the nodal void fraction perturbation as;

$$\delta\hat{\rho} = C_\rho\ \delta\langle\hat{\alpha}_{av}\rangle \tag{60a}$$

$$\delta\hat{\Lambda} = C_\Lambda\ \delta\langle\hat{\alpha}_{av}\rangle \tag{60b}$$

$$\delta\hat{\beta}_k = C_{\beta_k}\ \delta\langle\hat{\alpha}_{av}\rangle \tag{60c}$$

$$\delta(\widehat{\nu\Sigma_f}) = C_{\nu\Sigma_f}\ \delta\langle\hat{\alpha}_{av}\rangle \tag{60d}$$

$$\delta\hat{v} = C_v\ \delta\langle\hat{\alpha}_{av}\rangle \tag{60e}$$

where,

$$\delta\langle\hat{\alpha}_{av}\rangle = \left(\frac{1}{L_H}\sum_{k=B}^{N} T_{1,k}\right)\delta\hat{j}_i + \frac{1}{L_H}\sum_{k=B}^{N}\sum_{i=1}^{k} T_{2,i}\,\delta\hat{q}'''_{t,i}$$

$$+ \frac{1}{L_H}\left(\sum_{k=B}^{N} T_{3,k}(s)\right)\delta\hat{h}_{in} \qquad (60f)$$

and, C_ρ, C_ν, C_{β_k}, $C_{\nu\Sigma_f}$, C_v are the void coefficient for each nuclear parameter considered. These parameters were calculated by utilizing 3-dimensional nodal neutronics data and power distributions (5). Thus the nuclear coupling transfer function, $\Phi(s)$, has the form of,

$$\delta\hat{q}'''_{av} = \Phi(s)\ \delta\langle\hat{\alpha}_{av}\rangle \qquad (61)$$

where,

$$\Phi(s) = \left[\frac{C_v}{v_o} + \frac{C_{\nu\Sigma_f}}{\nu\Sigma_{f,o}} + \frac{C_\rho - s\sum\limits_{k=1}^{6}\frac{C_{\beta_\ell}}{s+\lambda_o} + s\frac{C_\Lambda}{\Lambda_o}\sum\limits_{k=1}^{6}\frac{\beta_{\ell,o}}{s+\lambda_\ell}}{\Lambda_o s + s\sum\limits_{\ell=1}^{6}\frac{\beta_{\ell,o}}{s+\lambda_\ell}}\right] q'''_{av,o}$$

A block diagram of the complete feedback system, including the thermal-hydraulic feedback, and the neutron-coupled feedback, is shown in Figure-6. Using the boundary conditions given by,

$$\delta(\Delta\hat{p}_{1\phi,total}) + \delta(\Delta\hat{p}_{2\phi,total}) = 0 \qquad (62)$$

the block diagram can be easily transformed into a closed loop diagram to permit application of conventional control theory methods.

To investigate the stability margin, we have used the well-known Nyquist criterion in the frequency domain in terms of the open-loop transfer function,

$$GH = \frac{\delta\hat{j}_{i,f}}{\delta\hat{j}_{i,t}}\ ,\quad \text{or,}\quad -\frac{\delta\hat{q}'''_F}{\delta\hat{q}'''_t}$$

For the application of the coupled kinetics/thermal-hydraulic model discussed above, the computer implementation of the model,

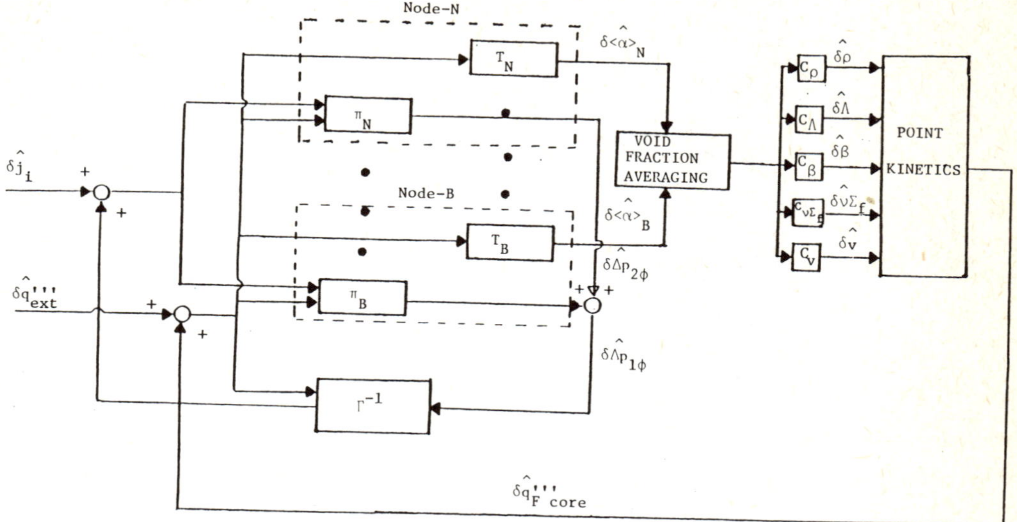

Fig. 6 The system block diagram.

NUFREQ-N, was used for the parametric analysis of a typical BWR/4. It was found (6) that nuclear coupling can have a significant effect on both BWR loop and parallel channel stability.

SUMMARY AND CONCLUSIONS

A detailed, state-of-the-art, BWR stability model has been derived and evaluated for conditions typical of BWR/4. It has been shown that, in order to obtain realistic estimates of system stability, the thermal-hydraulic and neutronic modelling must be as accurate as possible.

In particular, simplyfying assumptions made concerning phasic slip, heated wall dynamics and neutronics can lead to conservative evaluations of the stability margin, and unnecessary operating restrictions.

NOMENCLATURE

A_{x-s} - Cross-sectional flow area

c_p - Specific heat

C_K - Velocity of kinematic wave ($=C_o<j>+V_{gj}$)

C_o - Void concentration parameter

D_H - Hydraulic diameter

f - Darcy-Weisbach friction factor

G - Mass flux

h - Enthalpy

H_g - Gap conductance in a fuel rod

$H_{1\phi}$ - Single-phase heat transfer coefficient

k - Thermal conductivity

K - Local pressure loss coefficient

L - Total number of single-phase nodes in the channel

M - Total number of grid spacers

N - Total number of nodes in the channel

p - Pressure

P_H - Heater perimeter

V_{gj} - Drift velocity

α - Thermal diffusivity

$<\alpha>$ - Cross-sectional average void fraction

ρ - Density

$<\rho>$ - Cross-sectional average density of the two-phase mixture

Ω - Characteristic frequency of the phase change

λ - Boiling boundary

$\phi^2_{\ell o}$ - Two-phase friction loss multiplier

Φ - Two-phase local loss multiplier

$\hat{\delta\zeta}$ - The Laplace transform of the perturbation, $\delta\zeta$

Subscripts

B - Inlet boundary node index of the two-phase region

DC - Downcomer

exit - Channel exit

f - Saturated liquid

fw - Feedwater

g - Saturated vapor

i - Inlet

LP - Lower plenum

P - Upper plenum

R - Riser

V - Reactor vessel

w - Wall

1ϕ - Single-phase region

2ϕ - Two-phase region

o - Steady-state

REFERENCES

1. Lahey, R.T., Jr. and F.J. Moody, "The Thermal Hydraulics of a Boiling Water Nuclear Reactor", ANS (1977).

2. Zuber, N. and F.W. Staub, "The Propagation of the Wave Form of the Vapor Volumetric Concentration in Boiling Forced Convection System Under Oscillatory Conditions", Int. J. Heat Mass Transfer, Vol. 9, pp. 871-895 (1966).

3. Yadigaroglu, G. and A.E. Bergles, "Fundamental and Higher-Mode Density-Wave Oscillations in Two-Phase Flow; The Importance of the Single-Phase Region", ASME paper 71-HT-13, American Society of Mechanical Engineers (1971).

4. Lahey, R.T., Jr. and G. Yadigaroglu, "NUFREQ, A Computer Program to Investigate Thermo-Hydrodynamic Stability", NEDO-13344, July (1973).

5. Park, J.K. and M. Becker, "Development of Point Kinetics Parameters from Nodal Power Distribution", Trans. Am. Nucl. Soc., 35, pp 581 (1980).

6. Park, G.C., M. Becker and R.T. Lahey, Jr., "The Effect of Radial Non-Uniformity on BWR Stability Margins", Trans. Am. Nucl. Soc., 35, 647 (1980).

7. Zuber, N. and J.A. Findlay, "Average Volumetric Concentration in Two-Phase Flow Systems", J. Heat Transfer (1965).

APPENDIX I

Under the assumption that $q_o'''(z)$ and $\delta\hat{q}'''(s,z)$ are piecewise constant, within axial nodes, we can perform the integration node-by-node of Eq. (25). In order to perform the analytical integration, we must derive each parameter appearing in the integrand which has axial dependence, e.g., $\delta\langle\hat{j}(z)\rangle$, $\delta\hat{G}(z)$, $\delta\langle\hat{\alpha}(z)\rangle$, $\delta\hat{\phi}^2_{\ell o}(z)$, etc.

The mass flux in the two-phase region can be expressed as,

$$G = [\rho_f - \Delta\rho C_o\langle\alpha\rangle]\langle j\rangle - \Delta\rho V_{gj}\langle\alpha\rangle \tag{I.1}$$

where,

$$\Delta\rho \stackrel{\Delta}{=} \rho_f - \rho_g \tag{I.2}$$

Perturbing and Laplace transforming Eq. (I.1),

$$\delta\hat{G} = [\rho_f - \Delta\rho C_o\langle\alpha_o\rangle]\delta\langle\hat{j}\rangle - \Delta\rho(C_o\langle j_o\rangle + V_{gj})\delta\langle\hat{\alpha}\rangle \tag{I.3}$$

The nodal velocity, $\delta\langle\hat{j}\rangle_k$, is determined from perturbing, Laplace transforming and integrating the mixture continuity equation. For convenience, we can transform the z-coordinate to a ξ-coordinate, defined by,

$$\xi_{o,k}(z) = \frac{\{C_{k,o}(z)\}_k}{(C_{k,o})_k^{in}} \tag{I.4}$$

such that $\delta\langle\hat{j}\rangle_k$ can be expressed as,

$$\delta\langle\hat{j}\rangle_k = \delta\langle\hat{j}\rangle_k^{in} + \frac{(C_{K,o})_k^{in}}{C_o\Omega_{o,k}}(\xi_{o,k}-1)\,\delta\hat{\Omega}_k \tag{I.5}$$

The void fraction perturbation, $\delta\langle\hat{\alpha}\rangle_k$, can be calculated by perturbing and Laplace transforming the void propagation equation, to obtain,

$$s\delta\langle\hat{\alpha}\rangle_k + \frac{d\langle\alpha_o\rangle_k}{dz}\delta(\hat{C}_K)_k + (C_{K,o})_k\frac{d\delta\langle\hat{\alpha}\rangle_k}{dz} = \left[\frac{\rho_f}{\Delta\rho} - C_o\langle\alpha_o\rangle_k\right]\delta\hat{\Omega}_k - C_o\Omega_{o,k}\,\delta\langle\hat{\alpha}\rangle_k \tag{I.6}$$

By solving this Bernoulli-type differential equation, we obtain,

$$\delta\langle\hat{\alpha}\rangle_k = -P_{2,k}C_o\Omega_{o,k}\left\{\xi_{o,k}^{-2} - \xi_{o,k}^{-\frac{(s+C_o\Omega_{o,k})}{C_o\Omega_{o,k}}}\right\}\delta\langle\hat{j}\rangle_k^{in}$$

$$+ P_{2,k}(C_{K,o})_k^{in}\left\{\xi_{o,k}^{-2} - \xi_{o,k}^{-\frac{(s+C_o\Omega_{o,k})}{C_o\Omega_{o,k}}}\right\}\delta\hat{\Omega}_k + \xi_{o,k}^{-\frac{(s+C_o\Omega_{o,k})}{C_o\Omega_{o,k}}}\delta\langle\hat{\alpha}\rangle_k^{in} \tag{I.7}$$

where,

$$P_{2,k} \triangleq \frac{\frac{\rho_f}{\Delta\rho} - C_o\langle\alpha_o\rangle_k^{in}}{(C_{K,o})_k^{in}(s-C_o\Omega_{o,k})} \tag{I.8}$$

Now combining Eqs. (I.3), (I.5) and (I.7), we obtain $\delta\hat{G}(s,\xi_{o,k})$, which is a function of $\delta\langle\hat{j}\rangle_k^{in}$, $\delta\hat{\Omega}_k$ and $\delta\langle\hat{\alpha}\rangle_k^{in}$.

Next, a generalized two-phase friction loss multiplier can be written as,

$$\phi^2_{\ell o} = [C_{2,k} + C_{1,k} \frac{v_{fg}}{v_f} <x>] \tag{I.9}$$

where $C_{1,k}$ and $C_{2,k}$ are empirical correction factors, which are both unity for homogeneous flow. In this study we obtained $C_{1,k}$ and $C_{2,k}$ from a Martinelli-Nelson fit (1). If we perturb and Laplace transform Eq. (I.9), we obtain,

$$\delta(\hat{\phi}^2_{\ell o}) = C_1 \frac{v_{fg}}{v_f} \delta\hat{<x>}_k \tag{I.10}$$

Using the perturbed form of the Zuber-Findlay void-quality relation (7) and Eq. (I.7), we can calculate the transfer function of two-phase friction loss multiplier in terms of $\delta<\hat{j}>^{in}_k$, $\delta\hat{\Omega}_k$, and $\delta\hat{<\alpha>}^{in}_k$.

The last term of the integrand of Eq. (25) is the spatial drift-flux gradient, which represents the net momentum flux with respect to the movement of the center-of-mass of the two-phase mixture. Perturbing and Laplace transforming,

$$\delta\hat{S}_{DG} = \rho_f\rho_g <V'_{gj}>^2_o \left[\frac{-\delta<\hat{\bar{\rho}}>(<\bar{\rho}_o>^2-\rho_g<\bar{\rho}_o>)(2<\bar{\rho}_o>\delta<\hat{\bar{\rho}}>-\rho_g\delta<\hat{\bar{\rho}}>)}{(<\bar{\rho}_o>^2 - \rho_f<\rho_o>)}\right]$$

$$+ 2 \left(\frac{\rho_f-<\bar{\rho}_o>}{<\bar{\rho}>-\rho_g}\right)\left(\frac{\rho_f\rho_g}{<\bar{\rho}_o>}\right) <V'_{gj}>_o (C_o-1)\delta<\hat{j}> \tag{I.11}$$

Since $<V'_{gj}> \overset{\Delta}{=} (C_o-1)<j> + V_{gj}$, this can be written more compactly as,

$$\delta\hat{S}_{DG} = - \bar{V}'_1\delta<\hat{\bar{\rho}}> + 2\bar{V}'_2 (C_o-1) \delta<\hat{j}> \tag{I.12}$$

Now if we insert the axially dependent parameters derived above into Eq. (25), the final form of the two-phase pressure drop perturbation in the kth node can be expressed as,

$$\delta(\Delta\hat{p}_{2\phi})_k = G_{1,k}(s)\ \delta\hat{<j>}_k^{in} + G_{2,k}\ \delta\hat{\Omega}_k + G_{3,k}\ \delta\hat{<\alpha>}_k^{in} \qquad (I.13)$$

where,

$$G_{1,k}(s) \stackrel{\Delta}{=} \frac{(C_{K,o})_k^{in}}{g_c C_o \Omega_{o,k}} [sg_{1,k} + \frac{fG_o}{D_H \rho_f} \{\bar{\phi}_{\ell o}^2\ g_{1,k} + \frac{G_o f_{2,k}}{2}\}] \ell n \xi_{o,k}^{out}$$

$$+ \frac{(C_{K,o})_k^{in}}{g_c (C_o \Omega_{o,k} - s)} [sg_{2,k} + \frac{fG_o}{D_H \rho_f} \{\bar{\phi}_{\ell o}^2\ g_{2,k}$$

$$+ \frac{G_o f_{3,k}}{2}\}]\{(\xi_{o,k}^{out})^{\frac{C_o \Omega_{o,k} - s}{C_o \Omega_{o,k}}} - 1\} + \frac{(C_{k,o})_k^{in}}{g_c} \{\frac{fG_o^2 f_{1,k}}{2D_H \rho_f C_o \Omega_{o,k}}$$

$$+ g\Delta\rho \frac{P_{2,k}}{\xi_{o,k}^{out}}\}(\xi_{o,k}^{out} - 1) + \frac{fG_o^2 (C_{K,o})_k^{in} f_{4,k}}{2g_c D_H \rho_f (2C_o \Omega_{o,k} - s)} \{(\xi_{o,k}^{out})^{\frac{2C_o \Omega_{o,k} - s}{C_o \Omega_{o,k}}} - 1\}$$

$$+ \frac{\gamma_{2,k}}{g_c} (\xi_{o,k}^{out})^{-2} + \frac{g}{g_c} \Delta\rho (C_{K,o})_k^{in} P_{2,k} \frac{C_o \Omega_{o,k}}{s} \{(\xi_{o,k}^{out})^{-\frac{s}{C_o \Omega_{o,k}}} - 1\}$$

$$- \frac{\gamma_{2,k}}{g_c} (\xi_{o,k}^{out})^{-\frac{s + C_o \Omega_{o,k}}{C_o \Omega_{o,k}}} + \frac{1}{g_c} \{B_1(s, z_k) - B_2(s, z_{k-1})\} \qquad (I.14)$$

$$G_{2,k}(s) \stackrel{\Delta}{=} \frac{(C_{K,o})_k^{in} f G_o \bar{\phi}^2_{\ell o}}{g_c C_o \Omega_{o,k} \rho_f D_H} g_{3,k} (\xi^{out}_{o,k} - 1)$$

$$\frac{(C_{K,o})_k^{in}}{g_c C_o \Omega_{o,k}} (s\, g_{4,k} + \frac{f G_o \bar{\phi}^2_{\ell o}}{\rho_f D_H} + \frac{f G_o^2}{2 D_H \rho_f} f_{6,k}) \ell n\, \xi^{out}_{o,k}$$

$$\frac{g (C_{K,o})_k^{in} \Delta \rho P_{2,k}}{g_c C_o \Omega_{o,k}} \{ (\xi^{out}_{o,k})^{-1} - 1 \} + \frac{(C_{K,o})_k^{in}}{g_c C_o \Omega_{o,k}} \{ s g_{3,k}$$

$$+ \frac{f G_o^2}{2 D_H \rho_f} f_{5,k} - \gamma_{3,k} \frac{C_o \Omega_{o,k}}{(C_{k,o})_k^{in}} \} (\xi^{out}_{o,k} - 1)$$

$$+ \frac{f G_o^2 (C_{K,o})_k^{in}}{2 g_c D_H C_o \Omega_{o,k} \rho_f} \times \frac{f_{7,k}}{2} \{ (\xi^{out}_{o,k})^2 - 1 \}$$

$$- \frac{g (C_{K,o})_k^{in} \Delta \rho P_{2,k}}{g_c s} \{ (\xi^{out}_{o,k})^{-\frac{s}{C_o \Omega_{o,k}}} - 1 \}$$

$$+ \frac{f G_o^2 (C_{K,o})_k^{in} f_{9,k}}{2 g_c D_H \rho_f (2 C_o \Omega_{o,k} - s)} \{ (\xi^{out}_{o,k})^{-\frac{2 C_o \Omega_{o,k} - s}{C_o \Omega_{o,k}}} - 1 \}$$

$$+ \frac{(C_{K,o})_k^{in}}{g_c (C_o \Omega_{o,k} - s)} [s g_{5,k} + \frac{f G_o \bar{\phi}_{\ell o}}{2 \rho_f D_H} g_{2,k} + \frac{f G_o^2 f_{8,k}}{2 \rho_f D_H}$$

$$- \frac{(C_o \Omega_{o,k} - s) \gamma_{4,k}}{(C_{K,o})_k^{in} (\xi^{out}_{o,k})^2}] \{ (\xi^{out}_{o,k})^{\frac{C_o \Omega_{o,k} - s}{C_o \Omega_{o,k}}} - 1 \} + \frac{1}{g_c} \{ B_2(s, z_k) - B_2(s, z_{k-1}) \}$$

(I.14b)

$$G_{3,k}(s) = \frac{(C_{K,o})_k^{in}}{g_c(C_o\Omega_{o,k}-s)}[s\,g_{6,k} + \frac{fG_o}{\rho_f D_H}\overline{\phi}^2_{\ell o}\{g_{6,k}$$

$$+ \frac{G_o f_{10,k}}{2}\}]\{(\xi^{out}_{o,k})^{\frac{C_o\Omega_{o,k}-s}{C_o\Omega_{o,k}}} -1\}$$

$$+ \frac{fG_o^2(C_{K,o})_k^{in} f_{11,k}}{2g_c\rho_f D_H(2C_o\Omega_{o,k}-s)}\{(\xi^{out}_{o,k})^{\frac{2C_o\Omega_{o,k}-s}{C_o\Omega_{o,k}}} -1\} +$$

$$+ \frac{g(C_{K,o})_k^{in}\,\Delta\rho}{g_c s}\{(\xi^{out}_{o,k})^{-\frac{s}{C_o\Omega_{o,k}}} -1\} +$$

$$\frac{\gamma_{5,k}}{g_c}\{(\xi^{out}_{o,k})^{-\frac{s+C_o\Omega_{o,k}}{C_o\Omega_{o,k}}} -1\} + \frac{1}{g_c}\{B_3(s,z_k)-B_3(s,z_{k-1})\} \qquad (I.14c)$$

By the assumption of thermodynamic equilibrium in the two-phase region,

$$\delta\hat{\Omega}_k = \frac{P_H v_{fg}}{A_{x-s} h_{fg}}\,\delta\hat{q}''_k \qquad (I.15)$$

Now using the heated wall dynamics in relation in Eq. (19),

$$\delta(\hat{\Delta p}_{2\phi})_k = G_{1k}(s)\delta\langle\hat{j}\rangle_k^{in} + \frac{P_H v_{fg} Z_{3,k}}{A_{x-s} h_{fg}} G_{2,k}\delta\hat{q}'''_{t,k}$$

$$+ G_{3,k}(s)\delta\langle\hat{\alpha}\rangle_k^{in} \qquad (I.16)$$

In order to evaluate the node inlet parameters, $\delta\langle\hat{j}\rangle_k^{in}$ and $\delta\langle\hat{\alpha}\rangle_k^{in}$, we can use Eqs. (I.5) and (I.7). From the channel inlet boundary conditions,

$$\delta\langle\hat{j}\rangle_k^{in} = \delta\hat{j}_i - \Omega_{o,B}\delta\hat{\lambda} + \delta_k(B)\,\frac{P_H v_{fg}}{A_{x-s} h_{fg}} \sum_{j=B}^{k-1} Z_{3,j}\Delta z_j \delta\hat{q}'''_{t,j} \quad (I.17)$$

where,

$$\delta_k(i) \overset{\Delta}{=} \begin{cases} 0, & \text{if } k=i \\ 1, & \text{if } k\neq i \end{cases}$$

and,

$$\delta\langle\hat{\alpha}\rangle_k^{in} = X_{1,k}\,\delta\hat{j}_i + X_{2,k}^{1\phi} \sum_{s=1}^{L} H'_s \delta\hat{q}'''_{t,s}$$

$$+ \sum_{i=B}^{k-1} X_{2,i}^{k}\,\delta\hat{q}'''_{t,i} + X_{3,k}\delta\hat{h}_{in} \quad (I.18)$$

where, $X_{i,k}(s)$ $(i=1,3)$, $X_{2,k}^{1\phi}(s)$, $X_{2,k}^{k}(s)$ are transfer functions having,

$$X_{1,k} = (1-\Omega_{o,B}\Lambda_1) \sum_{i=B}^{k-1} a_i - \frac{C_k \Omega_{o,B}\Lambda_1}{\rho_g v_{fg} j_{i_o}} \quad (I.19a)$$

$$X_{2,k}^{1\phi} = -\left\{ \sum_{i=B}^{k-1} a_i + \frac{C_k}{\rho_g v_{fg} j_{i_o}} \right\} \Omega_{o,B}\Lambda_2 \quad (I.19b)$$

$$X_{2,i}^{k} = \begin{cases} \dfrac{P_H v_{fg}}{A_{x-s} h_{fg}} \left(\sum_{j=i+1}^{k-1} a_i \right) Z_{3,i}\,\Delta z_i + b_i & (\text{for, } i<k=1 \\ b_i & (\text{for, } i=k-1) \end{cases} \quad (I.19c)$$

$$X_{3,k} = -\left\{ \sum_{i=B}^{k-1} a_i + \frac{c_k}{\rho_f v_{fg} j_{i_o}} \right\} \Omega_{o,B}\Lambda_3 \quad (I.19d)$$

$$a_i(i<k-1) = \left(\prod_{i=1}^{k-1-i} \Theta_3^{k-j}\right) \Theta_1^i$$

$$a_{k-1} = \Theta_1^{k-1} , \; b_{k-1} = \Theta_2^{k-1}$$

$$b_i(i<k-1) = \left(\prod_{i=1}^{k-1-i} \Theta_3^{k-j}\right) \Theta_2^i$$

$$c_k = \left(\prod_{i=B}^{k-1} \Theta_3^i\right)$$

and, $\Theta_i(i=1,2,3)$ are defined in Eq. (I.7) as,

$$\delta\langle\hat{\alpha}\rangle_k = \Theta_1^k \; \delta\langle\hat{j}\rangle_k + \Theta_2^k \; \delta\hat{\Omega}_k + \Theta_3^k \; \delta\langle\hat{\alpha}\rangle_k^{in} \tag{I.20}$$

Thus, from Eqs. (I.16) - (I.18),

$$\delta(\Delta\hat{p}_{2\phi})_k = \{G_{1,k}(1-\Omega_{o,B}\Lambda_1) + G_{3,k}X_{1,k}\}\delta\hat{j}_i$$

$$+ \{X_{2,k}^{1\phi} \; G_{3,k} - \Omega_{o,B}\Lambda_2 G_{1,k}\Lambda_2 G_{1,k}\} \sum_{s=1}^{L} H'_s \delta\hat{q}'''_{t,s}$$

$$+ \sum_{s=1}^{k-1} \{\frac{P_H v_{fg}}{A_{x-s} h_{fg}} G_{1,k} \; Z_{3,i} \; \Delta z_i + G_{3,k} \; X_{2,i}^k\} \; \delta\hat{q}'''_{t,i}$$

$$+ \frac{P_H v_{fg} Z_{o,k}}{A_{x-s} h_{fg}} G_{2,k} \; \delta\hat{q}'''_{t,k} + (-G_{1,k}\Lambda_3\Omega_{o,B} + G_{3,k} \; X_{3,k})\delta\hat{h}_{in} \tag{I.21}$$

For the inlet node of the two-phase region, the additional term due to the boiling boundary perturbation can be added to Eq. (I.21).

The total pressure drop perturbation in the two-phase region is obtained by summing up all the nodes in the two-phase region. The final form of $\delta(\hat{\Delta p}_{2\phi,total})$ is,

$$\delta(\hat{\Delta p}_{2\phi}) = \Pi_1 \delta\hat{j}_i + \sum_{k=1}^{N} \Pi_2^k \,\delta\hat{q}'''_{t,k} + \Pi_3 \,\delta\hat{h}_i \qquad \text{(I.22a)}$$

where,

$$\Pi_1 = \sum_{k=B}^{N} G_{1,k}(1-\Omega_{o,B}\Lambda_1) + \sum_{k=B+1}^{N} (G_{3,k}X_{1,k})$$

$$- \frac{\Omega_{o,B}\Lambda_1}{\rho_f v_{fg} (C_{K,o})_B^{in}} G_{3,B} - R\Lambda_1 + \sum_m^M K_{1,m} \qquad \text{(I.22b)}$$

$$\Pi_2^k = \begin{cases} [\sum_{j=B+1}^{N} \{X_{2,j}^{1\phi} G_{3,j} - \Omega_{o,B}\Lambda_2 G_{1,j}\} - (G_{1,B} + \frac{G_{3,B}\Omega_{o,B}}{\rho_f v_{fg}(C_{K,o})_B^{in}} \\ + R)\Lambda_2]H'_k + \sum_{i=B}^{N} \delta_{n_{sp(i)}}(0)\{\sum_{j=1}^{N_{sp(i)}} K_{2,j}^k(z'_\ell)\} \quad (\text{for } k \leq L) \\ \text{or,} \\ \sum_{i=k+1}^{N} \{\frac{P_H v_{fg}}{A_{x-s} h_{fg}} G_{1,i} Z_{3,k}\Delta z_k + G_{3,i}X_{2,k} + \frac{P_H v_{fg} Z_{3,k}}{A_{x-s} h_{fg}}\} \\ + \delta_{k,B} \frac{P_H v_{fg}}{A_{x-s} h_{fg}} Z_{3,B} G_{3,B} + \sum_{i=k}^{N} \{\delta_{n_{sp(i)}}(0) \sum_{n=1}^{N_{sp(i)}} K_{2,i}^k(z'_\ell)\} \\ \qquad (\text{for } k \geq B) \end{cases} \qquad \text{(I.22c)}$$

$$\Pi_3 = - \Lambda_3 \Omega_{o,B} \sum_{k=B}^{N} G_{1,k} + \sum_{k=B+1}^{N} G_{3,k} X_{3,k}$$

$$+ \left(\frac{G_{3,B} \Omega_{o,B}}{\rho_f v_{fg} (C_{K,o})_B^{in}} + R\right)\Lambda_3 + \sum_{m}^{M} K_{3,m} (z_\ell') \tag{I.22c}$$

APPENDIX II

The profile of the void fraction perturbation in the k-th node, $\delta\hat{<\alpha>}_k$, is given in Eq. (I.7). We can obtain the node-average void fraction perturbation as,

$$\delta\hat{\overline{<\alpha>}}_k = \frac{1}{\Delta z_k} \int_{z_k}^{z_{k+1}} \delta<\hat{\alpha}(z)>_k \, dz \tag{II.1}$$

By changing the coordinate system, in accordance with Eq. (I.4),

$$\delta\hat{\overline{<\alpha>}}_k = \frac{(C_{K,o})_B^{in}}{\Delta z_k C_o \Omega_{o,k}} \int_1^{\xi_{o,k}^{out}} \delta<\hat{\alpha}(\xi_{o,k})>_k \, d\, \xi_{o,k} \tag{II.2}$$

thus $\delta\hat{\overline{<\alpha>}}$ has the form,

$$\delta\hat{\overline{<\alpha>}}_k = \psi_{1,k}(s)\, \delta<\hat{j}>_k^{in} + \psi_{2,k}(s)\, \delta\hat{\Omega}_k + \psi_{3,k}\, \delta<\hat{\alpha}>_k^{in} \tag{II.3}$$

From Eqs. (I.15), (I.17) and (I.18) we obtain,

$$\delta\hat{\overline{<\alpha>}}_k = T_{1,k}(s)\, \delta\hat{j}_i + \sum_{i=1}^{k} T_{2,i}^{k}(s)\, \delta\hat{q}_{t,i}''' + T_{3,k}(s)\, \delta\hat{h}_{in}$$

where,

$$T_{1,k} \stackrel{\Delta}{=} \{\psi_{1,k} (1-\Omega_{o,B}\Lambda_1) + \psi_{3,k} X_{1,k}\}$$

$$(X_{2,k}\,\psi_{3,k} - \Omega_{o,B}\,\Lambda_2\,\psi_{1,k})\; H_i'(s) \qquad (1\leq k<B)$$

$$T_{2,i}^{k} \overset{\Delta}{=} \begin{cases} \dfrac{P_H v_{fg}}{A_{x-s} h_{fg}}\,\psi_{1,k}\, Z_{3,i}\,\Delta z_i + \delta_k(2)\,\psi_{3,k}\, X_{2,i}^{k} & (i\leq k-1) \\[2ex] \dfrac{P_H v_{fg} Z_{3,k}}{A_{x-s}\, h_{fg}}\;\psi_{2,k} & (i=k) \end{cases}$$

$$T_{3,k} \overset{\Delta}{=} \psi_{1,k}\,\Lambda_3\,\Omega_{o,B} + \psi_{3,k}\, X_{3,k}$$

$$\psi_{1,k} \overset{\Delta}{=} \frac{P_{2,k}(C_{k,o})_k^{in}}{\Delta z_k}\left[\left(\frac{1}{\xi_{o,k}^{out}} - 1\right) - \frac{C_o\Omega_{o,k}}{s}\left\{(\xi_{o,k}^{out})^{-\frac{s}{C_o\Omega_{o,k}}} - 1\right\}\right]$$

$$\psi_{2,k} \overset{\Delta}{=} \frac{P_{2,k}(C_{k,o})_k^{in}}{\Delta z_k s}\left\{(\xi_{o,k}^{out})^{-\frac{s}{C_o\Omega_{o,k}}} - 1\right\} - \frac{P_{2,k}(C_{k,o})_k^{in}}{\Delta z_k C_o \Omega_{o,k}}\left(\frac{1}{\xi_{o,k}^{out}} - 1\right)$$

$$\psi_{3,k} = -\frac{(C_{k,o})_k^{in}}{\Delta z_k s}\left\{(\xi_{o,k}^{out})^{-\frac{s}{C_o\Omega_{o,k}}} - 1\right\}$$

THE MODES OF FLOW OSCILLATION IN MULTI-CHANNEL TWO-PHASE FLOW SYSTEMS

Shigeyasu Nakanishi, Mamoru Ozawa, and Seikan Ishigai

Department of Mechanical Engineering
Faculty of Engineering
Osaka University
Suita, Japan

Graduate School of Science and Technology
Kobe University
Kobe, Japan

Osaka University of Industry
Daitoh, Japan

ABSTRACT

In this paper, some results of the experimental investigations on oscillation modes of multi-channel two-phase flow systems are reported. In the first part, phase relation between oscillating channels is studied using a Freon-113 test rig; the number of parallel channels is varied up to five. It is found that, if the thermohydraulic characteristics of all channels are identical, they oscillate with the same amplitude and equally out of phase with each other, and that if the characteristics are not identical various modes become possible. In the second part, interaction between oscillating channels is studied in detail at a two-channel air-water test rig. In this rig, it is indispensable to equip every air and water supply lines with compressible capacities in order to realize flow oscillation, and only the oscillations originated from Ledinegg instability, i.e. some kinds of pressure drop oscillation, can be observed. Effects of the compressible capacities and the hydraulic states of the channels on the modes of oscillation are examined. Other than the types of oscillation observed in single channel systems, one inherent to multi-channel systems is observed which is named as modulation wave one. It is clear that this stems from channel interaction.

1. INTRODUCTION

It has been well established for boiling two-phase flow systems comprising few parallel channels that the threshold for thermohydraulic instability is fairly close to that of single channel systems. On the other hand, there are many ambiguous points about the behaviors of multi-channel systems after instability occurs, although a number of works have been devoted to clarify them. We considered them very interesting from both practical and academic points of view, and have investigated for years; a part of our earlier results was reported briefly in [1]. In this paper, we present more details of our experimental investigations about two topics; phase relation of flow oscillation in individual channels and interaction between oscillating channels.

Firstly, consider boiling systems without compressible capacities. If the system comprises two identical channels, two channels oscillates in phase or 180° out of phase with each other and, of course, only the latter case can be observed where the total flow rate is kept constant. In the systems comprising more than two channels, situation becomes much more complicated because of high degrees of freedom, so various arguments can be found in literatures. About the three channel systems, Akagawa et al. [2] reported that two channels oscillated in phase with different amplitudes and the third oscillated 180° out of phase with the former. D'Arcy [3] observed the following three modes; i) two channels oscillate with about the same amplitude and 180° out of phase and the third does not oscillate; ii) two channels oscillate with about the same amplitude and in phase and the third oscillates 180° out of phase, and iii) all three channels oscillate in a kind of three-phase type of oscillation. Harvie [4] argued that the modes i) and ii) predominated in the neighborhood of the thresholds for instability and only the mode iii) was observed at highly unstable condition. Koshelov et al. [5] conducted experiments with a test rig comprising four channels and reported that there existed oscillation modes other than those cited above and the mode observed varied accidentally from run to run. In the first part of this paper, we present experimental results obtained at a Freon-113 test rig comprising five channels.

Interactions between oscillating channels are expected to vary according to the system conditions. In the second part of this paper, they are experimentally investigated using an air-water test rig comprising two horizontal channels. Air-water system is employed for the reason that this system can be easily controlled to the desired unstable flow conditions. All experiments are conducted in the regime of pressure drop oscillation. It should be noted here that single channel gas-liquid systems have two types of pressure drop oscillation, i.e. relaxation type and quasi-static type [6]. The former is identical to that encountered in boiling systems, while the latter is considered as pecular to gas-liquid systems.

2. PHASE RELATION BETWEEN OSCILLATING CHANNELS

2.1 Experimental Apparatus and Procedures

Experimental works are conducted at a Freon-113 test rig installed in the Steam Power Engineering Laboratory of Faculty of Engineering, Osaka University, whose descriptions can be found in [1] or [7]. Figure 1 schematically shows the test section used here. It is composed of five identical channels connected in parallel by two headers and arranged vertically. The coolant, Freon-113, flows vertically upward in it. Each channel is made from stainless steel tube with 6.03 mm I.D./7.00 mm O.D. and 2.8 m long (heated length), and equipped with a Venturi-tube at each inlet for flow measure ment. Figures in parentheses in the figure indicate the numbers of the channels. Heating is made by supplying A.C. current directly to the individual channels and is uniform both along channel length and channel by channel. Flow distribution is carefully adjusted by inserting some suitable orifice plates just downstream of the Venturi-tubes. The test conditions are as follows; pressure at exit is about 0.4 MPa, flow rate per channel is

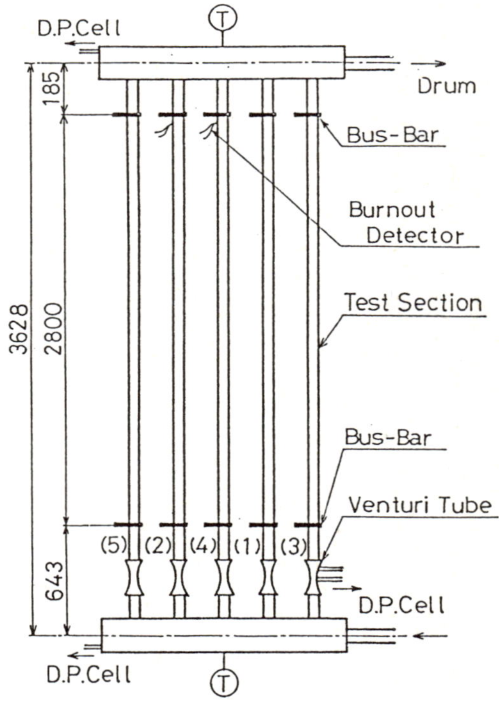

FIGURE 1 TEST SECTION FOR FREON-113 TEST RIG

30 - 40 kg/h for the three channel system and about 20 kg/h for the four and five channel systems, heating power is up to 2.0 kW per channel, and inlet subcooling is up to 90 K.

Experimental procedures are as follows; after adusting flow rate in each channel within 3 - 4 % discrepancy under adiabatic condition, total flow rate and inlet subcooling of the test section are set to the specified values and then heating is started. Heating power is gradually increased until flow instability occurs. Then, the power is adjusted to realize the desired oscillation condition in the test section, and fixed to record and collect the data required. Throughout the run, pressure at the exit header is maintained at the specified value by controlling heating and/or cooling of the vapor-liquid drum to which the delivery line from the test section is connected. For the runs with three channels, the channels Nos. 1, 2 and 4 are used; for the runs with four channels, the channels Nos. 1, 2, 4 and 5, and for the runs with five channels, all channels.

2.2 Results and Discussions

Results concerning instability threshold, types of instability and oscillation frequency have been published in [1]; the reader interested in them might refer it. In the following, only the results concerning phase relation of oscillating flows are shown.

2.2.1 Three channel system. If the flow is uniformly distributed among three channels, all channels oscillate with the same ampli-

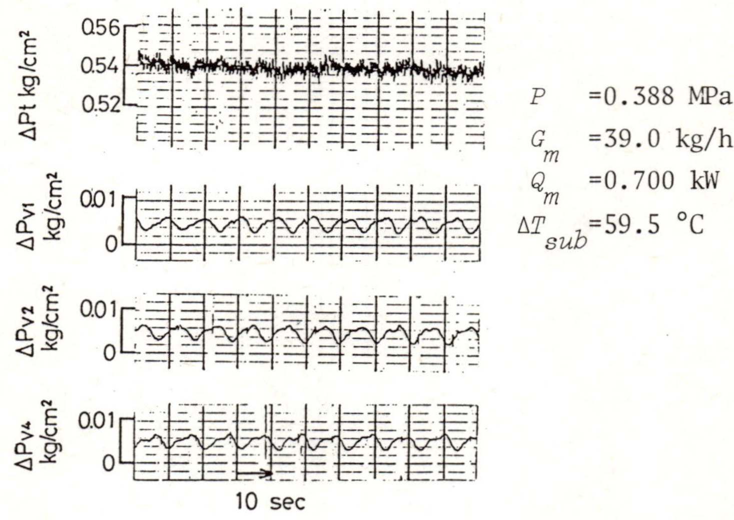

FIGURE 2 RECORDING OF OSCILLATION WITH THREE CHANNEL SYSTEM (CASE OF UNIFORM FLOW DISTRIBUTION)

tude and 120° out of phase with each other. Figure 2 shows an example of recordings of such runs, where P is system pressure; G_m - flow rate per channel; Q_m - heating power per channel; ΔT_{sub} - inlet subcooling; ΔP_t - pressure difference between the inlet and exit headers, and ΔP_{vi} - pressure difference signal from the Venturi-tube of the channel No.i. The instability observed is density wave oscillation as in other examples of runs cited in this section.

To examine effect of flow distribution, gagging of channels is readjusted and experiments are conducted in the same manner as the above. Figure 3 shows some examples of oscillation in such runs; the flow distrbution ratio among the channels is $G_1 : G_2 : G_4 =$ 35.5 : 30.4 : 34.1 under adiabatic and steady condition, where G_i is flow rate through the channel No. i. Under this setting, various modes of flow oscillation are observed from run to run. In the case of Figure 3, a, the channels Nos. 1 and 4 oscillate with about the same amplitude and 180° out of phase with each other, while the channel No. 2 much more weakly oscillates. In the case of Figure 3, b, the channels Nos. 1 and 2 oscillate with about the same amplitude but not the same wave form and in phase, while the channel No. 4 oscillates with about twice amplitude and 180° out of phase. In the case of Figure 3, c, three channels oscillate 120° out of phase with each other, but their wave forms are not similar. The

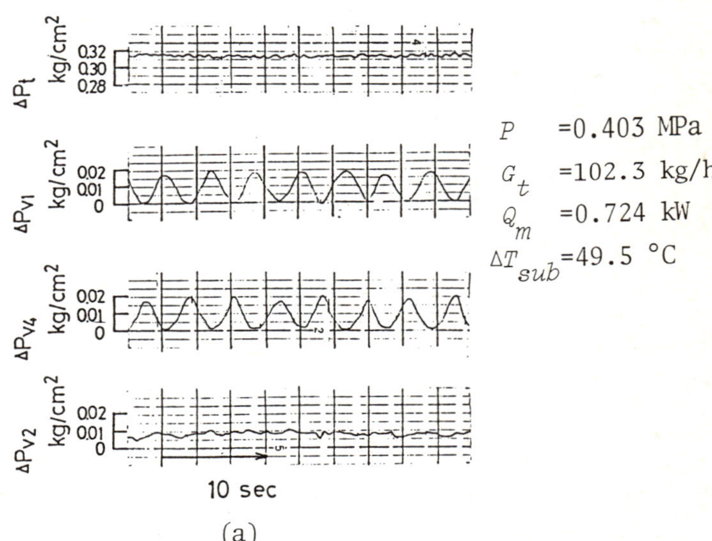

(a)

FIGURE 3 RECORDINGS OF OSCILLATION WITH THREE CHANNEL SYSTEM (CASES OF NON UNIFORM FLOW DISTRIBUTION)

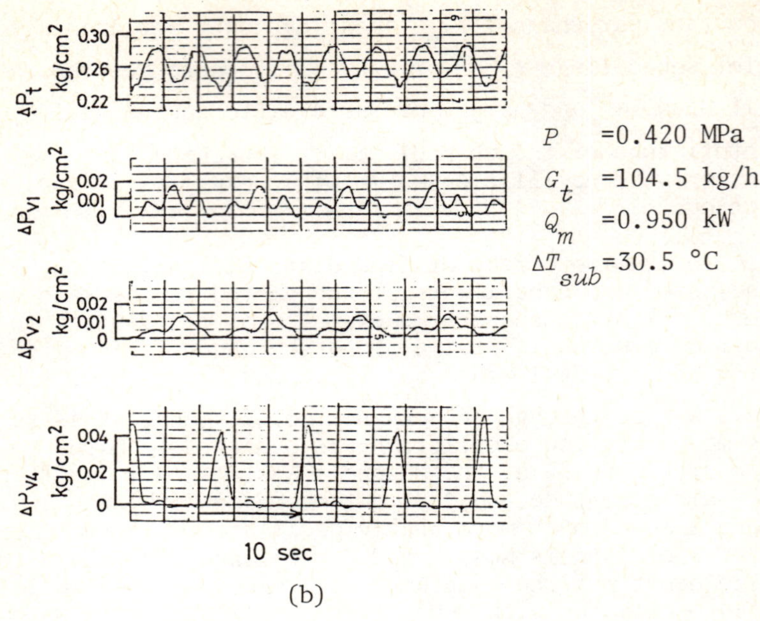

(b)

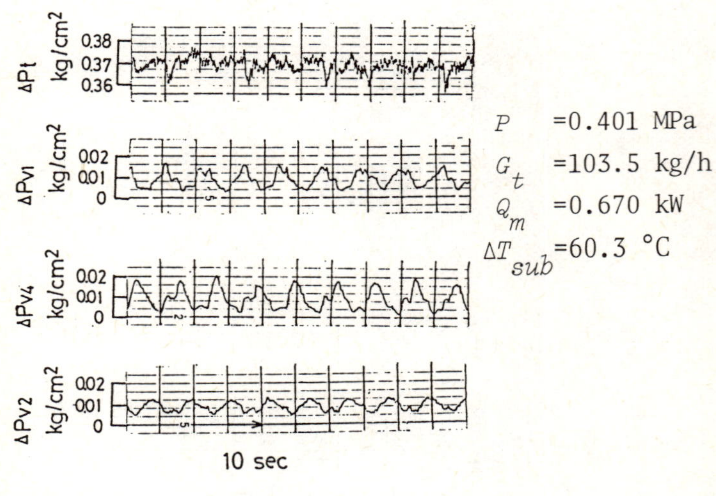

(c)

FIGURE 3 RECORDINGS OF OSCILLATION WITH THREE CHANNEL SYSTEM (CASES OF NON UNIFORM FLOW DISTRIBUTION) (CONTINUED) mode observed has no clear connection with the experimental condition and moreover different modes may occur at different runs even under almost the same experimental condition. Thus it is impossible to forecast the mode previously. However, it is clear that this variety of modes of oscillation is introduced by the nonuniform flow distribution, though its mechanism has not yet been identified.

If a compressible capacity is connected with the system just upstream of the inlet header, pressure drop oscillation can occur at a certain condition. It is a neccesary but not adequate condition for pressure drop oscillation of multi-channel system that the flow rate per channel corresponds to the negative slope region of the pressure drop characteristics of the boiling channel. For systems with no compressible capacity, it is known [2] that, when the state of the system falls into the negative slope region, flow redistribution occurs and flow rates of the individual channels spontaneously change to stabilize the whole systems. In this case, if one channel remains in the negative slope region, it is possible to stimulate pressure drop oscillation by connecting a compressible capacity with the system. Under this situation, the rest of channels essentially do not oscillate while the channel in the negative slope region shows pressure drop oscillation coupled with the compressible capacity. Namely, the neccesary and adequate condition for pressure drop oscillation of multi-channel boiling systems is that one channel rests on the negative slope region after flow redistribution. This condition is experimentally ascertained in our three channel system.

2.2.2 Four and five channel systems. For these systems, experiments are conducted at about 20 kg/h per channel due to pump capacity of the test rig. So, patterns of oscillation might be somewhat different from those of the three channel system, but the qualitatative conclusions must be not affected. Figure 4 shows a typical recording of oscillation with the four channel system. Four channels are divided into two pairs and channels in the same pair oscillate with about the same amplitude and in phase. Two pairs of channels oscillate 180° out of phase with each other. Members of the pairs are not fixed and may vary from run to run. Figure 5 shows an example of recordings of oscillation with the five channel systems. It is seen from the figure that all channels oscillate with the same amplitude and equal phase lag.

3. INTERACTIONS BETWEEN TWO OSCILLATING CHANNELS

3.1 Experimental Apparatus and Procedures

Figure 6 shows schematically the flow diagram of the air-water test rig comprising two horizontal channels. The test section is

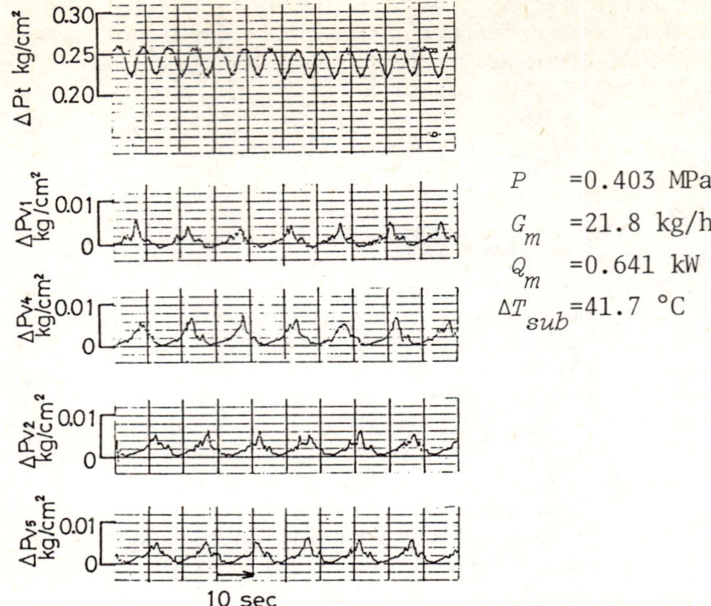

FIGURE 4 RECORDING OF OSCILLATION WITH FOUR CHANNEL SYSTEM

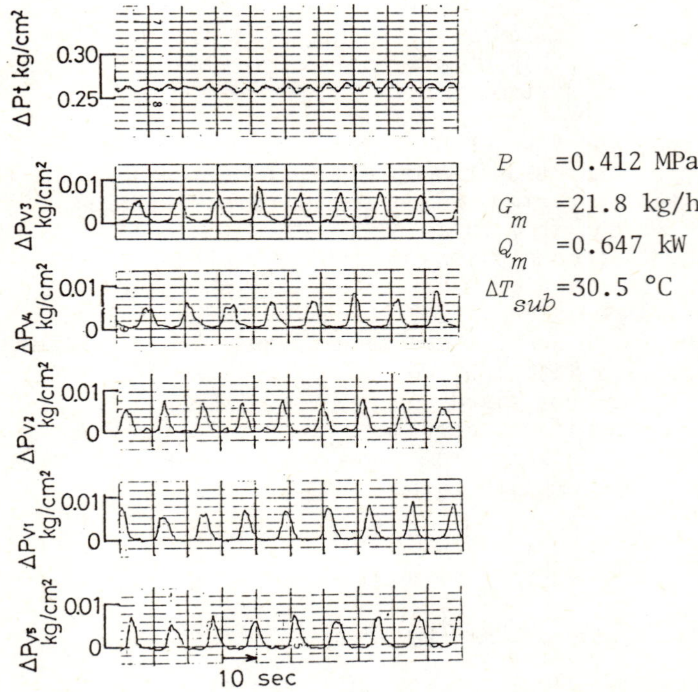

FIGURE 5 RECORDING OF OSCILLATION WITH FIVE CHANNEL SYSTEM

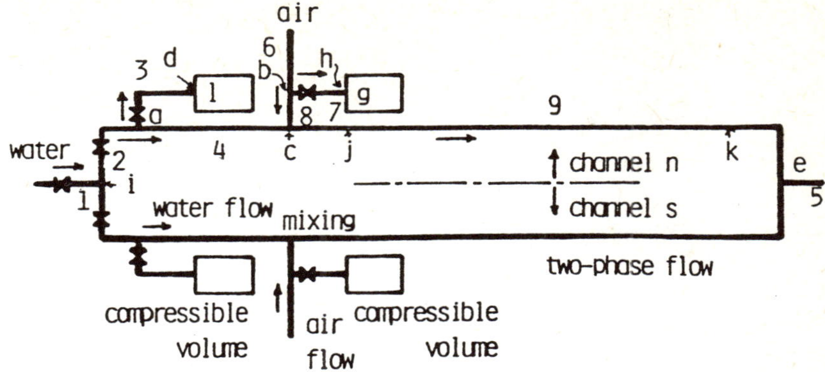

FIGURE 6.FLOW DIAGRAM OF AIR-WATER TEST RIG

TABLE 1 MAIN DIMENSIONS OF TEST SECTION

ELEMENT	LENGTH, mm	INNER DIAMETER, mm
1(-i)	---	7.7
2(i-a)	260	7.7
3(a-d)	900	7.7
4(a-c)	400	3.05
5(e-)	36	4.4
6(-b)	---	7.7
7(b-h)	100	7.7
8(b-c)	150	7.7
9(c-j)	43	3.05
(j-k)	3100	3.05
(k-e)	416	3.05

composed of two identical channels named the channels n and s, respectively. Each channel is made from glass tube with 3.05 mm I.D. and 3.359 m long, and equipped respectively with own compressible capacities for air, g, and for water, l. Pressure difference across the two-phase region is measured between the points j and k, whose distance is 3.100 m; main dimensions of the test section are shown in Table 1. Total water supply to the test section is kept constant, and air supply to each channel is also kept constant at the point b by using a critical flow nozzle and a pressure reducing valve.

Experimental procedures are follows; firstly, total water flow rate, air flow rates and volumes of the compressible capacities are set to the specified values. In the course of flow setting, all compressible capacities are disconnected from the main system by closing their connecting valves. After measuring statical values of pressure drops across the whole system (between the points i and e), ΔP_t, and across each channel, ΔP_n and ΔP_s, the system is brought pressure drop oscillation by opening the lines connected with the compressible capacities. Data are collected and recorded when the stationary state is established.

The test conditions are as follows; superficial water velocity in channel, w_{l0} = 0.187 m/s, superficial air velocity in channel, w_{g0} = 0.12 - 1.3 m/s, and volumes of the compressible capacities for water, V_l = 0 - 4 l and for air, V_g = 0.014 - 5 l.

3.2 Results and Discussions

The type of flow instability obtainable at the test rig described above is pressure drop oscillation excited by Ledinegg instability. In gas-liquid single channel systems, the pressure drop oscillation has two variants, i.e. relaxation type and quasi-static types [6]. The former is similar to that encountered in boiling systems and normally observed when the compressible capacity for liquid is small (Figure 7, a). The latter is pecular to gas-liquid system and normally observed when the compressible capacities both for liquid and for gas are relatively large (Figure 7, b); the name, quasi-static, comes from the fact that its pressure drop oscillates obeying the static flow characteristics. In multichannel systems, flow oscillation in some channel may interact with other channels and thus there may exist complex modes of oscilla-

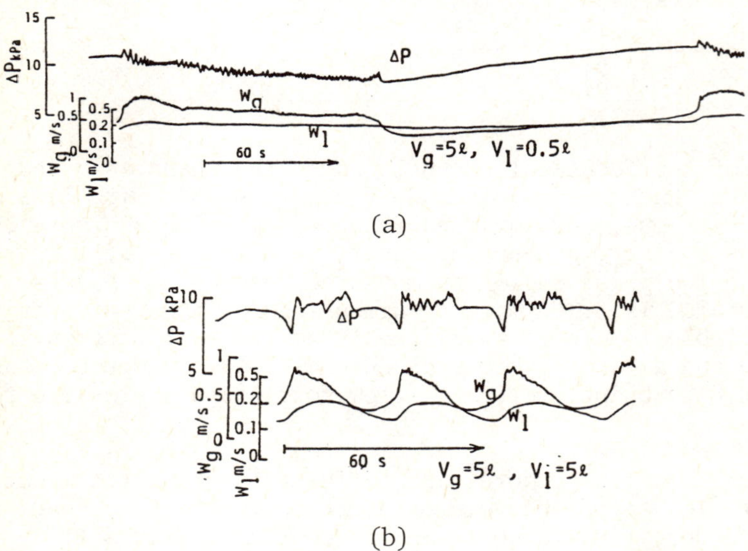

(a)

(b)

FIGURE 7 FLOW OSCILLATION IN AIR-WATER SYSTEM (SINGLE CHANNEL SYSTEM) tion. This interaction is discussed here for the simplest case, i.e. for the two channel system.

3.2.1 Modes of oscillation. Flow oscillations observed at our two channel test rig can be classified into the following four modes.

A) Single channel mode. Two channels oscillate with the same wave form and in phase. The type of oscillation is relaxation one. This mode is observed when the compressible capacities for air are large and those for water are very small (Figure 8, a).

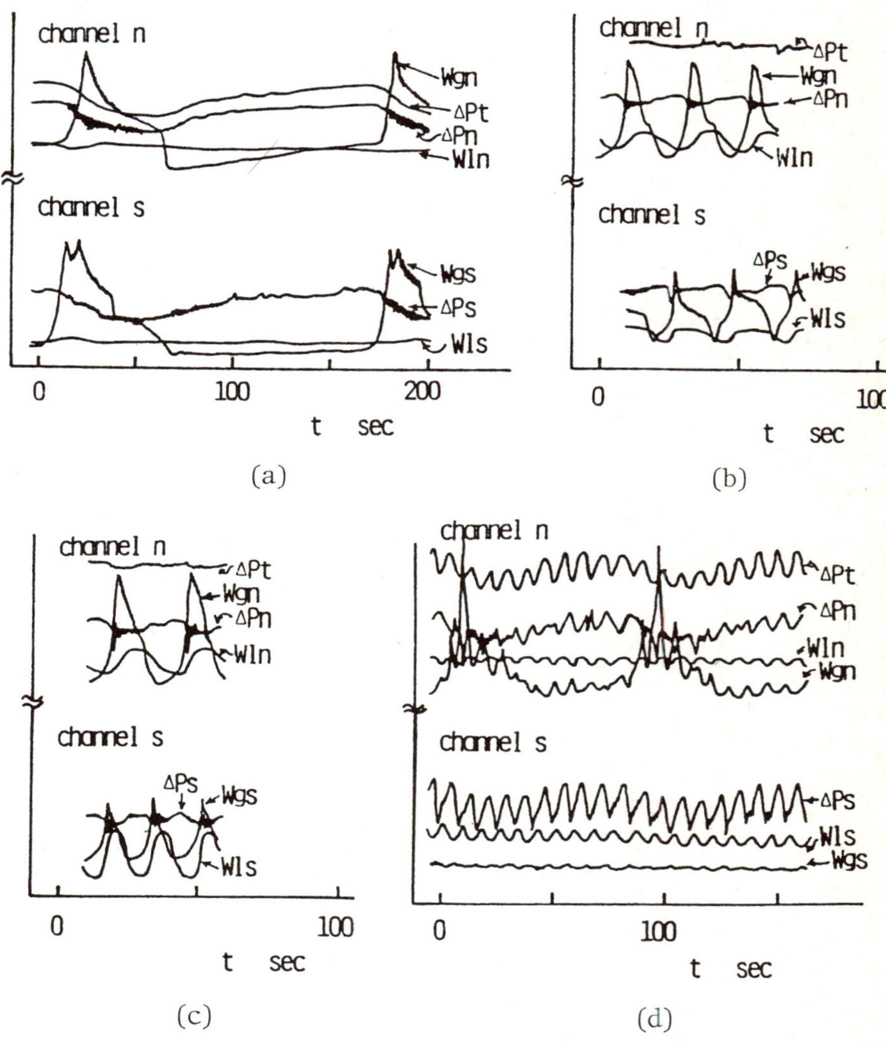

FIGURE 8 MODES OF OSCILLATION IN AIR-WATER SYSTEM (TWO CHANNEL SYSTEM)

B) U-tube mode. Two channels oscillate 180° out of phase with each other, and the type of oscillation is quasi-static one. This mode is observed when the compressible capacities for water are larger than in the case of A) (Figure 8, b).

C) Multi-channel mode. When the compressible capacities for water are further increased in the case of B), two channels start to oscillate independently of each other. The type of oscillation is quasi-static one. Intensity of oscillation becomes relatively weak (Figure 8, c).

D) Modulation-wave mode. When the compressible capacities for water are small and difference between volumes of two compressible capacities for air is large, one channel oscillates in the form of long-period fluctuation superposed with short-period fluctuation (Figure 8, d).

Between the modes A and B, a transition mode exists which takes the form of irregular alternation of these two modes. Overall pressure drop across the test section, ΔP_t, is kept essentially constant in the modes B and C, while it oscillates with large amplitude in the modes A and D. Figure 9 shows a pattern map in terms of volumes of the compressible capacities for the case where the flow rates and the volumes of the compressible capacities are identical in two channels.

To elucidate the effect of different volumes of the compressi-

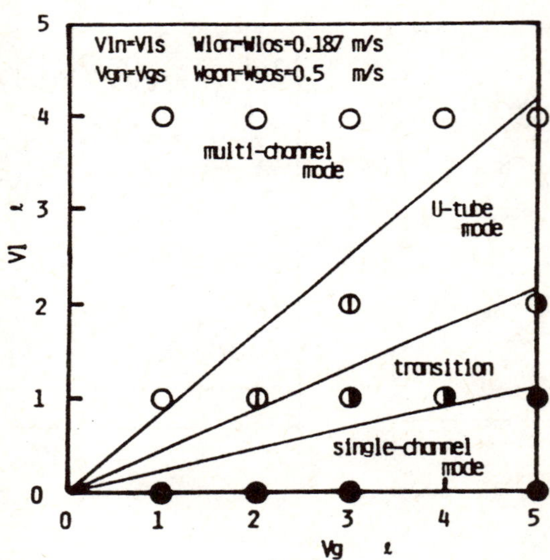

FIGURE 9 PATTERN MAP OF OSCILLATION MODES OF AIR-WATER SYSTEM (TWO CHANNEL SYSTEM)

ble capacities on the mode of oscillation, a series of experiments are made various combinations of volumes of the capacities. At runs with $V_{ls} = V_{ln} = 0\ l$, $V_{gn} = 0.014\ l$ and $V_{gs} = 1 - 5\ l$, the mode D is observed, i.e. the channel s oscillates in the form of modulation wave and the channel n shows high frequency oscillation. When V_{gn} increases to about $1\ l$, the mode changes to the mode A. At the runs with $V_g = 1 - 5\ l$, $V_l = 1 - 5\ l$, the mode C is observed.

3.2.2 Effect of hydraulic states of channels. The mode of oscillation observed in the individual channels is determined by volumes of the compressible capacities as well as the hydraulic states of each channel. Among the factors characterizing the hydraulic states, one of the most important is the slope of the pressure drop curve on which the operation point is located. Table 2 summarizes the results about the effect of combination of the states of channels at various couples of volumes of the compressible capacities. The roman numeral indicates the region of the pressure drop characteristics on which the channel state is located; I, II and III correspond to positive slope region in low flow rate, negative slope region and positive slope region in high flow rate, respectively.

TABLE 2 MODE OF OSCILLATION AND SLOPE OF PRESSURE DROP CURVE

$V_{ln} = V_{ls}$, l		0	0	1	1	4	4
$V_{gn} = V_{gs}$, l		1	5	1	5	1	5
STATE OF CHANNELS	I I	S S	S S	S S	S S	S S	S S
	I III	S S	S S	S S	S S	S S	S S
	III III	S S	S S	S S	S S	S S	S S
	I II	Q R	L R	S Q	S R	S Q	S Q
	II II	R R	R R	Q Q	R R	Q Q	Q Q
	III II	R R	R R	G Q	G R	S Q	G Q

S means stable; R - relaxation type; Q - quasi-static type; G - case where only gas flow oscillate, and L - case where only liquid flow oscillates. The combination of II and II is the case studied at the above.

Clearly, if at least one channel is located at the region II, the system becomes unstable. In the case where only one channel is located in the region II, the unstable channel oscillates in relaxation or quasi-static mode according to volumes of the compressible capacities and the stable one oscillates with the same frequency and in phase with the former. The latter becomes at rest as the volume of the capacity for liquid increases. The oscillation frequency is much higher than in the case where two channels are located in the region II.

4. CONCLUDING REMARKS

In the first part of this paper, it is established that for boiling systems comprising few channels that, if flow is distributed uniformly among all channels, they oscillate with the same amplitude and equal phase lag. So, confusions about phase relation in multi-channel systems are clearly settled. In the second part, it is shown using air-water system that various modes of oscillation are possible in multi-channel systems. Although the results obtained at air-water system should not be directly applied to the actual boiling systems and their interpretation has not been obtained as yet, they must provide clues for investigation of interaction between oscillating channels of boiling systems.

The authors are deeply indebted to Professor K. Akagawa and Associate Professor T. Fujii of Kobe University for their permission to use their valuable materials.

REFERENCES

1. Ozawa, M., S. Nakanishi and S. Ishigai. "Flow Instabilities in Multi-Channel Boiling Systems." ASME Paper 79-WA/HT-55 (1979).
2. Akagawa, K., T. Sakaguchi, M. Kono and M. Nishimura. "Study on Distribution of Flow Rate and Flow Stabilities in Parallel Long Evaporators." Bulletin of the JSME 14 (1971) 837-848.
3. D'Arcy, D.F. "An Experimental Investigation of Boiling Channel Flow Instability." Proceedings on Two Phase Flow Dynamics, Eindhoven Vol. 2 (1967) 1173-1223.
4. Harvie, J.D. "An Experimental Investigation of Flow Insta-

bility in Freon-12 and Comparison with Water Data." Inst. of Chem. Engrs. Symp. Ser. 38 (1974) E1.

5. Koshelov, I.I., A.V.Surnov and L.V.Nikitina. "Usloviya vozniknoveniya pul'satsii na modeli vertikal'nykh ekranov." Energomashinostroenie 15-10 (1969) 3-5 (in Russian.).
6. Ozawa, M.,K. Akagawa, T. Sakaguchi and T. Suezawa. "Oscillating Instabilities in Gas-Liquid Two-Phase Flow Systems." Preprint of 1980 International Seminar, Nuclear Reactor Safety Heat Transfer, Dubrovnik (1980).
7. Ozawa, M., S. Nakanishi, S. Ishigai, Y. Mizuta and H. Tarui. "Flow Instabilities in Boiling Channels. Part 1 Pressure Drop Oscillation." Bulletin of the JSME 22 (1979) 1113-1118.

ACKNOWLEDGMENT

The authors are deeply indebted to Professor K. Akagawa and Associate Professor T. Fujii of Kobe University for their permission to use their valuable materials.

A LINEARIZED STABILITY ANALYSIS OF FORCED VAPOR FLOW CONDENSATION

M.Ünsal

Department of Mechanical Engineering
Middle East Technical University
Gaziantep Campus, Gaziantep, Turkey

ABSTRACT

The hydrodynamic stability of thin condensate films adjacent to flowing vapors is considered. The conditions representing mass, momentum, and energy conservation across the liquid/vapor interface are formulated for the case of negligibly small vapor hydrodynamic boundary layer thickness. The problem formulation is analyzed by means of a perturbation method yielding the base flow problem and the linearized stability problem. The linearized stability problem is analyzed yielding the neutral stability curves and closed form approximate expressions for the wave amplification rates and the wave velocities. The results of the stability analysis are discussed with reference to forced vapor flow condensation on vertical and horizontal surfaces. The effects of vapor velocity, density ratio, Prandtl number, acceleration effect parameter and the surface tension parameter on the stability characteristics of infinitesimal disturbances are identified.

1. INTRODUCTION

Interfacial stability characteristics have been utilized in the past when modelling the physical behaviour in some two-phase flows. Estimation of a fraction of the pressure drop in two-phase annular mist flow via a critical wavelength concept [1] and derivation of expressions for boiling heat transfer coefficients [2] are examples. The interfacial stability characteristics are usually deduced from hydrodynamic stability analyses which are also essential for the estimation of the strict validity regions of previous theoretical research based on the steady flow assumption.

The linearized stability of liquid films undergoing heat transfer and interfacial phase change has been studied in the past. Bankoff [3], Marschall and Lee [4], Ünsal and Thomas [5] have presented solutions based on the long wave perturbation method. Recently, Spindler [6] analyzed the linearized stability problem by three different procedures. The parallel flow approximation was relaxed in this latter study and a quasi-parallel treatment of the stability problem was presented. The effect of vapor motion on the stability characteristics of condensate films which has been neglected in all of these previous investigations is the subject matter of the present study. It is expected that results from the present analytical study will aid identification of some of the instability mechanisms in forced vapor flow condensation. A linearized stability analysis of steady condensate films adjacent to flowing vapors such as those treated in References [7-9] is considered. The stability problem is analyzed by the long wave perturbation method.

2. FORMULATION

Stability of the free interface of a condensate film flowing down an inclined surface under the action of gravity and flowing vapor is considered (Fig.1). The velocity and temperature in the condensate film are to be determined from the two-dimensional incompressible mass, momentum and energy equations. The wall temperature is assumed to be constant. The liquid satisfies the no-slip condition at the surface. The liquid/vapor interface is assumed to be at the saturation temperature of the vapor. Mass, momentum and energy conditions must be satisfied across the liquid/vapor interface. Although these interfacial conditions have been previously formulated [10,11], a different formulation is presented here in order to render the stability problem tractable by a straightforward perturbation solution procedure. The vapor boundary layer is neglected in the present formulation which limits the applicability of the present results to cases with large condensation rates and small vapor viscosity. It is natural that disturbances originating at the liquid/vapor interface will bring about coupled disturbed motions in the liquid and the vapor phases. A stability analysis of the flow system consisting of the condensate film, the vapor hydrodynamic boundary layer and the free stream vapor flow will not be attempted in the present study. Instead, the stability problem will be simplified by limiting the analysis to cases with strong vapor suction. When the condensation mass transfer rate is large the vapor boundary layer thickness will be negligibly small. Following this observation, the interfacial mass energy and momentum conditions will be written for a thin interfacial control volume (Fig.2). The vapor boundary layer is contained within the control volume. The mass,momentum and energy flows across the surfaces of the control volume which are perpendicular to the liquid/vapor interface are neglected as well as the capacitive effects inside the control volume. The mass conservation condition is

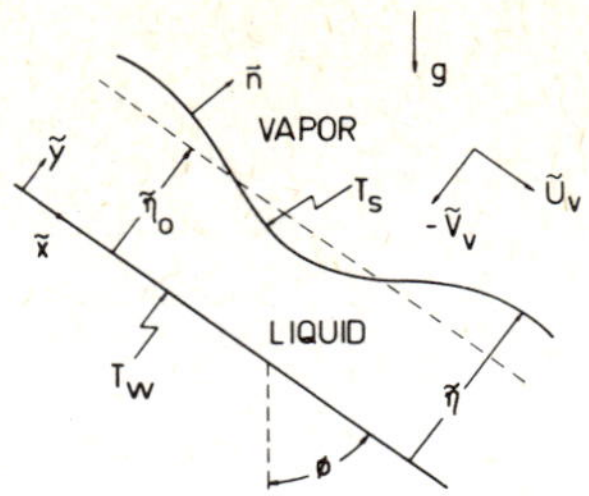

Fig. 1 Coordinate system for vapor and condensate flow.

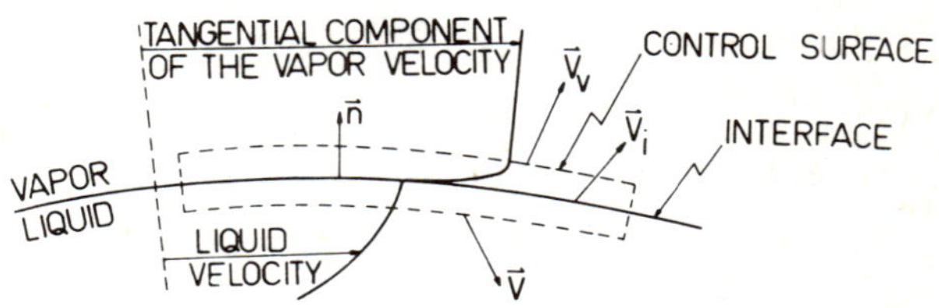

Fig. 2 The interfacial control volume.

$$\rho(\vec{V} - \vec{V}_i).\vec{n} = \rho_v(\vec{V}_v - \vec{V}_i).\vec{n}\ , \quad \text{at } \tilde{y} = \tilde{\eta} \tag{1}$$

The heat flux in the vapor phase will be zero for a saturated vapor and usually negligible for a superheated vapor. The energy interfacial condition is given by

$$\vec{q}.\vec{n} - h_{fg}\rho(\vec{V} - \vec{V}_i).\vec{n} = 0\ , \quad \text{at } \tilde{y} = \tilde{\eta} \tag{2}$$

The interfacial momentum condition is obtained by equating the forces to the momentum generation rate :

$$\rho\vec{V}(\vec{V} - \vec{V}_i).\vec{n} - \rho_v\vec{V}_v(\vec{V}_v - \vec{V}_i).\vec{n} - \tilde{p}_v.\vec{n} - \vec{n}.\Pi - \frac{\sigma}{r}\vec{n} = 0$$
$$\text{at } \tilde{y} = \tilde{\eta} \tag{3}$$

In the interfacial conditions given above, $\vec{V}_v$ is the free stream vapor velocity and, as a consequence, the viscous boundary layer has been neglected. The velocity of the interface and the unit normal to the interface are given by

$$I_{\tilde{t}} + \vec{V}_i\ .\ \nabla I = 0, \qquad \vec{n} = \nabla I/\ |\nabla I| \tag{4 ,5}$$

A constant free stream $\tilde{x}$-component vapor velocity, $\tilde{U}_{vo}$, has been considered in the previously reported studies of steady forced vapor flow condensation. The same vapor velocity condition is used in the base flow of the present stability analysis. This condition gives a a uniform vapor $\tilde{y}$-component velocity in the $\tilde{y}$-direction for the base flow. Additionally, the parallel flow approximation will be utilized. Thus, $\tilde{V}_{vo}$ will be given a locally constant treatment and it will be possible to represent the base flow vapor motion by a potential function. Since the viscous effects are neglected inside the vapor phase, the time dependent motion of the vapor will also be expressed by a potential function. Consequently, the components of the vapor velocity will be written in terms of a potential function,

$$\tilde{U}_v = \tilde{\Phi}_{\tilde{x}}, \qquad \tilde{V}_v = \tilde{\Phi}_{\tilde{y}} \tag{6}$$

and $\tilde{\Phi}$ will satisfy :

$$\tilde{\Phi}_{\tilde{x}\tilde{x}} + \tilde{\Phi}_{\tilde{y}\tilde{y}} = 0 \tag{7}$$

The vapor pressure will be determined by :

$$\tilde{p}_v + \frac{1}{2}\rho_v(\tilde{\Phi}^2_{\tilde{x}} + \tilde{\Phi}^2_{\tilde{y}}) + \rho_v\tilde{\Phi}_{\tilde{t}} = \tilde{p}_{vo} + \frac{1}{2}\rho_v(\tilde{U}^2_{vo} + \tilde{V}^2_{vo}) \tag{8}$$

Since $\vec{V}_v$ is the velocity outside the vapor boundary layer, the tangential component of $\vec{V}_v$ will not be equal to the tangential component of the liquid velocity at the liquid/vapor interface. It will be assumed that the free stream vapor flow is not subjected to disturbances other than those originating at the phase interface. Also, the disturbances must decay with distance from the interface. This latter condition yields :

$$\tilde{U}_v = \tilde{U}_{vo}, \qquad \tilde{V}_v = \tilde{V}_{vo}, \qquad \text{at } \tilde{y} = \infty \tag{9}$$

In order to simplify formulation, the velocity of the interface will be eliminated from the interfacial conditions by equation (4). The tangential component of the interfacial momentum condition will be simplified using equations (1) and (2). The normal component of equation (3) will be put into a simpler form by means of equations (1), (2), and (8). Finally, the left hand side of equation (1) will be expressed in terms of the heat flux by substitution from equation (2). Expressing the liquid velocity by a stream function,

$$\tilde{U} = \tilde{\Psi}_{\tilde{y}}, \qquad \tilde{V} = -\tilde{\Psi}_{\tilde{x}}$$

and introducing dimensionless variables,

$$\Psi = \tilde{\Psi}/\tilde{U}_o\tilde{\eta}_o, \qquad \Phi = \alpha\tilde{\Phi}/\tilde{U}_o\tilde{\eta}_o \tag{10 ,11}$$

$$x = \alpha\tilde{x}/\tilde{\eta}_o, \qquad y = \tilde{y}/\tilde{\eta}_o \tag{12, 13}$$

$$z = \alpha y, \qquad t = \alpha\tilde{U}_o\tilde{t}/\tilde{\eta}_o \tag{14, 15}$$

$$p_{vo} = \tilde{p}_{vo}/\rho\tilde{U}^2_o, \qquad p = \tilde{p}/\rho\tilde{U}^2_o \tag{16, 17}$$

$$\theta = (\tilde{T} - \tilde{T}_w)/(\tilde{T}_s - \tilde{T}_w), \qquad \eta = \tilde{\eta}/\tilde{\eta}_o \tag{18 ,19}$$

the problem formulation is expressed by the following dimensionless equations, boundary and interfacial conditions. Differential Equations :

$$\Phi_{xx} + \Phi_{zz} = 0 \tag{20}$$

$$\Psi_{yyy} + n = \alpha R(p_x + \Psi_{yt} + \Psi_y\Psi_{xy} - \Psi_x\Psi_{yy}) - \alpha^2\Psi_{xxy} \tag{21}$$

$$p_y - mR^{-1} = -\alpha R^{-1}\Psi_{xyy} + \alpha^2(\Psi_y\Psi_{xx} - \Psi_x\Psi_{xy} + \Psi_{xt}) - \alpha^3R^{-1}\Psi_{xxx} \tag{22}$$

$$\theta_{yy} = \alpha PrR(\Psi_y\theta_x - \Psi_x\theta_y + \theta_t) - \alpha^2\theta_{xx} \tag{23}$$

Boundary conditions at y = 0 :

$$\Psi_x = 0, \qquad \Psi_y = 0, \qquad \theta = 0 \tag{24 ,25 ,26}$$

Interfacial conditions at $y = \eta$:

$$\theta = 1 \tag{27}$$

$$\begin{aligned}\Psi_{yy} &= F(\theta_y - \alpha^2\theta_x\eta_x)(\Phi_x - \Psi_y)(1 + \alpha^2\eta^2{}_x)(1 - \alpha^2\eta^2{}_x)^{-1} \\ &\quad - \alpha(1-\gamma)N_dR^{-1}(\theta_y - \alpha^2\theta_x\eta_x)^2(1 - \alpha^2\eta^2)^{-1}\eta_x + \alpha^2\Psi_{xx} \\ &\quad + \alpha^2 4\eta_x\Psi_{xy}(1 - \alpha^2\eta^2{}_x)^{-1}\end{aligned} \tag{28}$$

$$\begin{aligned}&p + (\gamma-1)N_dR^{-2}(1 + \alpha^2\eta^2{}_x)^{-1}(\theta_y - \alpha^2\theta_x\eta_x)^2 + \gamma(\Phi_x^2 + \Phi_z^2)/2 \\ &+ \gamma\Phi_t + N_bR^{-5/3}\eta_{xx}(1 + \alpha^2\eta_x^2)^{-3/2} + \alpha 2R^{-1}\,[\Psi_{xy}(1 - \alpha^2\eta_x^2) + \eta_x\cdot \\ &(\Psi_{yy} - \alpha^2\Psi_{xx})](1 + \alpha^2\eta_x^2)^{-1} = p_{vo} + \gamma(U^2 + V^2)/2\end{aligned} \tag{29}$$

$$R\gamma(\alpha\Phi_x\eta_x - \Phi_z + \alpha\eta_t) = F(\theta_y - \alpha^2\theta_x\eta_x) \tag{30}$$

$$\alpha R(\Psi_y\eta_x + \Psi_x + \eta_t) = F(\theta_y - \alpha^2\theta_x\eta_x) \tag{31}$$

Boundary conditions at $y = \infty$

$$\Phi_x = U, \qquad \Phi_z = -V \tag{32 ,33}$$

It is noted that the tangential component of the interfacial momentum condition, equation (28), is a generalized version of the "asymptotic shear stress interfacial condition" utilized in References [7] and [8].

3. DERIVATION OF THE BASE FLOW AND THE LINEARIZED STABILITY PROBLEMS

A long wave asymptotic solution of the problem posed by equations (21)-(23), boundary conditions (24)-(26), and interfacial conditions (27)-(29) is sought by assuming

$$\Psi(x,y,t) = \Psi_o(x,y,t) + \alpha\Psi_1(x,y,t) + 0(\alpha^2) \quad (34)$$

$$p(x,y,t) = p_o(x,y,t) + \alpha p_1(x,y,t) + 0(\alpha^2) \quad (35)$$

$$\theta(x,y,t) = \theta_o(x,y,t) + \alpha\theta_1(x,y,t) + 0(\alpha^2) \quad (36)$$

Substitution of expansions (34)-(36) into the problem and solving the resulting perturbation problems yields the solutions for Ψ_o θ_o, p Ψ_1, and θ_1. The algebraic details are omitted.

$$\Psi_o = - ny^3/6 + [c_1\eta^{-1} \Phi_x(x,\alpha\eta,t) + c_2\eta] y^2 \quad (37)$$

$$\theta_o = y/\eta \quad (38)$$

$$p_o = mR^{-1}(y-\eta) - \gamma [\frac{1}{2} (\Phi_x^2 + \Phi_z^2) + \Phi_t]_{z = \alpha\eta} -N_b R^{-5/3}\eta_{xx} + p_{vo} + \frac{\gamma}{2} (U^2 + V^2) + (1-\gamma)N_d R^{-2}\theta^2_{oy} \quad (39)$$

$$\theta_1 = PrR [-\eta^{-2}\eta_t(y^2-\eta^2)/6 -(2C\eta^{-2}\eta_x + C_x\eta^{-1})(y^3-\eta^3)/12 + n\eta^{-2}\eta_x(y^4-\eta^4)/40] y \quad (40)$$

$$\Psi_1 = R[p_{ox}y^3/6 + C_t y^4/12 + CC_x y^5/30 + Dy^2] \quad (41)$$

$$c_1 = F/2(1+F), \qquad c_2 = n(1+F/2)/2(1+F) \quad (42\ ,43)$$

$$C = c_1\eta^{-1}\Phi_x(x,\alpha\eta,t) + c_2\eta \quad (44)$$

$$D = - \frac{1}{2} (1+F)^{-1}[(1+F/2)p_{ox}\eta + (1+F/3)C_t\eta^2 + (2/3 +F/6)CC_x\eta^3 + (1-\gamma)N_d R^{-2}\eta^{-2}\eta_x] + FPr(1+F)^{-2}[-\eta_t/6 - (2C\eta\eta_x + \eta^2 C_x)/8 \; n\eta^2\eta_x/20] [\Phi_x(x,\alpha\eta,t) - n\eta^2/2] \quad (45)$$

Substitution of the first order solutions for Ψ and θ into conditions (30) and (31) now yields the interfacial mass conservation condition,

$$\Phi_z + j_o\eta^{-1} - j_1\eta_t - j_2\eta^2\eta_x - j_3\eta_x\Phi_x - j_4\eta\Phi_{xx} + 0(\alpha^2) = 0 \quad \text{at} \quad z = \alpha\eta \quad (46)$$

and the interfacial energy conservation condition.

$$k_o\eta^{-1} - k_1\eta_t - k_2\eta^2\eta_x - k_3(\eta_x\Phi_x + \eta\Phi_{xx}) - k_4\eta\eta_x\Phi_{xz} = \frac{\partial}{\partial x}$$

$$[k_5\eta^3\eta_x - k_6\eta^3\Phi_x\Phi_{xx} + k_7\eta^3\Phi_z\Phi_{xx} - k_8\eta^3\Phi_{xt} + k_9\eta^3\eta_{xxx} + k_{10}\eta_x$$
$$+ k_{11}\eta^2\eta_t\Phi_x - k_{12}\eta^4\eta_t - k_{13}\eta^2\eta_x\Phi_x^2 - k_{14}\eta^5\Phi_{xx} - k_{15}\eta^6\eta_x$$
$$- k_{16}\eta^6\eta_x\Phi_x] + O(\alpha^3) \qquad \text{at} \quad z = \alpha\eta \tag{47}$$

The constants coefficients in Equations (46) and (47) are listed in the Appendix. The transformation of conditions (46) and (47) from $y = \alpha\eta$ to $y = 0$ by expansions like (48) greatly simplifies the

$$\Phi_x(x,\alpha\eta,t) = \Phi_x(x,o,t) + \alpha\eta\Phi_{xz}(x,o,t) + O(\alpha^2) \tag{48}$$

subsequent algebraic operations. The problem is then reduced to that consisting of the transformed form of conditions (46)-(47), conditions (32)-(33) and differential equation (20). In order to analyze this problem within the framework of the theory of hydrodynamic stability, the following transformations are introduced

$$\eta \to \overline{\eta}(x) + \varepsilon\,\eta(x,t) \tag{49}$$

$$\Phi \to Ux - Vz + \varepsilon\,\Phi(x,z,t) \tag{50}$$

In the remaining analysis $\eta(\Phi)$ will represent the disturbance component of the film thickness (the vapor potential function). The problem is next separated into base flow and disturbance components. This is accomplished by substituting (49)-(50) into (20), (32), (33), and the transformed form of conditions (46) and (47). The steady parts of the film thickness and the potential function constitute the steady solution of the problem. Under the parallel flow approximation, the x-derivatives of $\overline{\eta}(x)$ and V are neglected from the disturbance equation and the coefficients of this equation are simplified by letting $\overline{\eta} = 1$. The base flow problem and the linearized stability problem are given below.

The Base Flow Problem :

a) Interfacial energy condition

$$k_o/\overline{\eta} - k_2\overline{\eta}^2\overline{\eta}_x - k_3U\overline{\eta}_x + O(\varepsilon^2) = \frac{d}{dx}\{k_5\overline{\eta}^3\overline{\eta}_x + k_9\overline{\eta}^3\overline{\eta}_{xxx} + k_{10}\overline{\eta}_x$$
$$+ k_{13}U^2\overline{\eta}^{-2}\overline{\eta}_x - k_{15}\overline{\eta}^6\overline{\eta}_x - k_{16}U\overline{\eta}^4\overline{\eta}_x\} + O(\alpha^3) \tag{51}$$

b) Interfacial mass condition

$$j_o/\overline{\eta} - j_2\overline{\eta}^2\overline{\eta}_x - j_3U\overline{\eta}_x - V + O(\varepsilon^2) + O(\alpha^2) = 0 \tag{52}$$

The linearized Stability Problem :

a) Disturbance vapor motion and vapor boundary conditions away from the interface

$$\Phi_{xx} + \Phi_{zz} = 0 \tag{53}$$

$$\Phi_x = \Phi_z = 0 \qquad \text{at} \quad z = \infty \tag{54}$$

b) Interfacial energy condition

$$k_0\eta + k_1\eta_t + k_{17}\eta_x + k_{18}\eta_{xx} + k_9\eta_{xxxx} + k_{19}\eta_{xt} + k_3\Phi_{xx} + k_{20}\Phi_{xxz}$$

$$- k_{21}\Phi_{xxx} - k_8\Phi_{xxt} = 0(\varepsilon) + 0(\alpha^3) \qquad \text{at} \quad z = 0 \tag{55}$$

c) Interfacial mass condition

$$j_0\eta + j_1\eta_t + (j_2 + Uj_3)\eta_x + j_4\Phi_{xx} - \Phi_z - \alpha\Phi_{zz} = 0(\varepsilon) + 0(\alpha^2)$$

$$\text{at} \quad z = 0 \tag{56}$$

4. SOLUTION OF THE LINEARIZED STABILITY PROBLEM

Equations (51) and (52) give the base flow film thickness $\overline{\eta}(x)$ and the base flow y-component velocity, -V. An asymptotic solution of the base flow problem is not presented here. It is noted however that the solution of the base flow problem will yield the following asymptotic expression for the base flow Reynolds number.

$$\text{Re} = (k_2R/3 + uk_3R^{1/3})/\alpha + 0(F) \tag{57}$$

A traveling wave solution is sought for the linearized stability problem. A disturbance film thickness of the form

$$\eta = a_1(e^{i\chi} + e^{-i\chi}) \tag{58}$$

gives the following corresponding solution for (53) and (54)

$$\Phi = a_2e^{-z}(e^{i\chi} + e^{-i\chi}) \tag{59}$$

The substitution of the disturbance film thickness and the disturbance potential function into (55) and (56) yields two linear homogeneous algebraic equations for a_1 and a_2. Setting the determinant to zero, one obtains

$$k_{28}c^2 + (-k_{26} + ik_{27})c - (k_{24} + ik_{25}) = 0 \tag{60}$$

Letting $c_i = 0$ in Equation (60), one finds the following neutral stability condition.

$$f \equiv k^2_{25}k_{28} - k_{25}k_{26}k_{27} - k_{24}k^2_{27} = 0 \tag{61}$$

Inspection of Equation (61) shows that the neutral stability condition is determined by eight dimensionless numbers

$$f(\alpha, R, u, Pr, N, F, \gamma, \phi) = 0 \tag{62}$$

The neutral stability curves are obtained by locating the zeros of equation (61) numerically. Letting $c = c_r + ic_i$ in (60) and equating the real and imaginary parts of the resulting expression to zero, one finds

$$c_r = (k_{25} + c_i k_{26})/(k_{27} + 2c_i k_{28}) \tag{63}$$

for the wavespeed and

$$B_1 + B_2 c_i + B_3 c_i^2 + B_4 c_i^3 + B_5 c_i^4 = 0 \tag{64}$$

for the linear wave amplification rate. The coefficients in equation (64) are listed in the Appendix. Noting that the wave amplification rate is zero on the neutral stability curve and small elsewhere, equation (64) will be written in the form,

$$B_1 + B_2 c_i + \varepsilon(B_3 c_i^2 + B_4 c_i^3 + B_5 c_i^4) = 0 \tag{65}$$

and an asymptotic solution, valid for small amplification rates, will be obtained by assuming

$$c_i = c_{i0} + \varepsilon c_{i00} + O(\varepsilon^2) \tag{66}$$

subsitition of (66) into (65), solving for c_{i0} and c_{i00}, then letting $\varepsilon = 1$ gives the following asymptotic expression valid about the neutral stability condition.

$$c_i \cong - B_1/B_2 - B_1^2 B_3/B_2^3 + B_1^3 B_4/B_2^4 + B_1^3 B_4/B_2^4 - B_1^4 B_5/B_2^5 \tag{67}$$

5. DISCUSSION

The solution presented in the previous section is valid for all values of the inclination angle, ϕ. In this section, the stability of thin condensate films will be discussed for the three flow configurations shown in Fig. 3a. These are condensation on a vertical surface (ϕ=0), condensation below a horizontal surface (ϕ=-90), and condensation above a horizontal surface (ϕ=90). Neutral stability curves are presented in this section to display the effects of Pr, N, F, γ, and u on the stability of the phase interface. These curves are obtained by locating the zeros of equation (61) numerically. The solution of equation (61) yields an (α,R) pair on the neutral stability curve. The corresponding Reynolds number is computed from (57).

The neutral stability curves for condensation on a vertical surface are shown in Figures 3b-3e. It is seen from Fig.3b that an increase of the vapor velocity from zero to 250 has a stabilizing effect when Pr=2.62, N=12347, F=0.0333, and γ=0.00061. These values of the flow parameters correspond to condensation of saturated steam at 100C

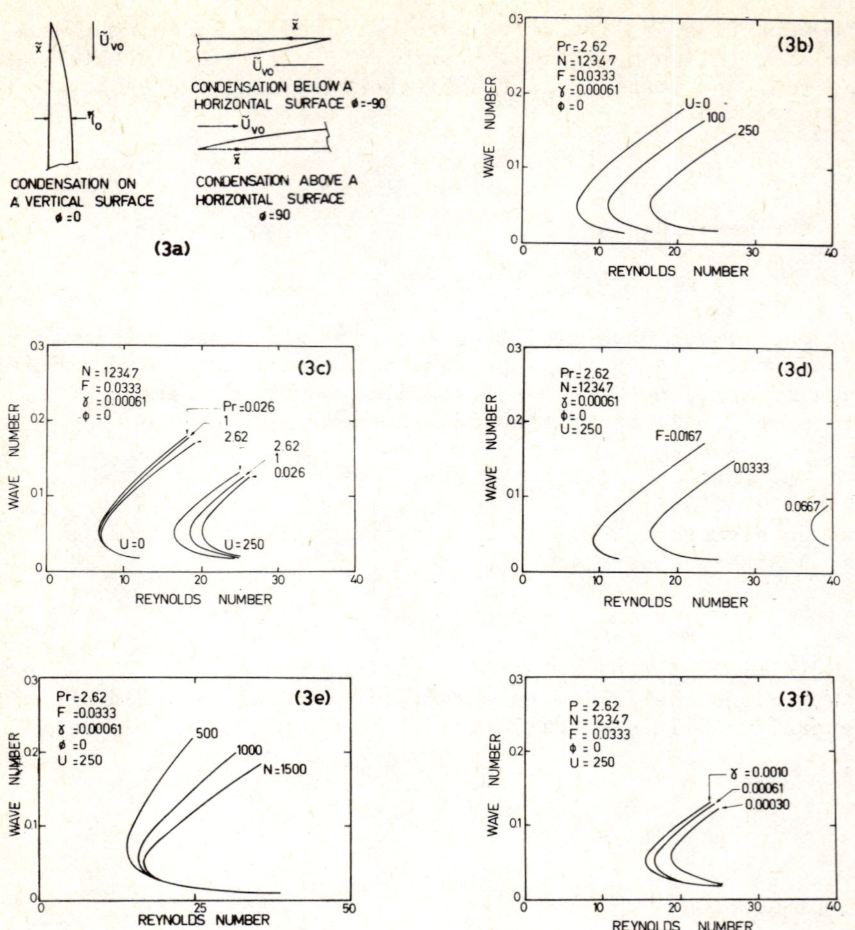

Fig. 3 3a: Flow configurations, 3b-3f: Neutral stability curves for condensation on a vertical surface.

onto a surface at 53C. An increase in the Prandtl number has a stabilizing effect when u=0 and a destabilizing effect when u=250 (Fig.3c). By inspection of Figs. 3d-3f, it is observed that increasing F or N is stabilizing while increasing γ is destabilizing when u=250.

Some of the instability mechanisms will be explored by inspection of the disturbance interfacial shear, the disturbance interfacial velocity and the disturbance liquid pressure. Substitution of transformations (49) and (50) into (37) and (39), differentiation of (37) twice with respect to y, evaluation of the resulting expressions at the phase interface, and simplification for small F yields the following expressions for forced vapor flow condensation on a vertical surface.

$$\Psi_y\Big|_{y=\bar{\eta}+\varepsilon\eta} = \bar{\Psi}_{oy} + \varepsilon(n\eta - Fj_o\eta_x) + 0(\alpha) + 0(\varepsilon^2) \quad (68)$$

$$\Psi_{yy}\Big|_{y=\bar{\eta}+\varepsilon\eta} = \bar{\Psi}_{oyy} + \varepsilon F(-\ n\eta/2 - U\eta - j_o\eta_x) + 0(\alpha) + 0(\varepsilon^2) \quad (69)$$

$$p = \bar{p} + \varepsilon[\,FR^{-1}(U - c_r)\eta_x + R^{-2}(N_b R^{1/3} - 3N_d)\eta] + 0(\alpha) + 0(\varepsilon^2) \quad (70)$$

The terms multiplying ε in equations (68)-(70) represent the disturbance parts of the interfacial x-component liquid velocity, the interfacial liquid tangential shear, and the liquid pressure. The $n\eta$ term in equation (68) represents the gravity induced disturbance interfacial liquid velocity. The direction of this disturbance velocity is depicted in Fig.4a by vectors labeled (1). This disturbance velocity has a positive sign at the wave crests and is destabilizing. The term containing the factor j_o in equation (68) appears as a consequence of the disturbance vapor motion. The direction of this term is depicted by vectors labeled (2) in Fig.4a and it has a stabilizing effect.

The disturbance interfacial liquid shear stress consists of three terms. The first one of these which contains the factor n in equation (69) is a gravity induced disturbance shear stress. An increase in the interfacial velocity at the wave crest induces a disturbance tangential shear which acts in the opposite direction (vectors labeled (1) in Fig.4b). The disturbance shear given by the $U\eta$ term in equation (69) has a negative sign at the wave crests. This component of the disturbance interfacial shear stress is in the same direction with the gravity induced disturbance shear stress. The direction of the last component of the disturbance shear stress given in equation (69) is shown by vectors labeled (2) in Fig.4b. This last term represents the disturbance vapor effect. All of the three terms, each representing a component of the disturbance shear stress, are stabilizing.

The first term in the expression for the disturbance pressure is destabilizing for $U>c_r$ and stabilizing for $U<c_r$. The term containing N_b is stabilizing and that containing the N_d is destabilizing.

The neutral stability curves for condensation below a horizontal surface are shown in Figs. 4c-4f. The effects of the dimensionless parameters are consistent with those for condensation on a vertical The particular value of ϕ is -90 for this case giving n=0 and m=2. Taking u=0, F=0 and c=0, the neutral stability condition for a motionless isothermal liquid film underside of a horizontal surface is

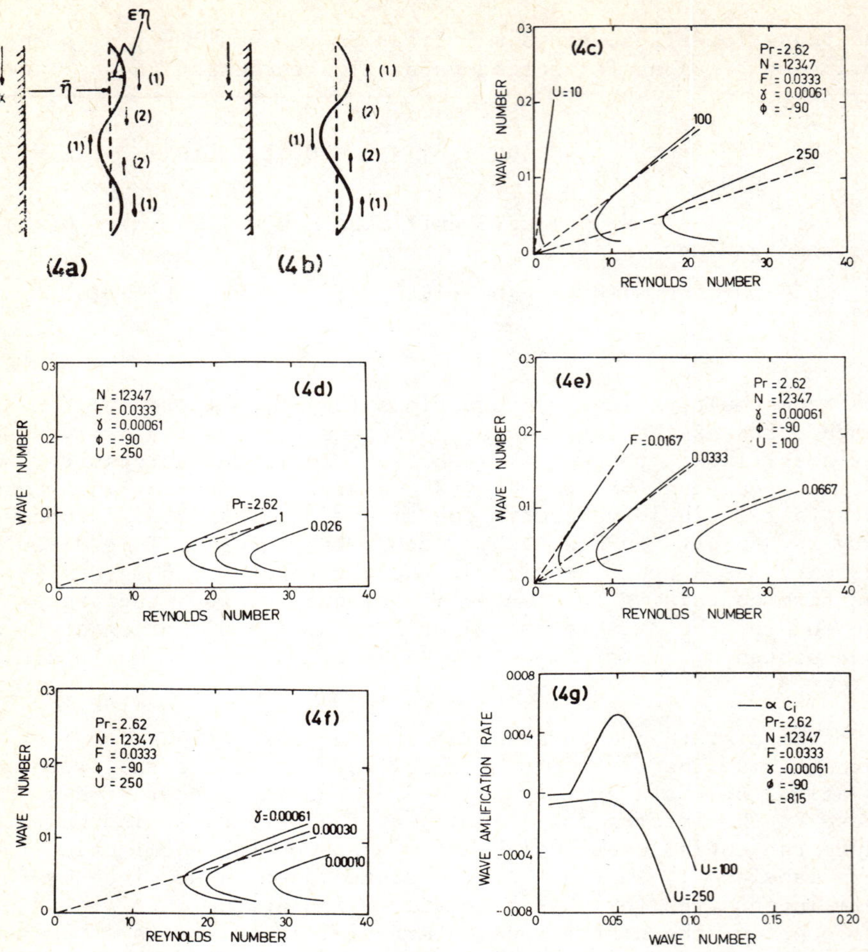

Fig. 4 4a: Disturbance interfacial velocity, 4b: Disturbance interfacial shear, 4c-4f: Neutral stability curves for condensation under a horizontal surface, 4g: Linear wave amplification rates for condensation under a horizontal surface.

obtained as

$$\alpha^2 = 2R^{2/3}/N \tag{71}$$

or, in dimensional form

$$(2\pi/\lambda)^2 = \rho g(1 - \gamma)/\sigma \tag{72}$$

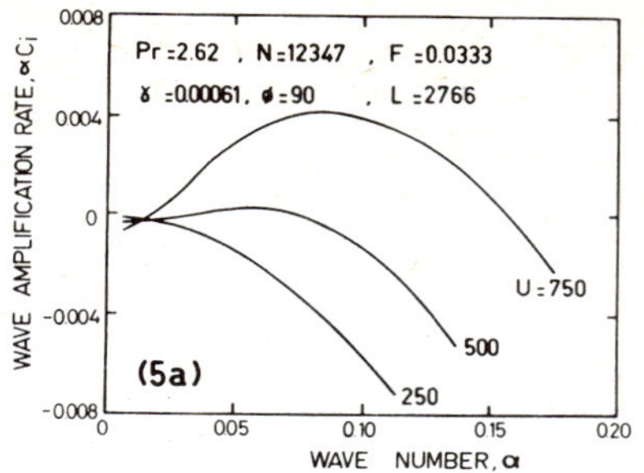

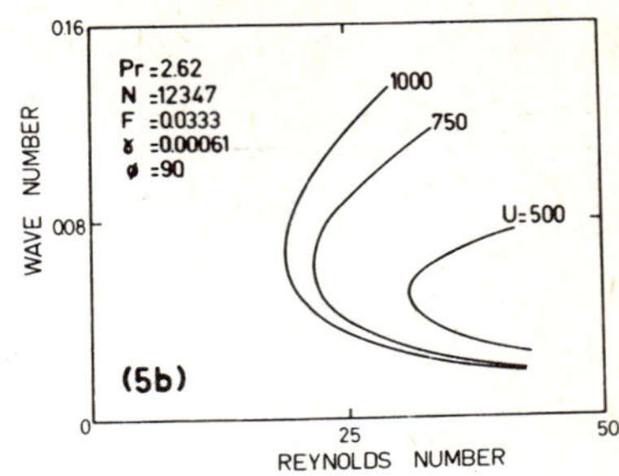

Fig. 5 5a: Linear wave amplification rates for condensation above a horizontal surface, 5b: Neutral stability curves for condensation above a horizontal surface.

This neutral stability condition represents the cut off wavenumber for Taylor instability and is in agreement with that given in [12]. Expressing R in terms of the Reynolds number by substitution from equation (57) yields

$$\alpha = 2(2/N)^{1/2}Re/uF \tag{73}$$

The dashed lines in Figures 4c-4f are the graphs of equation (73). These coincide with the upper branches of the neutral stability curves whenever u is small. The strong stabilizing effect of the vapor velocity is depicted in Fig. 4g. An increase of the dimensionless vapor velocity from 100 to 250 is sufficient to completely stabilize the condensate film for the particular physical situation corresponding to Fig. 4g.

Evaluation of equation (65) for small vapor velocities and condensation above a horizontal surface results in large critical Reynolds numbers which may fall beyond the validity region of the present solution. The linear wave amplification rates are depicted in Fig. 5a for one location on the horizontal surface and for three vapor velocities. Corresponding neutral stability curves are given in Fig. 5b. Increasing the vapor velocity brings about a destabilizing effect for this particular example.

An experimental investigation of the interfacial stability characteristics in forced vapor flow condensation was not found in the open literature. Therefore, a comparison with an experimental data for vertical isothermal film flow [13] is presented in this section. Equation (67) may be simplified for a vertical isothermal liquid film by letting γ=u=F=m=0, n=2. Setting the derivative of the resulting expression for αc_i to zero, one obtains the following expression for the most unstable linear disturbance

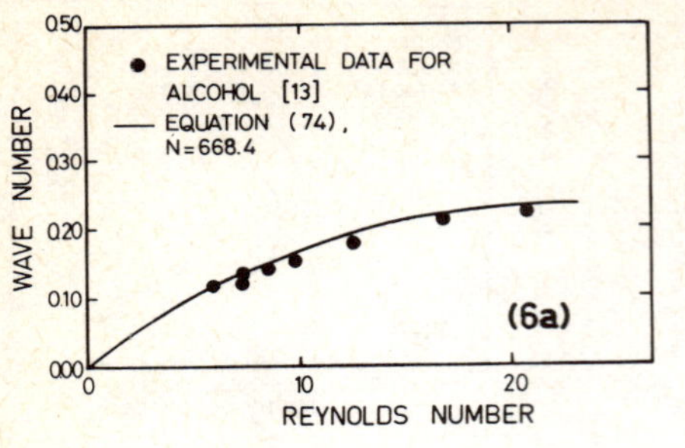

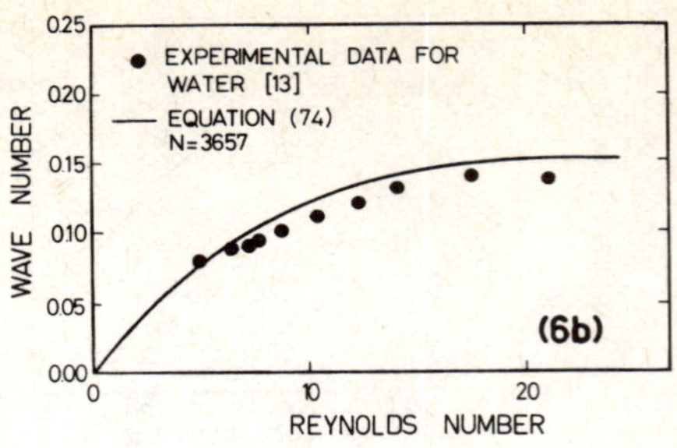

Fig. 6 Comparison of predicted wavenumbers with experimental data for isothermal liquid films.

$$\alpha = [-144/25R^2 + ((144/25R^2)^2 + 3456/375NR^{1/3})^{1/2}]^{1/2} \quad (74)$$

The dimensionless number R is related to the Reynolds number by R=3Re/2. The wavenumber predictions from equation (74) are compared with the experimental data of Reference [13] in Fig.6.

6. CONCLUSIONS

The linearized stability of forced vapor flow condensation on plane surfaces of arbitrary inclination has been investigated by the longwave perturbation method. The formulation and the solution presented in this study constitute an asymptotic analysis of the stability problem valid about the strong suction limit.

The results of the present investigation show that gravity is destabilizing and that low speed cocurrent vapor flow has a stabilizing effect for condensation on vertical and below horizontal surfaces. Another effect of cocurrent vapor flow is observed when the vapor velocity exceeds the wavespeed. This latter effect tends to destabilize the interface. The interface of a condensate film above a horizontal surface is found to be stable at low cocurrent vapor velocities. Such an interface tends to be unstable with increasing vapor velocity.

The results indicate that increasing the Prandtl number may have a stabilizing or a destabilizing effect depending on the value of the vapor velocity. It is found that decreasing γ and increasing F or N have stabilizing effects for forced vapor flow condensation on vertical and below horizontal surfaces.

NOMENCLATURE

a_1, a_2	Constants
$B_1,\ldots,B_5$	Multipliers given in the Appendix
c	c_r+ic_i, Complex wave velocity
C	Defined in equation (44)
c_1, c_2	Defined in equations (42), (43)
c_p	Specific heat of the liquid
D	Defined in equation (45)
f	Defined in equation (61)
F	$c_p\Delta T/h_{fg}Pr$, acceleration effect parameter
g	gravitational acceleration
h_{fg}	enthalpy of phase change
I	$\tilde{y} - \tilde{\eta}$
$j_0,\ldots,j_6$	Multipliers given in the Appendix
k	Thermal conductivity of the liquid
$\underset{\sim}{k}_0,\ldots,k_{28}$	Multipliers given in the Appendix
$\tilde{L}$	Distance from leading edge
L	$\tilde{L}/(2\nu^2/g)^{1/3}$
m	$-2\sin\phi$
$\dot{m}'$	Mass flow rate per unit length
n	$2\cos\phi$
$\vec{n}$	Unit normal to interface
N	$2^{1/3}\sigma/\rho\nu^{4/3}g^{1/3}$
N_b	$\alpha^2 N$
N_d	F^2/γ
p	Dimensionless liquid pressure
$\overline{p}$	Dimensionless base flow liquid pressure
$\tilde{p}$, $\tilde{p}_v$	Liquid pressure, vapor pressure
$\tilde{p}_{vo}$, p_{vo}	Dimensional and dimensionless base flow vapor pressure
Pr	Prandtl number
$\vec{q}$	Liquid heat flux
r	Radius of curvature

R	$g\tilde{\eta}_0{}^3/2\nu^2$
Re	m'/μ, Reynolds number
$\tilde{t}$, t	Time, dimensionless time
$\tilde{T}$, $\tilde{T}_s$, $\tilde{T}_w$	Liquid, vapor, surface temperatures
ΔT	$\tilde{T}_s - \tilde{T}_w$
u	$\tilde{U}_{vo}/(g\nu/2)^{1/3}$
$\tilde{U}$, $\tilde{U}_v$	Liquid, vapor $\tilde{x}$-component velocity
$\tilde{U}_o$	$g\tilde{\eta}_0{}^2/2\nu$, A reference velocity
$\tilde{U}_{vo}$	Base flow x-component vapor velocity
U	$\tilde{U}_{vo}/\tilde{U}_o$
$\tilde{V}$, $\tilde{V}_v$	Liquid, vapor $\tilde{y}$-component velocity
$\tilde{V}_{vo}$	Base flow $\tilde{y}$-component vapor velocity
V	$-\tilde{V}_{vo}/\tilde{U}_o$
$\vec{V}$, $\vec{V}_v$, $\vec{V}_i$	Liquid, vapor, interface vector velocity
$(\tilde{x},\tilde{y})$, (x,y)	Dimensional, dimensionless coordinates
z	αy
α	$2\pi\tilde{\eta}_o/\lambda$, Wavenumber
γ	ρ_v/ρ
ε	A small parameter
$\tilde{\eta}$	Film Thickness
$\tilde{\eta}_0$	Base flow film thickness
$\bar{\eta}$	Dimensionless base flow film thickness
η	Dimensionless film thickness
θ	Dimensionless liquid temperature
$\bar{\theta}$	Dimensionless base flow liquid temperature
λ	wavelength
μ	Liquid viscosity
ν	Liquid kinematic viscosity
Π	Liquid stress tensor
ρ, ρ_v	Liquid, vapor density
σ	Surface tension
ϕ	Angle of inclination
$\tilde{\Phi}$	Vapor potential function
Φ	Dimensionless vapor potential function

χ x-ct

$\tilde{\psi},\psi$ Dimensional, dimensionless liquid stream function

$\overline{\psi}$ Dimensionless base flow stream function

Subscripts :

$x,y,t,\tilde{x},\tilde{y},\tilde{z},t$ Partial differentiation with respect to the subscript

REFERENCES

1. Friedel, L. "Momentum exchange and pressure drop in two-phase flow." Proceedings of the ASI on two-phase flows and heat transfer, Istanbul (1976) 239-303.

2. Collier, J.G. "Post-dryout heat transfer- A review of the current position." Proceedings of the ASI on two-phase flows and heat transfer, Istanbul (1976) 769-813.

3. Bankoff, S.G. "Stability of liquid flow down a heated inclined plane." Int.J.Heat and Mass Transfer 14(1971) 337-385.

4. Marschall, E. and Lee C.Y. "Stability characteristics of condensate films." Wärme-und Stoffübertragung 1(1973) 32-37.

5. Ünsal, M. and Thomas W.C. "Linearized stability analysis of film condensation." J.Heat Transfer 100(1978) 629-634.

6. Spindler, B."Linear stability of liquid films with interfacial phase change." Int.J.Heat and Mass Transfer 25(1982) 161-172.

7. Shekriladze, I.G.and Gomelauri, V.I. "Theoretical study of laminar film condensation of flowing vapour." Int.J.Heat and Mass Transfer 9(1966) 581-591.

8. Denny, V.E.and Mills, A.F. "Nonsimilar solutions for laminar film condensation on a vertical surface." Int.J.Heat and Mass Transfer 12(1969) 965-979.

9. Koh, J.C.Y."Film Condensation in a forced convection boundary layer flow." Int.J.Heat and Mass Transfer 5(1962) 941-954.

10.Hsieh, D.Y. "Effects of heat and mass transfer on Rayleigh-Taylor instability." J.Basic Engineering 94(1972) 156-162.

11. Delhaye, J.M. "Local Instantaneous Equations". Two-Phase Flows and Heat Transfer, eds S.Kakac and F.Mayinger, Hemisphere Publishing Corporation, Washington, 1(1977) 59-79.

12.Bellman, R.and Pennington, R.H. "Effects of surface tension and viscosity on Taylor instability." Quart.J.Appl.Math. 12(1954) 151-162.

13.ter Haar, D.(Editor)" Collected Papers of P.L.Kapitza Pergamon Press, Oxford(1965) 662-709.

APPENDIX

$$j_0 = F/\gamma R$$

$$j_1 = \alpha(1+FPr/3\gamma)$$

$$j_2 = \alpha nFPr(22+7F)/80\gamma(1+F)$$

$$j_3 = \alpha(1+F^2Pr/8\gamma(1+F))$$

$$j_4 = \alpha F^2Pr/8\gamma(1+F)$$

$$j_5 = j_2 + Uj_3$$

$$j_6 = j_4 - 1 + \alpha$$

$$k_0 = F/R$$

$$k_1 = \alpha(1+FPr/3)$$

$$k_2 = \alpha n(1+F/4+FPr(22+7F)/80)/(1+F)$$

$$k_3 = \alpha F(1+FPr/4)/2(1+F)$$

$$k_4 = \alpha k_3$$

$$k_5 = \alpha^2 m(4+F)/12(1+F)$$

$$k_6 = \alpha^2 R(F^2(6+F+5Pr)/80(1+F)^3 - \gamma(4+F)/12(1+F)$$

$$k_7 = \alpha^2 R\gamma(4+F)/12(1+F)$$

$$k_8 = \alpha^2 R(F(5+F)/24(1+F)^2 - \gamma(4+F)/12(1+F))$$

$$k_9 = \alpha^2 N_b R^{-2/3}(4+F)/12(1+F)$$

$$k_{10} = \alpha^2(1-\gamma)N_d/6R$$

$$k_{11} = \alpha^2R(5+F)F/24-FPr/6)/(1+F)^2$$

$$k_{12} = \alpha^2Rn(5+F)(1+F/2)/24 - FPr/12)/(1+F)^2$$

$$k_{13} = \alpha^2RF^2((6+F)/80 - Pr/16)/(1+F)^3$$

$$k_{14} = \alpha^2RnF((6+F)(1+F/2)/80 - FPr/32)/(1+F)^3$$

$$k_{15} = \alpha^2Rn^2((6+F)(1+F/2)^2/80 - FPr(22+7F)/320)/(1+F)^3$$

$$k_{16} = \alpha^2RnFPr(11+F)/80(1+F)^3$$

$$k_{17} = k_2 + Uk_3$$

$$k_{18} = k_5 + k_{10} + U^2k_{13} - k_{15} - Uk_{16}$$

$$k_{19} = Uk_{11} - k_{12}$$

$$k_{20} = \alpha k_3 - j_0k_7$$

$$k_{21} = Uk_6 + k_{14}$$

$$k_{22} = k_0 + k_9 - k_{18}$$

$$k_{23} = k_3 - k_{20}$$

$$k_{24} = j_6k_{22} - j_0k_{23} - j_5k_{21}$$

$$k_{25} = j_6k_{17} + j_0k_{21} - j_5k_{23}$$

$$k_{26} = j_6k_{19} + j_1k_{21} + j_5k_8$$

$$k_{27} = j_6k_1 + j_0k_8 - j_1k_{23}$$

$$k_{28} = j_1k_8$$

$$B_1 = k_{24}k_{27}^2 + k_{25}k_{26}k_{27} - k_{25}^2k_{28}$$

$$B_2 = 4k_{24}k_{27}k_{28} + k_{26}^2k_{27} + k_{27}^3$$

$$B_3 = 4k_{24}k_{28}^2 + k_{26}^2k_{28} + 5k_{27}^2k_{28}$$

$$B_4 = 8k_{27}k_{28}^2$$

$$B_5 = 4k_{28}^3$$

A COMPARISON OF FLOODING MODELS FOR AIR-WATER AND STEAM-WATER FLOW

S.G. Bankoff and S.C. Lee

Chemical Engineering and Mechanical and Nuclear Engineering Departments
Northwestern University
Evanston, Illinois 60201, USA

ABSTRACT

A number of models have been proposed over the years for flooding, or countercurrent flow limitation (CCFL), for gas-liquid flow. These may be roughly classified as surface wave stability theories, static equilibrium theories, envelope or limiting operating condition theories, and semi-empirical correlations. The scatter of the data is rather large, partly because of the different definitions of flooding which have been employed, and partly because of the strong influence of the entrance and exit conditions. Some progress has been made recently in distinguishing between various critical conditions and locations for flooding; these effects tend to distinguish steam-water flooding from air-water flooding. The various theories are reviewed briefly and comparisons made with representative data. Directions for future research are indicated.

1 INTRODUCTION

When the gas velocity reaches a certain value in countercurrent flow to a liquid film flowing down along the tube wall, large amplitude waves eventually appear on the interface, so that the flow becomes chaotic, with continuous entrainment of liquid droplets torn from the tips of the waves. At this point neither the gas flow rate, nor the liquid flow rate can be further increased without altering the flow pattern. This phenomenon, representing a limit of the attainable gas-liquid two-phase countercurrent flow region, is known as "flooding".

The flooding problem has attracted much attention since the pioneering work on the design of packed columns by Sherwood, et al. [1]. In recent years, interest in the flooding phenomenon has grown in conjunction with the safety analysis of nuclear reactors. Flooding may limit the penetration of emergency coolant into the hot reactor core, during a postulated loss of coolant accident (LOCA) in either a boiling-water reactor (BWR) equipped with a spray system, or a pressurized-water reactor (PWR) with an upper head injection system [2,3]. Much work has been performed in order to improve the understanding of the flooding phenomenon in connection with evaluation of the performance of these safety systems.

A number of investigators [4-13] have attempted to derive theoretical criteria for the onset of flooding in a vertical annular flow. However, a purely theoretical approach is very difficult, and most investigators have therefore set up simplified equations with extensive idealizations and/or employ some empirical equations for the countercurrent flow parameters, such as mean film thickness, interfacial friction factor, etc. Generally, they show that these analytical models agree well with their own experimental data or selected data in their range of validity. A number of small scale experiments have been performed in order to investigate flooding under various conditions and flow variables. However, a wide spread of data is found even under ostensibly the same flow conditions, probably because of the differences in entrance and exit conditions, and also because of the differences in the definition of what constitutes flooding. However, these extensive experiments have produced a number of useful empirical correlations, particularly for vertical adiabatic two-phase countercurrent flow.

Flooding in an inclined countercurrent flow may also be of importance in connection with the injection of emergency core cooling water into the hot leg of a PWR [2]. Some investigators performed flooding experiments for a nearly horizontal stratified flow [17] and for inclined annular flow [25]. The results indicate that various models for slug formation in a horizontal stratified flow may be applicable to prediction of onset of flooding in a nearly horizontal countercurrent flow with some modification [17].

Despite these efforts the flooding problem is not completely resolved. For example, a completely satisfactory scaling law has not yet been established, and a consistent effect of the fluid properties, such as viscosity and surface tension, has not been synthesized from the available data. Hence, it may be useful to compare the existing analytical and empirical models for flooding in order to evaluate the state of the art, and to recommend future studies in this area. Tien and Liu [14] have recently published a survey report on vertical two-phase countercurrent flooding, consisting principally of a brief introduction to existing models for onset of flooding in vertical annular flow. However, the present reviews existing analytical and empirical models in inclined stratified flow as well as vertical annular flow, and discusses validity of these models under various flow conditions.

The paper consists of six sections. Section Two discusses the fundamentals of two-phase countercurrent flow. The most important empirical correlations for countercurrent flow parameters are discussed. In the following two sections the existing analytical and empirical models are classified and briefly described. Finally, these models are compared with extensive experimental data, and their validity and limitations of applicability are discussed.

2 FUNDAMENTALS OF COUNTERCURRENT FLOW

Some basic characteristics of countercurrent flow must be understood in order to investigate flooding, partly because some empirical information is essential, and partly because sudden changes in the flow parameters represent a transition of the flow regime which in many cases can be identified with flooding. It is also important to consider the geometry of the countercurrent flow, such as vertical annular, inclined or vertical stratified, perforated horizontal plate (such as the upper tie plate of a PWR), packed bed or packed tower. We shall not consider the latter two of these possibilities. In any case some of the key parameters are pressure drop, void fraction, and interfacial wave structure.

2.1 Pressure Drop

The pressure drop measured in countercurrent flow has always been a good indicator for predicting flow pattern change either in adiabatic or in condensing flows [16-18]. Figure 1 shows typical pressure drops for the gas phase in a vertical annular flow [16] and in an inclined steam-water flow [17], respectively. For adiabatic flow, as seen in Figure 1a, the pressure drop slowly increases before the flooding point as the gas flow rate is increased under constant liquid flow rate. In this region, the film flow exhibits two- or three-dimensional waves at the interface. As the gas flow rate is further increased, roll waves eventually appear on the pebbly interfacial structure, eventually resulting in formation of liquid slugs and bridging within the tube. As a preliminary to the formation of slugs, a rough thick liquid film, causing considerable augmentation of the interfacial shear stress, develops. The pressure drop therefore begins to rise sharply with small increases in gas flow rate, once roll waves have appeared. The onset of flooding following formation of slugs generally corresponds to the point where maximum pressure drop is attained, and leads to unsteady conditions.

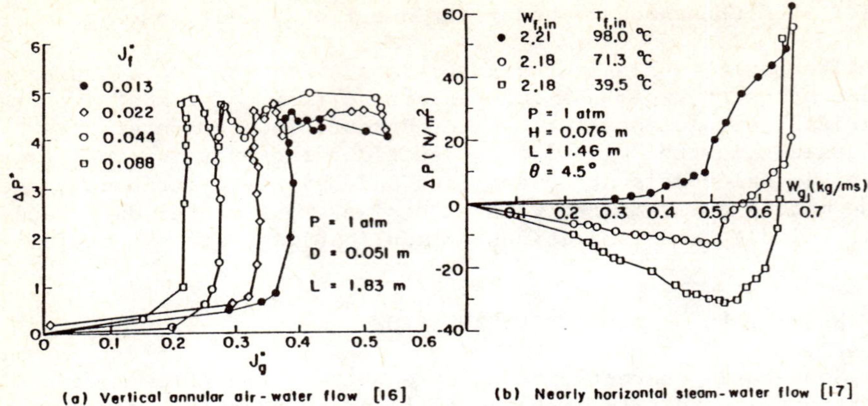

Figure 1. Typical pressure drops in countercurrent two-phase flow.

The expression for pressure drop in a condensing flow consists of three terms, describing the effects of gravity, friction, and momentum transfer due to condensation. Of these, the gravity term is usually negligible compared to the other two terms. There is an increase in pressure due to the loss of vapor phase momentum by condensation, which monotonically increases to an asymptotic value as the vapor flow rate is increased. Therefore, the pressure drop is at first negative, decreasing in the low steam flow region with increasing steam flow rate because the pressure gain term is dominant in this region. However, when the steam flow rate approaches the flooding velocity, the frictional loss begins to rise sharply because of slug formation in the same manner as for air-water flow. Thus, the pressure drop exhibits a sharp increase near the flooding point, becoming strongly positive. The steam flow rate at the minimum pressure drop corresponds to the point of inception of roll waves on the interface. From these experimental results one may conclude that the measurement of pressure drop in countercurrent flow predicts the onset of flooding quite well for both adiabatic and condensing flows.

2.2 Mean Film Thickness

The mean film thickness, which is directly related to either the liquid fraction or the void fraction, plays an important role. For instance, several investigators, such as Imura, et al. [4], Richter [13], and Grolmes, et al. [18] employed empirical correlations of the mean film thickness in their flooding analyses.

Many earlier experiments on mean film thickness in vertical

annular flow without an upwards gas flow showed that the Nusselt equation holds quite well in the laminar flow regime [20-22]

$$\delta^* = 1.442 \, Re_f^{1/3} \qquad Re_f \leq 400 \tag{1}$$

where the dimensionless film thickness and liquid Reynolds number are defined by

$$\delta^* = \delta[g\sin\theta(\rho_f - \rho_g)\rho_f/\mu_f^2]^{1/3} \tag{2}$$

$$Re_f = W_f/\mu_f \tag{3}$$

For turbulent films, Feind [22] correlated his experimental data with the following equation:

$$\delta^* = 0.532 \, Re_f^{\frac{1}{2}} \qquad 400 \leq Re_f \leq 2{,}000 \tag{4}$$

An alternative correlation for vertical turbulent films, based on the extensive experimental data of Belkin, et al. [21], is due to Wallis [8]:

$$\delta^* = 0.159 \, Re_f^{2/3} \qquad 250 \leq Re_f \leq 7{,}500 \tag{5}$$

We note that the dimensionless mean film thicknesses calculated by means of equations (4) and (5) agree quite well up to $Re_f = 2{,}000$, as indicated by the experimental range for equation (4). This equation should not be extrapolated beyond this point, however, because it deviates significantly from the results given by equation (5).

Measurements of mean film thickness in vertical two-phase countercurrent flow have been carried out by several investigators, such as Feind [22], Hewitt and Wallis [23], and Hawley and Wallis [16]. Their results indicate that the mean film thickness is not significantly altered by the upwards gas flow, although no quantitative expressions were presented. This result seems to agree with the experimental result of Bharathan, et al. [9] who suggested that the mean film thickness for the smooth film regime with countercurrent flow may be approximated by that for a falling film with no gas flow.

Recently, Lee and Bankoff [17] found three different empirical correlations on the mean film thickness in an inclined countercurrent steam-saturated water flow for three distinct film regimes, as shown in Figure 2: smooth film regime, three-dimensional wave regime,

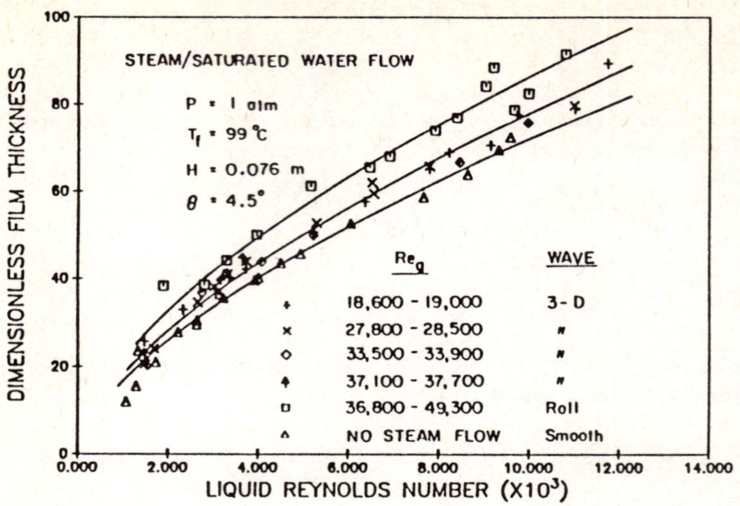

Figure 2. Dimensionless mean film thickness in a nearly horizontal countercurrent flow.

roll wave regime.
For smooth film regime,

$$\delta^* = 0.208\ Re_f^{0.635} \qquad 600 \leq Re_f \leq 9{,}600 \qquad (6)$$

For three-dimensional wave regime,

$$\delta^* = 0.229\ Re_f^{0.633} \qquad 1{,}400 \leq Re_f \leq 12{,}000 \qquad (7)$$

For roll wave regime

$$\delta = 0.334\ Re_f^{0.603} \qquad 1{,}800 \leq Re_f \leq 11{,}000 \qquad (8)$$

These results indicate that the mean film thicknesses with countercurrent gas flow increase by approximately 10% and 20%, respectively, for the rough three-dimensional wave and roll wave regimes over the smooth film thicknesses without upward gas flow. Hence, it seems reasonable to apply equations (1), (4) and (5) to flooding analysis of vertical countercurrent flow without large error.

2.3 Interfacial Shear Stress

Interfacial shear stress is closely tied to the flow regime transitions in two-phase flow systems. Detailed analyses of the local behavior have been performed, and the shear stress depends upon both the large and small scale interfacial wave structure. However, these are not easy to measure, so it is usually preferable to express the interfacial shear stress in terms of macroscopic properties such as void fraction, gas and liquid flow rates and fluid physical properties. Bharathan, et al. [9] considered countercurrent flow to fall into three distinct film regimes: smooth film regime, transition regime, and rough film regime. Assuming that the wall shear stress is small compared to the interfacial shear stress in the rough film regime where flooding occurs, an empirical correlation for the friction factor was derived, using time- and space-averaged data on the void fraction and pressure drop. According to the experimental results, the interfacial friction factor, f_i, in vertical annular countercurrent flow is of the form:

$$f_i = 0.005 + a(D^*/2)^b(1 - \sqrt{\alpha})^b \tag{9}$$

where

$$f_i = 2\tau_i/\rho_g\overline{U}_g^{\,2} \tag{10}$$

and

$$\log_{10} a = -0.56 + 9.07/D^* \tag{11}$$

$$b = 1.63 + 4.74/D^* \tag{12}$$

The dimensionless diameter, D^*, defined by

$$D^* = D[(\rho_f - \rho_g)g/\sigma]^{\frac{1}{2}} \tag{13}$$

may be looked upon as the square root of a Bond number. The viscosity is not present in equation (9), so that an extension to rather viscous liquids is not recommended.

Lee and Bankoff [17] recently measured interfacial shear stresses in a nearly horizontal stratified steam/saturated-water flow. It was found that the interfacial friction factor for the three-dimensional wave regime is of the same order of magnitude as for the smooth film regime, but that it is strongly dependent upon the gas Reynolds number as well as the liquid Reynolds number for

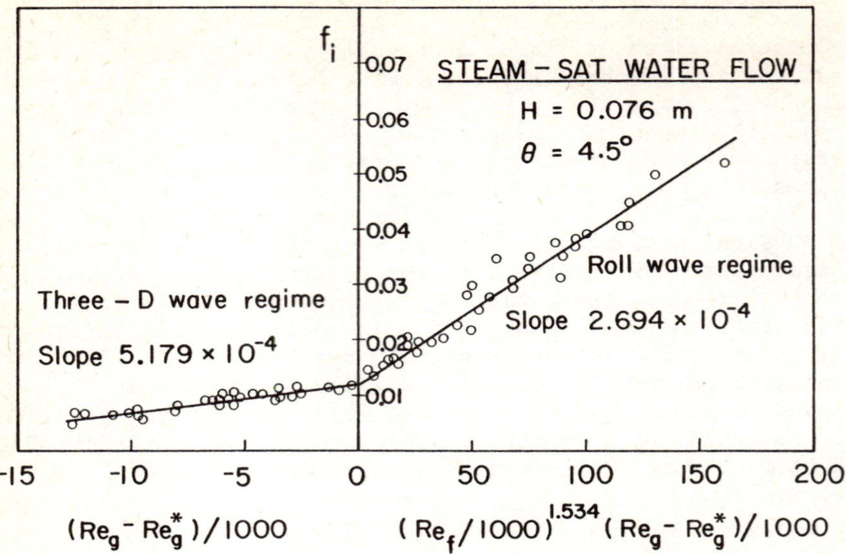

Figure 3. Interfacial friction factors in a nearly horizontal steam/saturated-water flow.

the roll wave regime (Figure 3). According to their results, the interfacial friction factor during the build-up of the roll waves, which seems to indicate a transition regime according to Bharathan, et al. [9] classification, can be expressed as follows:

$$f_i = 0.012 + 2.694 \times 10^{-4}(Re_f/1000)^{1.534}(Re_g - Re_g^*)/1000 \quad (14)$$

where

$$Re_g^* = 1.837 \times 10^5 Re_f^{-0.184} \quad (15)$$

Interfacial shear stress for fully developed roll waves is difficult to measure because of the significant entrainment of water droplets torn from the tips of the roll waves. Nevertheless, it is thought that the interfacial friction factor may be described by equation (14) reasonably well, since it is observed that the frequency of the roll waves continues to increase up to the flooding point as the gas flow rate is increased beyond the inception of entrainment.

3 CLASSIFICATION OF FLOODING MODELS

A number of analytical and empirical models have been produced for vertical annular flow limitation. Several different viewpoints have been expressed for the analytical models. Some models consider flooding to be the result of interfacial instability between two superposed fluids flowing at different velocities. Various assumptions, such as potential flow [4-5], viscous laminar flow [6], and finite-amplitude surface waves [7], are then employed to develop models. From another viewpoint, flooding is a limiting condition for countercurrent flow, which suggests an envelope theory based on the steady hydrodynamic equations. Usually the envelope is taken to be the locus of tangents to the operating lines in the (J_f*, J_g*) plane for constant void fraction, and hence represents a limiting curve separating the operating region from an unattainable region for countercurrent flow. Models based on this approach include the separate-cylinders model [8], the drift flux model [8] and the separated-flow model [9-10]. Still another viewpoint develops the flooding equations based on postulated interfacial shapes. As examples one can note the stationary wave model [11], hanging film model [12] and roll wave model [13]. A summary of the analytical models for a vertical two-phase countercurrent flow is given in Table 1.

As mentioned above, flooding in a nearly stratified flow is of interest in connection with the injection of emergency core cooling water into the hot legs of a PWR. Lee and Bankoff [17] studied flooding and waterhammer in a nearly-horizontal, stratified steam-water countercurrent flow. It was found that slug formation models for co-current horizontal flow can be successfully modified to predict onset of flooding in a nearly-horizontal countercurrent flow. These will be discussed later. The slug formation models to be considered include:

A. Mishima and Ishii model [26]
B. Gardner model [27]
C. Wallis and Dobson model [28]
D. Kordyban and Ranov model [29]
E. Taitel and Dukler model [30]

Details on the slug formation models are presented in Table 2.

4 EXPERIMENTAL INVESTIGATIONS

4.1 Experimental System

The basic experimental system for studying flooding in vertical annular flow consists of an upper plenum, test section, and lower plenum. Liquid is introduced into the test section either through a porous injector or top flood entry. It is well-known that the liquid entry condition has a definite effect on the flooding. The

Table 1. Summary of the analytical models for vertical countercurrent flooding

Model	Basic Eq.	Empirical Equation	Fluids Compared	Flooding Equation
Potential Flow [4]	Laplace Bernoulli	Mean Film Thickness Critical Wave Length	Air/Water /Ethyl Alcohol /n-Heptane	$\bar{U}_g + \bar{U}_f = \frac{\sigma}{\rho_g}\left(k - \frac{1}{R - \delta}\right)^{1/2}$
Viscous Flow [6]	Orr-Sommerfeld		Air/Water /Glycerol	$\bar{U}_g = \bar{U}_g(Re,\ kc_i,\ \text{fluid properties})$ $\bar{U}_{g,crit} = \min(\bar{U}_g)$ when $kc_i > 0$
Finite-Amplitude Wave [7]	Momentum	Interfacial Shear	Air/Water /Glycerol /Silicone oil	$J_g^*/\alpha + (1/c^*)^{1/2} J_f^*/(1 - \alpha) = \bar{U}_{cr}[\rho_g/(\rho_f - \rho_g)gD]^{1/2}$ $2f_i J_g^{*2}/\alpha^{2.5} + 2f_w J_f^{*2}/(1 - \alpha)^2 = (1 - \alpha)$
Separate Cylinders [8]	Continuity Momentum		Air/Water	$J_g^{*2/(n+1)} + J_f^{*2/(m+1)} = 1$
Separated Flow [9]	Momentum	Interfacial Shear	Air/Water	$F(\alpha, J_g^*, J_f^*) = 2f_i J_g^{*2}/\alpha^{2.5} - 2f_w J_f^{*2}/(1 - \alpha)^2 - (1 - \alpha) = 0$ $G(\alpha, J_g^*, J_f^*) = \frac{\partial F}{\partial \alpha} = 0$
Stationary Wave [11]	Wave Profile	Pressure Distribution	Air/Water	$A = A(\bar{U}_f, \bar{U}_g, \text{fluid properties})$ $\frac{dA}{d\bar{U}_g} \to \infty$
Hanging Film [12]	Bernoulli		Air/Water	$K_{g,crit} = 1.87$
Roll Wave [13]	Momentum	Mean Film Thickness Interfacial Shear	Air/Water	$\frac{f_w}{4} Bo J_g^{*4}(Bo J_g^{*2} J_f^{*2} + 4) + 150 f_w J_g^{*2} = 1$

Table 2. Summary of various slug formation models

Authors	Basic Eq.	Wave	Stability Criterion	Modified Flooding Equation
Kordyban Ranov [29]	Laplace Bernoulli	Finite-amplitude	$\bar{U}_{cr} = \rho^{*\frac{1}{2}} (\frac{g}{k})^{\frac{1}{2}} G(k\delta_g)$ $G(k\delta_g) = 1/[\coth(k\delta_g - 0.9) + 0.45\coth^2(k\delta_g - 0.9)]^{\frac{1}{2}}$	$J_{mg}^*/\alpha + (1/\rho^*)^{\frac{1}{2}} J_{mf}^*/(1-\alpha)$ $= (\cot\theta/2Hk)^{\frac{1}{2}} G(k\delta_g)$
Wallis Dobson [28]	Laplace Bernoulli	Small-amplitude	$\bar{U}_{cr} = 0.5[(\rho_f-\rho_g)g\delta_g/\rho_g]^{\frac{1}{2}}$	$J_{mg}^*/\alpha + (1/\rho^*)^{\frac{1}{2}} J_{mf}^*/(1-\alpha)$ $= 0.354\ (\alpha\cot\theta)^{\frac{1}{2}}$
Taitel Dukler [30]	Bernoulli	Solitary-roll	$\bar{U}_{cr} = (\delta_g'/\delta_g)[(\rho_f-\rho_g)g\delta_g\cos\theta/\rho_g]^{\frac{1}{2}}$	$J_{mg}^*/\alpha + (1/\rho^*)^{\frac{1}{2}} J_{mf}^*/(1-\alpha)$ $= (\delta_g'/\delta_g)\ (\alpha\cot\theta/2)^{\frac{1}{2}}$
Gardner [27]	Bernoulli Energy Continuity	Large-amplitude	$\bar{U}_{cr} = \Gamma[(\rho_f-\rho_g)gH/\rho_g]^{\frac{1}{2}}$ $1-(1-1/\rho^*)(1-\alpha) =$ $[4\Gamma - 3(3+\rho^*)\Gamma^2 + 4(1+\rho^*)\Gamma^3]/$ $[2-2(3+\rho^*)\Gamma + 3(1+\rho^*)\Gamma^3]$	$J_{mg}^*/\alpha + (1/\rho^*)^{\frac{1}{2}} J_{mf}^*/(1-\alpha)$ $= \Gamma(\cot\theta/2)^{\frac{1}{2}}$
Mishima Ishii [26]	Laplace Bernoulli	Finite-amplitude	$\bar{U}_{cr} = 0.487\ (\rho_f g\delta_g/\rho_g)^{\frac{1}{2}}$	$J_{mg}^*/\alpha + (1/\rho^*)^{\frac{1}{2}} J_{mf}^*/(1-\alpha)$ $=0.344\ (\alpha\cot\theta)^{\frac{1}{2}}$

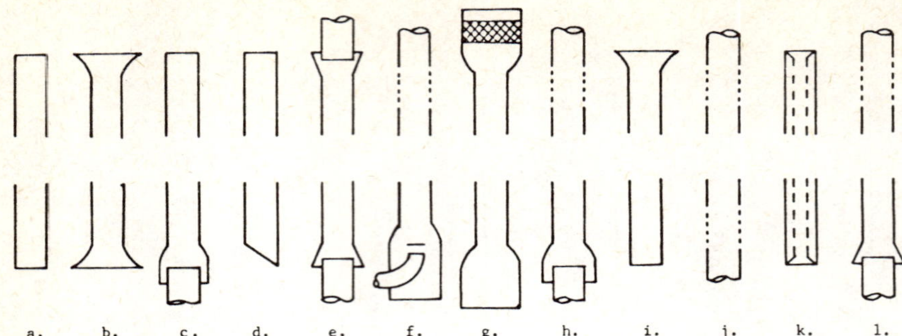

Figure 4. Geometries of the tube ends: (a) Wallis [42]; (b) Wallis [42]; (c) Kamei, et al. [45]; (d) Hewitt [25]; (e) Feind [22]; (f) Clift, et al. [40]; (g) Grolmes, et al.[18]; (h) Pushkina and Sorokin [43]; (i) Pushkina and Sorokin [43]; (j) Hewitt and Wallis [23]; (k) Chung [5]; (l) Dukler and Smith [44].

gas flooding velocities with porous injection are higher than those with top flood entry, even under the same conditions, since the former excludes the possibility of an initially-imposed disturbance at the water entrance. The tube end conditions used by various investigators are listed in Figure 4. Gas is supplied into the tube either directly through a pipe aligned with the tube axis or indirectly via the lower plenum.

The flooding condition inside the tube can be reached either by gradually increasing the gas flow rate for a fixed liquid flow rate or vice versa. Clift, et al. [40] reported that the flooding points determined by both methods are identical, but Wallis, et al. [41] found that they are not usually identical. In any case, most investigators have not made this distinction in their flooding experiments. The onset of flooding was determined by visual observations in most studies, but a few investigators [16-18] employed gas pressure drop measurements as a second criterion.

4.2 Empirical Correlations

Due to the complicated nature of flooding, it is often convenient to evaluate flooding characteristics, such as the dependence on geometrical conditions, the effect of fluid properties, and the tube diameter effect, by means of empirical correlations. A number of useful empirical correlations by various investigators have been produced over the past twenty years.

Two important dimensionless parameters have emerged: the

Wallis number [42] and the Kutateladze number [43]. Both may be looked upon as a ratio of inertial force to hydrostatic force, but the length scale in the Wallis number is a characteristic diameter, while in the Kutateladze number it is the Laplace capillary length:

$$J_k^* = j_k \left[\frac{\rho_k}{gD(\rho_f - \rho_g)}\right]^{1/2} \tag{16}$$

$$K_k = \bar{U}_k \left[\frac{\rho_k^2}{g\sigma(\rho_f - \rho_g)}\right]^{1/4} \tag{17}$$

where $k = g$ for gas phase
$k = f$ for liquid phase

Note that the Kutateladze number K is often defined by using the superficial velocity instead of the actual velocity [5].

One of the most frequently used correlations was given by Wallis [42], in the following form:

$$J_g^{*1/2} + mJ_f^{*1/2} = C \tag{18}$$

where m = 0.8 - 1.0, and C = 0.7 - 1.0, mainly depending upon the tube end conditions. Other investigators, such as Hewitt and Wallis [23], Clift, et al. [40], Hewitt [25], and Dukler and Smith [44], also correlated their flooding data with equation (18). Several investigators [43,5] suggested a similar correlation, based on the Kutateladze number.

$$K_g^{1/2} + mK_f^{1/2} = C \tag{19}$$

Pushkina and Sorokin [43] gave m = 0, $C = 3.2^{1/2}$ for the critical gas velocity, while Chung [5] found empirically that m = 0.65 - 0.80, with C a function of the dimensionless diameter, D*. Equations (18) and (19) are similar, in view of the relationship between the Wallis parameter and the Kutateladze number:

$$K_k = D^{*1/2} J_k^* \tag{20}$$

Several correlations [45-50] for adiabatic flow have also developed which are based on the Reynolds number, Froude number, or Weber number, rather than the conventional Wallis or Kutateladze numbers. These are listed in Table 3.

A few correlations [24,31,51] have emerged for steam-water

Table 3. Summary of various empirical correlations for vertical countercurrent flooding

Authors	Tube size(m)	Test fluids	Experimental range	Flooding equation
Wallis [42]	D=0.0127,0.019, 0.0254,0.051 L=1.22	Air/Water		$J_g^{*\frac{1}{2}} + J_f^{*\frac{1}{2}} = C$ $C = 0.725 - 0.875$
Kamei, et al. [45]	D=0.019,0.0316 0.0418 L=2.5	Air/Water Air/Millet-jelly Air/Soap solution	$50 < Re_f < 700$	$\frac{\rho_g j_g}{\rho_f j_f} = 36.2 \frac{Bo^{0.23} N_f^{0.462}}{Re_f^{1.225} \mu^{*0.71} \rho^{*0.13}}$
Feind [22]	D=0.02,0.05 L=0.96,2.5	Air/Water Air/Diethylene-glycol solution	$0.15 < Re_f < 3000$	$m \frac{Re_g}{Re_f^n} \frac{\rho^{*0.4}}{\mu^{*0.75}} + 1.4 \times 10^4 = 1300(\frac{R}{\delta})^{1.25}$ $m = 58.2,\ n = 0.33$ if $Re_f \le 400$ $m = 157.7, n = 0.5$ otherwise
Hewitt & Wallis [23]	D=0.0318	Air/Water	$J_f^* < 0.30$	$J_g^{*\frac{1}{2}} + J_f^{*\frac{1}{2}} = 1$
Clift, et al. [40]	D=0.0318 L=1.83	Air/Water Air/Glycerol	$0.01 < J_f^* < 0.09$	$J_g^{*\frac{1}{2}} + 0.34\ J_f^{*\frac{1}{2}} = 0.79 \quad K_{g,crit} = 2.24$
Tobilevich, et al. [46]	D=0.0327,0.0525 L=3.0	Air/Water Air/Sugar solution Steam/Water	$Re_f < 1800$	$Fr = a \exp(b\ \rho^{*-0.2} K_g)$ $a = 0.129,\ b = -14.14$ if $Fr > 0.012$ $a = 0.065,\ b = -10.12$ otherwise
Pushkina & Sorokin [43]	D=0.31,0.031,0.012 L=2.5	Air/Water	$0\ 005 < J_f^* < 0\ 064$	$K_{g,crit} = 3.2$

Table 3. (continued)

Authors	Tube size(m)	Test fluids	Experimental range	Flooding equation
Alekseev, et al. [48]				$K_g = 0.32\left[\frac{\sigma^{3/4}}{Q_f g^{1/4}(\rho_f-\rho_g)^{3/4}}\right]^{-0.22} Bo^{0.26}$
Grolmes, et al. [18]	D=0.004-0.025 L=0.04-1.30	Nitrogen/Water	$0.4 < Re_f < 140$	$\overline{U}_g = 1.15\left[\rho^* \frac{g\delta}{0.006+200\delta^2(\mu_R/\mu)^{0.44}}\right]^{1/2}$
Dukler & Smith [44]	D=0.051 L=3.94	Air/Water	$77.5 < Re_f < 822.8$	$J_g^{*1/2} + J_f^{*1/2} = 0.88$
Hewitt [25]	D=0.0127,0.0318 L=1.22,2.44	Air/Water Air/Silicone oil Air/Glycerol	$0.01 < J_f^* < 0.275$	$J_g^{*1/2} + J_f^{*1/2} = C$ C depends upon fluid property
Diehl & Koppany [47]	D=0.0266,0.0525 0.0158	Hydrogen/Diesel oil Air/Water Steam/Water		$j_g = F_1 F_2(\sigma/\rho_g)^{0.5}$ $F_1 = [D/(\sigma/80)]^{0.4}$ if $F_1 < 1$ $F_1 = 1$ otherwise $F_2 = (\rho_g j_g/\rho_f j_f)^{0.25}$
Suzuki & Ueda [50]	D=0.029,0.01,0.018 L=0.5 — 2.0	Air/Water Air/Glycerol		$[\rho_g(\overline{U}_g+\overline{U}_f)^2/\rho_f g\delta]^2 = a \log X + b$ $X = 0.63\, Re_f^{-1/3} Bo^{1/2} \mu^{*2/3}$ a and b are tube length-depedent
Chung [5]	D=0.0159,0.0318 0.046,0.070 L=0.914	Air/Water Air/Silicone oil Air/White oil	$0.128 < K_f < 2.265$	$K_g^{1/2} + mK_f^{1/2} = C_1 \tanh(C_2 D^{*1/4})$ $m = 0.65 - 0.80$ $C_1 = 1.79-2.1$ $C_2 = 0.8-0.9$

condensing flow in a single channel or in multiple paths. The vapor condensation effect is generally incorporated by calculating the steam flow reduction if the exit temperature of the inlet subcooled water were raised to the saturation temperature, which corresponds to maximum condensation.

Block and Crowley [51] found that the flooding equation (18) remains valid for steam-water flow in a simulated PWR downcomer geometry, provided that the effective dimensionless steam flow rate, J_{ge}^*, is substituted for the dimensionless steam inlet flow rate:

$$J_{ge}^* = J_g^* - f(T_{sat} - T_f)\left(\frac{\rho_f}{\rho_g}\right)^{1/2} \frac{C_{p\ell}}{i_{fg}} J_{f,in}^* \tag{21}$$

where f is the condensation efficiency, defined as the ratio of the actual condensation rate to the maximum attainable condensation rate.

Bankoff, et al. [24] introduced a new dimensionless group, H_k^*, defined by:

$$H_k^* = j_k \left[\frac{\rho_k}{gw(\rho_f - \rho_g)}\right]^{1/2} \tag{22}$$

where the interpolation scaling, w, is given by the following relation:

$$w = D^{\gamma} \left[\frac{\sigma}{g(\rho_f - \rho_g)}\right]^{(1-\gamma)/2} \qquad 0 \le \gamma \le 1 \tag{23}$$

In equation (23), the empirical exponent, γ is a dimensionless geometry-dependent constant. Employing the effective steam flow rate, the flooding data in a vertical countercurrent flow through horizontal perforated plates were successfully correlated by the following equation:

$$\left[H_g^* - f(T_{sat} - T_f)\left(\frac{\rho_f}{\rho_g}\right)^{1/2} \frac{C_{p\ell}}{i_{fg}} H_f^*\right]^{1/2} = C \tag{24}$$

Tien [31] suggested a similar correlation based on the Kutateladze number for flooding with condensation in a single channel:

$$[K_g - f(T_{sat} - T_f)(\frac{\rho_f}{\rho_g})^{\frac{1}{2}} \frac{C_{p\ell}}{i_{fg}} K_f]^{\frac{1}{2}} + K_f^{\frac{1}{2}} = 1.79 \tag{25}$$

According to equation (25), the subcooling of the liquid phase exhibits in the (K_g, K_f) plane a minimum point and an unstable positive-slope flooding curve, which results in a hysteresis effect. However, equation (25) has not been examined against any flooding data as yet.

Recently, Wallis et al [32] carried out an experiment on the limiting countercurrent flow characteristics of steam-water annular flow in a vertical tube for various water temperatures. It was found that the boundary on which water penetration is limited by the upward steam flow can be described by equation (18) with $C = 0.69$ to 0.8, depending on the tube end condition, and usually $m \approx 1$.

5 COMPARISON AND DISCUSSION

The comparisons of various models described earlier with the experimental data will be made separately for both flooding with phase change and flooding without it. For analytical models, comparisons are made with flooding data in various two-phase flows, including air-water flow, in order to examine the applicability of those models to various flow conditions. Their validities for the dependence of tube diameter are also investigated. Similarly, empirical correlations for flooding without and with phase change will be compared with flooding data in air-water flow and in steam-water flow, respectively, because most correlations have been developed from flooding data in these flows. In addition, as mentioned above, some discussion on slug formation models in a nearly horizontal stratified flow will be presented.

5.1 Flooding without Phase Change

Figure 5 shows comparisons of the viscous flow [6] and stationary wave [11] theoretical models with flooding data in a vertical air-water flow. Unlike other models, these two models imply a laminar liquid film, since both assume a parabolic velocity distribution in the liquid film.

It is shown that the viscous flow model generally appears to be in good agreement at moderate liquid Reynolds numbers, considering the wide spread of the experimental data by various investigators. This model can account for the effect of the liquid entrance condition by varying the interfacial friction factor. It has been shown that the gas flooding velocity will be substantially lowered

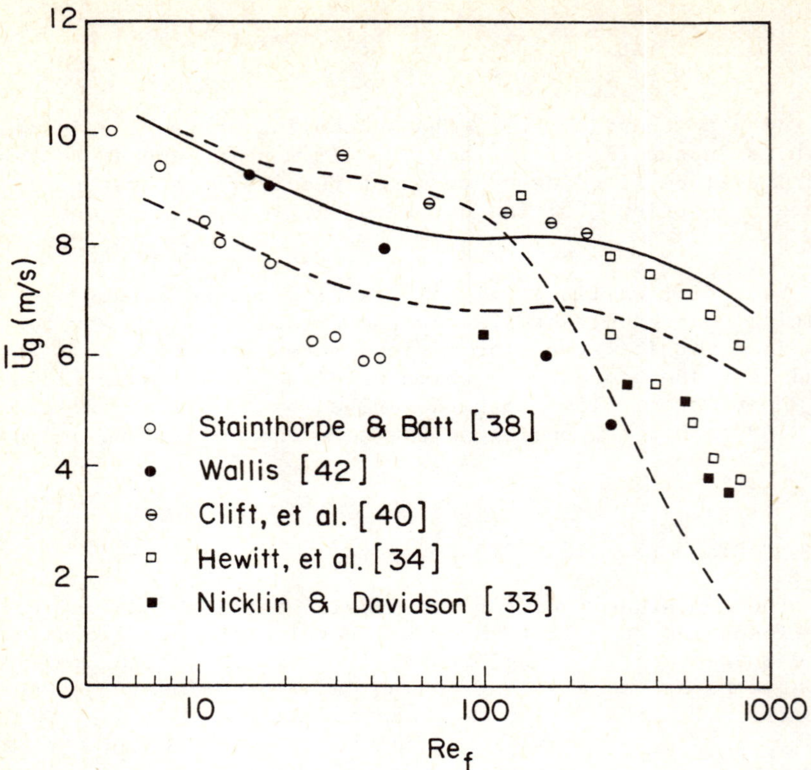

Figure 5. Comparison of two theoretical models with the flooding data for air-water flow: ——— viscous flow model [6] with f_i = 0.01; — — — viscous flow model with f_i = 0.014; – – – – stationary wave model [11]

unless a smooth liquid film is introduced [42,49]. This is because the higher interfacial friction factor due to initially-imposed disturbances at the sharp liquid entry is more likely to cause the formation of unstable waves. The stationary wave model indicates good agreement with flooding data for Reynolds numbers less than approximately 400. Beyond this Reynolds number, large discrepancies are found, attributed by the authors to the onset of turbulence in the liquid film. However, this argument appears to be doubtful because the assumption on the velocity distribution in this model has a negligible effect on the windward part of the wave profile. This is shown by referring to the differential equation developed by these authors for the shape of the front of the wave, based upon a momentum and mass balance. It turns out that a change from a laminar velocity profile to a uniform (turbulent) velocity profile

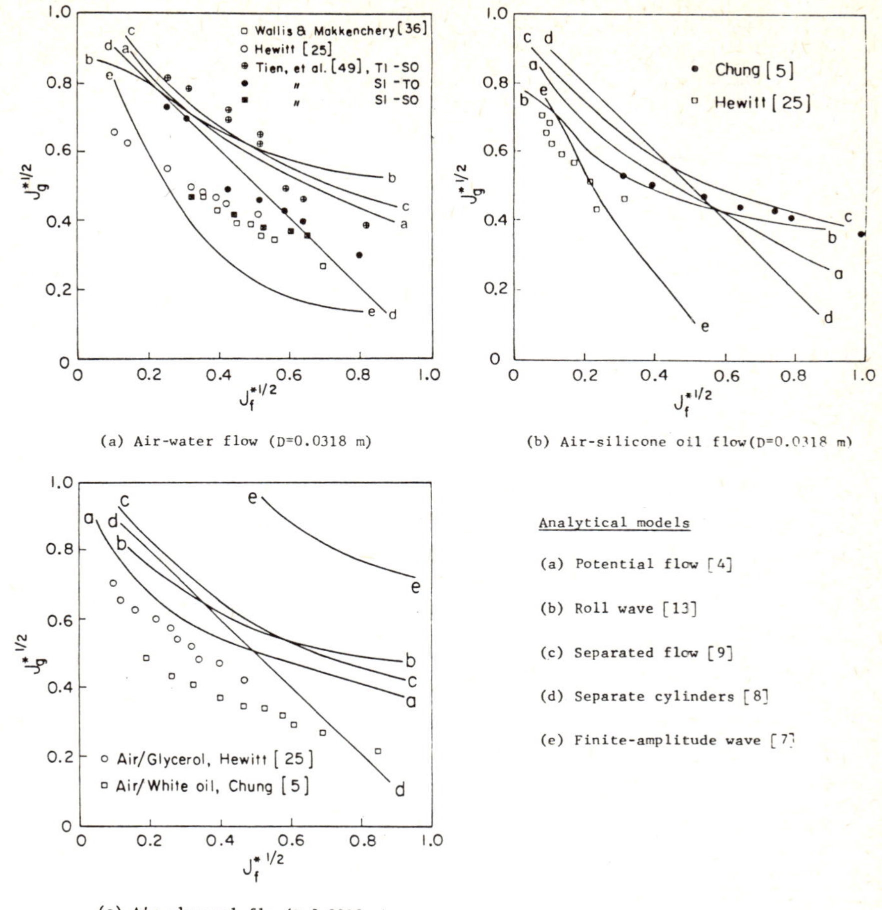

(a) Air-water flow (D=0.0318 m)

(b) Air-silicone oil flow (D=0.0318 m)

(c) Air-glycerol flow (D=0.0318 m)

Figure 6. Comparisons of various analytical models with three different two-phase flow systems: TI; tapered inlet, SO; sharp outlet, SI; sharp inlet, TO; tapered outlet.

assumption has an insignificant effect on the calculated wave shape. Indeed, it is therefore likely that the empirical pressure distribution over the windward part of a stationary wave no longer holds for the turbulent film because it does not represent well the behavior of the wave near the crest.

Although both models give successful predictions for the onset of flooding in the laminar liquid film region, they suffer a serious drawback, in that the numerical solutions which have been generated are dimensional, rather than dimensionless. Hence, at the present time, the governing equations and boundary conditions

Table 4. Properties of various two-phase flow systems

Fluids	Surface tension (N/m)	Liq. Density (kg/m^3)	Liq. Viscosity (Ns/m^2)	Gas Density (kg/m^3)	Gas Viscosity (Ns/m^2)
Air/ Water	0.0730	1000	1.1×10^{-3}	1.3	1.8×10^{-5}
Air/ Silicone oil	0.0175	820	0.82×10^{-3}	1.3	1.8×10^{-5}
Air/ Glycerol	0.0669	1170	18.0×10^{-3}	1.3	1.8×10^{-5}
Air/ White oil	0.0310	840	32.0×10^{-3}	1.3	1.8×10^{-5}

must be solved again for each new application.

Comparisons of other analytical models for vertical annular countercurrent two-phase flow with various two-phase countercurrent flooding data are shown in Figure 6. The three two-phase flow systems compared are air-water, air-silicone oil, and air-glycerol flows. The fluid properties are tabulated in Table 4. For these comparisons, the Wallis parameters were chosen as the basic variables, partly because extensive flooding data are available, and partly because most analytical models can also be easily expressed in terms of these parameters.

For air-water flow it appears that the upper region (low J_f^*) of flooding data can be predicted reasonably well by three analytical models: potential flow [4], roll wave [13] and separated flow [9]. The reason for this agreement is not clear. None of these models take into account liquid entrance effects, which tend to be very important at high gas velocities (low J_f^*). This is shown by the divergence of the data at small J_f^* for smooth and abrupt end conditions (Figure 6a), while the two sets of points tend to converge as J_f^* is increased. The finite-amplitude model predictions are significantly lower than all the other models, and are indeed below the smooth-end condition data. This indicates that the Jeffreys analysis[35] of the critical relative velocity for capillary wave growth is not applicable to vertical annular flow without modification, since the critical relative velocity is very sensitive to the mean liquid depth for thin films (for which the capillary wave concept is valid), and possibly also orientation. Another point is that in the Jeffreys analysis the waves are traveling in the same direction as the wind, but in countercurrent vertical flow the waves travel in the opposite direction to the gas flow. The separate-cylinders model differs from the other two in two respects: 1) the predicted curve is a straight line in the $J_g^{*\frac{1}{2}}$ vs. $J_f^{*\frac{1}{2}}$ plane of slope - 1, while all other models yield concave curves. 2) fair agreement is shown with the Tien, et al. data with one smooth and one sharp end condition, but not with the other data. Although the straight line formula is widely used, it seems reasonable that at very low J_f^* (very thin films), the flooding velocity must be very high. Conversely, there is a maximum J_f^* for gravity drainage in a pipe or channel, in the absence of any gas flow, which has nothing to do with the flooding mechanism, and hence may be expected to be larger than the straight line plot would indicate.

The effects of viscosity and surface tension on the flooding curve are illustrated in Figures 6b and 6c. Previous experimental results on the effect of fluid properties on flooding indicate that surface tension tends to stabilize the liquid film, while viscosity has generally a destabilizing effect [5,19,25,40]. The analysis by Cetinbudaklar and Jameson [6] also shows that viscosity tends to destabilize the liquid film.

The separated-flow and roll-wave models take into account the effect of surface tension through the dimensionless diameter, D* (which is also the square root of the Bond number, Bo). As shown in Figure 6b, surface tension has a stabilizing effect. As Bo decreases, J_g^* increases at fixed J_f^*. However, these models do not contain the viscosity explicitly, and hence cannot indicate general trends for the viscosity effect. If the empirical correlations for the interfacial shear stress employed in these models were further developed for various two-phase flows, they could be thus generalized. For the potential flow model, the viscosity and surface tension terms are included in the empirical correlations for the mean film thickness and the dimensionless critical wave length. It is predicted that lower surface tension causes lower gas flooding velocities, in agreement with experimental data. However, at higher liquid flow rates, the gas flooding velocities tend to be underestimated. Figure 6c shows, surprisingly enough, that the potential flow model seems to incorporate the effect of viscosity on flooding well. Although the effect of viscosity is not considered in the stability analysis of a traveling wave, it is reflected in the evaluation of the dimensionless critical wave number. For the finite-amplitude wave model, the surface tension effect is contained in the calculation of the critical relative velocity, which is proportional to a fractional power of the surface tension. Hence, higher surface tension results in higher gas flooding velocity at fixed J_f^*, in agreement with experimental trends. However, as shown in Figures 6a and 6b, J_g^* is strongly underpredicted for large J_f^*, and the effect of decreased surface tension is to make the deviation larger. On the other hand, from Figures 6a and 6c, one sees that J_g^* is greatly overpredicted, so that the effect of increased viscosity seems too strong in the model.

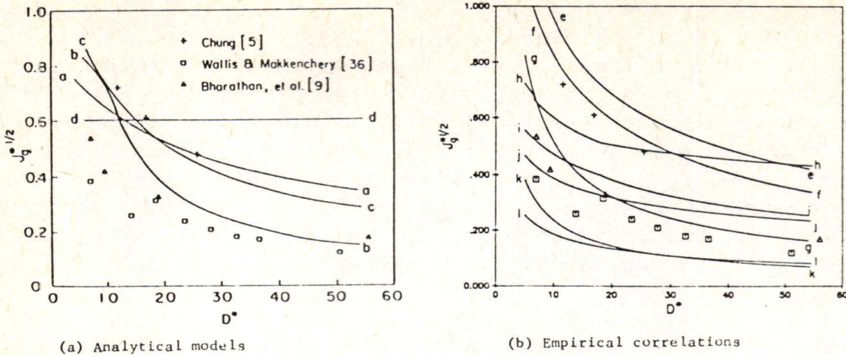

Figure 7. Predictions of gas flooding velocity with variation of tube diameter for $J_f^* = 0.05$: (a) potential flow [4] (b) roll wave [13]; (c) separated flow [9]; (d) separated cylinders [8]; (e) Pushkina and Sorokin [43]; (f) Chung [5]; (g) Kamei, et al. [45]; (h) Suzuki and Ueda [50]; (i) Feind [22]; (j) Alekseev, et al. [48]; (k) Grolmes, et al. [18]; (i) Tobilevich, et al. [46].

This probably comes about from the use of the Jeffreys sheltering theory, together with the formula for small capillary waves in determining the wave speed. This gives the critical relative velocity $\overline{U}_{cr} \sim \nu^{\frac{1}{2}}$ whereas other models predict that increased viscosity decreases J_g^* (destabilizes the film). Finally, the separate cylinders model gives no information on the effect of fluid properties, since neither surface tension nor viscosity are contained in the Wallis parameters.

The effect of D^* on J_g^*, when the dimensionless liquid velocity is fixed at $J_f^* = 0.05$, is shown in Figures 7a and 7b for the analytical models and empirical correlations, respectively. One sees that J_g^* monotonically decreases as D^* increases at fixed J_f^*. However, the separate cylinders model does not predict this effect. A better representation allows the constant, C, (and possibly m) to be a function of the Bond number, and possibly geometric parameters. This has been done for BWR tube bundle data by Sun [15], and for perforated horizontal plates by Bankoff, et al. [24]. One also sees from equations (18) and (19) that for fixed J_f^*, $J_g^* \propto (D^*)^n$, where n = 0 for the Wallis correlation, (small diameters) and $n = -\frac{1}{2}$ for the Pushkina and Sorokin correlation (constant K at large diameters). To bridge this gap, another flooding parameter (H^*) has been introduced, (Bankoff, et al. [24], following earlier work by Collier et al. [37]). This employs an interpolative length scale between the tube diameter (Wallis) and the Laplace capillary length (Kutateladze). However, this parameter was introduced for

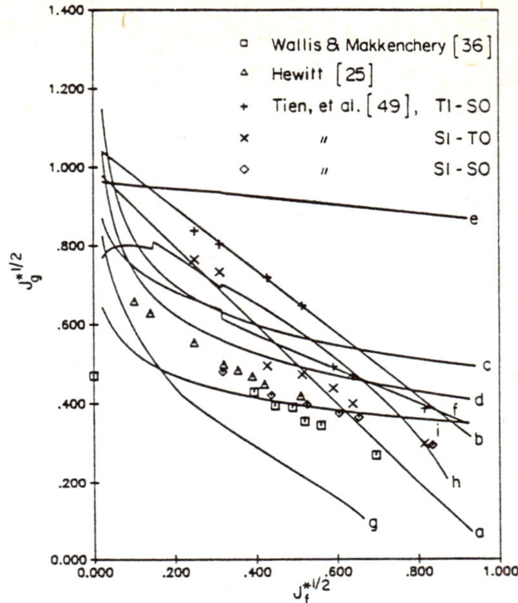

Figure 8. Comparison of various empirical correlations with the flooding data for air-water flow: (a) Wallis [42]; (b) Chung [5]; (c) Kamei, et al [45]; (d) Alekseev, et al. [48]; (e) Pushkina and Sorokin [43]; (f) Feind [22]; (g) Tobilevich, et al. [46]; (h) Suzuki and Ueda [50]; (i) Grolmes, et al. [18]

perforated-plate flooding, and has not yet been applied to vertical tubes. Figure 7b also shows the strong effect of smooth (Chung [5]) vs. abrupt end conditions, which tends to make quantitative judgments concerning the relative merits of the various correlations difficult.

Figure 8 shows the comparison of various empirical correlations with the same experimental data used in Figure 6a. Among these, the correlation by Pushkina and Sorokin [43] is unique in that it correlates data for the critical gas velocity at which the liquid flow starts to climb up along the tube wall. The predictions of most of the correlations lie within the relatively wide band data. Some correlations, such as those of Feind [22], Grolmes, et al. [18], and Alekseev, et al. [48], used the actual velocity, and hence the conversion into superficial velocity was made by using equations (1) and (4). It is useful to note that the correlations of Kamei, et al. [45], Feind [22], and Tobilevich, et al. [46] may be used

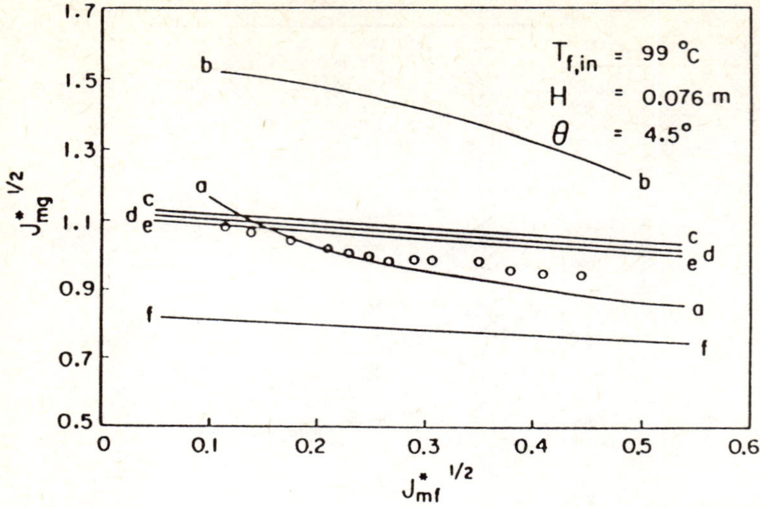

Figure 9. Comparison of various models with the flooding data for nearly horizontal countercurrent flow: (a) Lee and Bankoff [17]; (b) Taitel and Dukler [30]; (c) Gardner [27]; (d) Wallis and Dobson [28]; (e) Mishima and Ishii [26]; (f) Kordyban and Ranov [29].

for two-phase flows other than air-water flow. The correlation by Suzuki and Ueda [50] is the only one to incorporate the effect of the tube length. This may be reasonable, since Hewitt, et al. [34] reported that a longer tube length acts to decrease the gas flooding velocity. However, attention has not been paid to this effect by most investigators.

Comparison of various analytical models with flooding data for nearly horizontal stratified countercurrent flow is shown in Figure 9. The modified Wallis parameters, defined by the following equation, were employed as the basic variables:

$$J_{mk}^{\ *} = j_k \left[\frac{\rho_k}{2gH\sin\theta(\rho_f - \rho_g)} \right]^{\frac{1}{2}} \tag{26}$$

Five slug formation models, as well as the separated flow model by Lee and Bankoff [17], are included in this comparison. The void fraction for these slug formation models was calculated from equation (7), which represents the mean film thickness for the three-dimensional wave region. The flooding curve by Lee and Bankoff represents the envelope generated by the locus of the steady hydrodynamic equation for stratified countercurrent flow with the following empirical correlation for the interfacial shear stress combined from equations (8) and (10):

$$f_i = 0.012 + 6.091 \times 10^{-8} N_{mf}^{0.848} [N_{mg} J_{mg}^* - 183.7 (N_{mf} J_{mf}^*)]^{-0.184} \cdot (1-\alpha)^{2.544} \quad (27)$$

where the modified two-phase Grashof number N_{mk} is defined by

$$N_{mk} = [2gH^3 \sin\theta \rho_k (\rho_f - \rho_g)/\mu_k^2]^{\frac{1}{2}} \quad (28)$$

It is shown in Figure 9 that the onset of flooding in a nearly horizontal flow can be successfully predicted by the separated flow model and by three slug formation models: Mishima and Ishii model [26], Wallis and Dobson model [28], and Gardner model [27]. The prediction by the Taitel and Dukler model [30] constitutes a much higher flooding curve. This discrepancy may be attributed to ignoring the frictional loss at the interface in the Bernoulli-type equation. If the interfacial shear were taken into consideration in calculating the gas pressure at the wave crest (which was estimated as much as 80 percent of the momentum change between the crest and the trough by Kordyban [39] in cocurrent horizontal flow), the prediction would be closer to the flooding data. The Kordyban-Ranov [29] theory predicts a considerably lower flooding curve compared to the experimental data. This seems to be partly because the critical wave length, which was estimated to be 0.025 - 0.05 m from their cocurrent-flow experimental observations, may change significantly in nearly horizontal countercurrent flow. However, it may be concluded from this comparison that the slug formation model in cocurrent horizontal flow can be successfully applied to the prediction for the onset of flooding in nearly horizontal countercurrent flow with some modifications.

5.2 Flooding with Phase Change

Countercurrent flow accompanying phase change at the interface exhibits two different flow instabilities which bound the normal condensation process. A typical map for a nearly horizontal countercurrent flow of steam and cold water is illustrated in Figure 10. The line of $R_T = 1$ represents the ideal complete condensation line that corresponds to the mass flow ratio when ideal mixing of steam and cold water takes place, and hence R_T is defined by

$$R_T = C_{p\ell}(T_{sat} - T_f) W_f / i_{fg} W_g \quad (29)$$

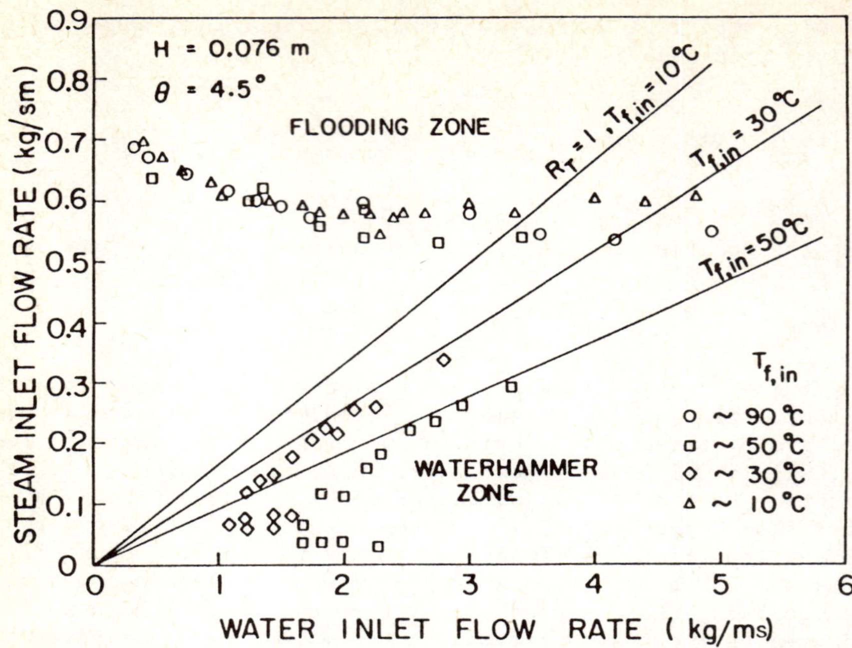

Figure 10. Stability map of nearly horizontal countercurrent flow of steam and cold water [17].

The condensation-induced waterhammer zone is characterized by significant pressure oscillations, probably due to unsteady collapse, or near-collapse, of local steam pockets. One sees that condensation-induced waterhammer cannot exist when the water flow rate is insufficient to condense all the incoming steam. Thus, the condensation-induced waterhammer zone is very closely related to the heat transfer mechanism in countercurrent flow. Flooding data for various degrees of liquid subcooling are also illustrated in Figure 10. The results indicated that the vapor flooding velocity does not depend significantly upon the water inlet temperature, since the water outlet temperature reaches nearly the saturation temperature, regardless of the inlet temperature.

Wallis, et al. [32] made a systematic study of air-water and steam-water countercurrent flow in a transparent vertical tube, which demonstrated some of the similarities and also differences of the two systems. The large spread of data by different investigators also becomes better understood from these results. Wallis, et al. found quite different results in the limiting curves of $J_g^{*1/2}$ vs. $J_f^{*1/2}$, based upon minor modifications of the steam supply and exit systems. In addition, the effect of using rounded vs. square ends on the tube was significant. Hysteresis effects were also found, as shown in Figure 11. These effects represent the possibility of bistable modes of operation. Referring to Figure

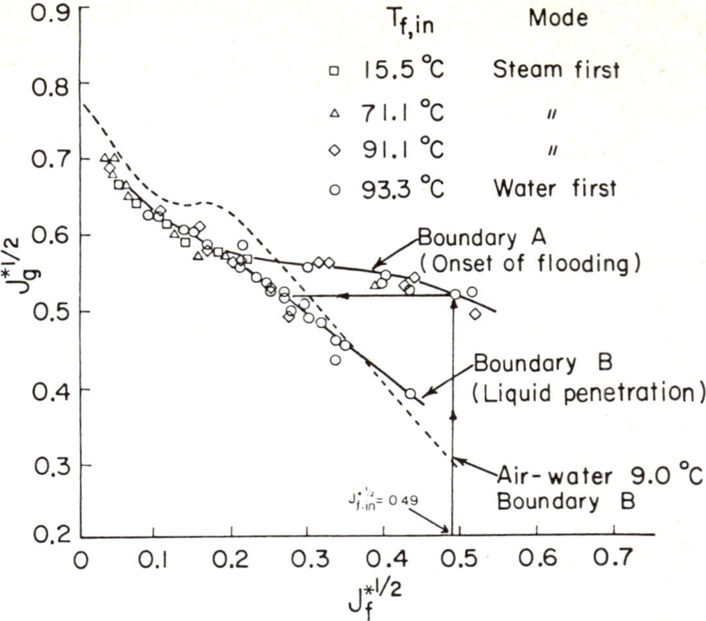

Figure 11. Flooding in vertical steam-subcooled water flow [32].

11, when the water flow rate is fixed at $J_f^{*\frac{1}{2}} = 0.49$ and the steam flow rate is gradually increased, the flooding curve is reached at the Boundary A when an instability manifests itself. At this point the upper plenum begins to fill, and small increases in steam flow rate then cause the water delivered downwards to jump to the Boundary B. Further increases or decreases of $J_g^{*\frac{1}{2}}$ cause the delivered water, $J_{f,d}^{*\frac{1}{2}}$, to follow Boundary B, excess water being carried off in the upper plenum overflow line. If, however, $J_g^{*\frac{1}{2}}$ is reduced to the point of intersection of Boundary B with the vertical line $J_{f,\ in}^{*\frac{1}{2}} = J_{f,d}^{*\frac{1}{2}}$, complete delivery begins and the upper plenum empties. The cycle can then repeat itself. The same phenomenon was earlier shown by Clift, et al.[40] for air/water, but the presence of steam condensation accentuates the possibility of instability and "chugging". With air-water the critical flooding locus can occur either at the liquid entry points or exit, depending upon the geometric details. With steam, however, the critical flooding location can shift sharply from top to bottom. In the water first mode all the steam is condensed up to a certain steam flow rate, at which point steam begins to penetrate the upper plenum. When this occurs, the flow rate of water downwards is reduced, perhaps at the same time as the upper plenum water begins heating up. Consequently, the temperature of the water reaching the lower plenum becomes nearly saturated, so that the critical flooding

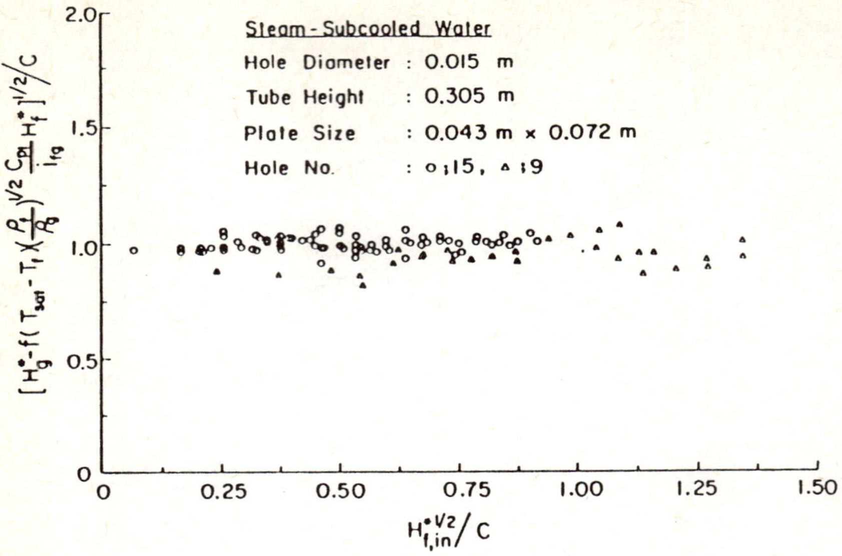

Figure 12. Comparison of 15 hole and 9 hole data with EOCB correlation by Bankoff, et al. [24].

locus is at the bottom, where the steam velocity is greatest. Nevertheless, these data indicate that in the "steam-first" mode the initial degree of subcooling of the water does not have a strong effect on these boundaries, so that the process can be considered to be hydrodynamic-limited, rather than heat transfer-limited, since the exit water becomes effectively saturated. Clearly, additional work is needed to explore these interesting possibilities.

Bankoff, et al. [24] studied the onset of penetration, or the end of complete bypass of water entering the upper plenum (EOCB), for a number of horizontal perforated plates with steam upflow. The data for a variety of perforation geometries, in the absence of condensation, could be correlated by a Wallis type equation, using, however, the dimensionless volumetric flux, H_g^* and H_f^*. With cold water, however, the onset of penetration, sometimes called "weeping" in the chemical engineering literature, was related to the delivery of cold water to the plate, either due to inlet jet water momentum or turbulent mixing in the upper plenum pool. An effective H_g^*, taking into account condensation of part of the steam in the boundary layer above the plate, then correlated the data quite well for onset of penetration (Figure 12). These investigators also found a band in the J_g^* vs. J_f^* plane (or equivalently in the $\dot{h}_g$ vs. $\dot{h}_f$ plane, where $\dot{h}_g$ and $\dot{h}_f$ are the enthalpy inlet flow rates for steam and water, respectively). Within the band oscillations took place in the liquid level in the upper plenum. When the steam flow rate was reduced to a critical level, complete delivery of the water downwards occurred. A hysteresis loop, similar to that observed by Wallis, et al. [32] in vertical tube flow, could thus be obtained.

6 SUMMARY AND CONCLUSIONS

Flooding models for condensing two-phase flow, as well as for adiabatic two-phase flow, are reviewed and compared with extensive data. A partial classification of these models would be: stability theory of a traveling wave, envelope theory, static equilibrium theory, and slug formation theory. Their ranges of validity are examined and discussed under various flow conditions. Most models, including some of the empirical correlations, exhibit a concave flooding curve on the ($J_g^{*\frac{1}{2}}$, $J_f^{*\frac{1}{2}}$) plane, as contrasted to the linear relationship of the widely-used, simple and convenient Wallis correlation.

The available evidence on the parametric dependence of the flooding curve has been discussed. The liquid entrance and exit conditions seem to have a strong effect upon flooding. In particular, the gas flooding velocity is substantially reduced if the liquid film is not introduced smoothly. This seems to be one of the important factors accounting for the wide spread of flooding data. More careful experimentation is recommended to investigate this effect. Most models incorporate a tube diameter effect, but the effect of the tube length, if any, is not clear yet. Viscosity and surface tension seem to exhibit opposite effects on the flooding velocity, as shown both experimentally and analytically.

Comparison of flooding models with steam-water flow data shows that the concept of an effective vapor flux seems to account for top flooding. However, there appears to be no significant difference between condensing flow and adiabatic flow if flooding takes place at the bottom of the test section, and the exit liquid reaches nearly the saturation temperature. Additional systematic work on condensing flow is recommended to investigate the condensing effect. Another area for further investigation is condensation-induced waterhammer, in addition to flooding, in order to obtain a firm theoretical and experimental basis for a stability map in steam-water flow.

NOMENCLATURE

A	amplitude of interface disturbance
a	empirical constant defined in equation (11)
Bo	Bond number, $D^2/[\sigma/(\rho_f - \rho_g)g]$
b	empirical constant defined in equation (12)
C	empirical constant
$C_{p\ell}$	specific heat of liquid (kJ/kgK)
c_i	imaginary part of complex wave velocity (m/s)
D	tube diameter (m)
D^*	dimensionless tube diameter defined in equation (13)
F	function
Fr	Froude number, $Q_f/(gD^3)^{\frac{1}{2}}$
f	condensation efficiency, or friction factor
g	gravitational acceleration (m/s^2)
H	channel height (m)
H^*	dimensionless flux defined in equation (22)
i_{fg}	latent heat (kJ/kg)
J^*	dimensionless parameter defined in equation (16)
J_m^*	modified J^*, parameter defined in equation (26)
j	superficial velocity (m/s)
K	dimensionless parameter defined in equation (17)
k	wave number (m^{-1})
L	tube length (m)
m	empirical constant
N_f	two-phase Grashof number $[gD^3\rho_f(\rho_f - \rho_g)/\mu_f^2]^{\frac{1}{2}}$

N_m modified two-phase Grashof number, defined in equation (28)

n constant

P pressure (N/m^2)

ΔP pressure drop (N/m^2)

Q flow rate per unit width (m^3/sm)

R radius (m)

R_T thermodynamic ratio defined in equation (29)

Re Reynolds number, W/μ

Re_g^* critical gas Reynolds number defined in equation (15)

T temperature (K)

U velocity in the flow direction (m/s)

W mass flow rate per unit width (kg/sm)

We Weber number $\sigma/(\rho_f \overline{U}_f^2 \delta)$

w length scale defined in equation (23)

X dimensionless parameter

α void fraction

Γ dimensionless parameter

γ experimental constant

δ mean film thickness (m)

δ_g depth of channel occupied by gas (m)

θ inclination angle

μ viscosity (Ns/m^2)

μ^* viscosity ratio, μ_f/μ_g

ν kinematic viscosity (m^2/s)

ρ density (kg/m^3)

ρ^* density ratio, ρ_f/ρ_g

σ surface tension (N/m)

τ shear stress (N/m^2)

Subscripts

cr critical relative

d delivery

e effective

f liquid

g gas

i interface

in inlet

k g or f

max maximum

min minimum

R reference

r relative

sat saturation

w wall

Superscripts

$-$ mean

$*$ dimensionless

$'$ disturbed

REFERENCES

1. Sherwood, T. K., Shipley, G. H., and Holloway, F. A. L. "Flooding velocities in packed columns." Ind. Eng. Chem, 30 (1938) 765-769.

2. Block, J. A. and Schrock, V. E. "Emergency cooling water delivery to the core inlet of PWR's during LOCA." ASME Symp. on Thermal and Hydraulic Aspects of Nuclear Reactor Safety 1 (1977) 109-132.

3. Lahey, R. T. "The status of boiling water nuclear reactor safety technology." ASME Symp. on Thermal and Hydraulic Aspects of Nuclear Reactor Safety 1 (1977) 151-171.

4. Imura, H., Kusuda, H., and Funatsu, S. "Flooding velocity in a countercurrent annular two-phase flow." Chem. Eng. Sci. 32 (1977) 79-87.

5. Chung, K. S. "Flooding phenomena in counter-current two-phase flow systems." Ph.D. Thesis, University of California, Berkeley (1978.

6. Cetinbudaklar, A. G. and Jameson, G. J. "The mechanism of flooding in vertical counter-current two-phase flow." Chem. Eng. Sci. 24 (1969) 1669-1680.

7. Zvirin, Y., Duffey, R. B., and Sun, K. H. "On the derivation of a countercurrent flooding theory." Symp. on Fluid Flow and Heat Transfer over Rod or Tube Bundles, ASME, New York (1979) 111-119.

8. Wallis, G. B. "One-dimensional two-phase flow." Chap. 11, McGraw Hill, New York (1969).

9. Bharathan, D., Wallis, G. B. and Richter, H. J. "Air-water countercurrent annular flow." EPRI NP-1165 (1979).

10. Dobran, F. "Condensation heat transfer and flooding in a countercurrent subcooled liquid and saturated vapor flow." Symp. on Thermal-Hydraulics in Nuclear Power Technology (1981) 9-20.

11. Shearer, C. J. and Davidson, J. F. "The investigation of a standing wave due to gas blowing upwards over a liquid film; its relation to flooding in wetted-wall columns." J. Fluid Mech. (1965) 321-336.

12. Wallis, G. B. amd Kuo, J. T. "The behavior of gas-liquid interfaces in vertical tubes." Int. J. Multiphase Flow 2 (1976) 521-536.

13. Richter, H. J. "Flooding in tubes and annuli." Int. J. Multiphase Flow 7 (1981) 647-658.

14. Tien, C. L. and Liu, C. P. "Survey on vertical two-phase counter current flooding." EPRI NP-984 (1979).

15. Sun, K. H. "Flooding correlations for BWR bundle upper tieplates and bottom side-entry orifices." Second Multiphase Flow and Heat Transfer Symp., Miami Beach, FL (1979).

16. Hawley, D. L. and Wallis, G. B. "Experimental study of liquid film fraction and pressure drop characteristics in vertical countercurrent annular flow." EPRI NP-2280 (1982).

17. Lee, S. C. and Bankoff, S. G. "Steam-subcooled water countercurrent flow in an inclined channel." Topical Report to Nuclear Regulatory Commission (1982), to be published.

18. Grolmes, M. A., Lambert, G. A., and Fauske, H. K. "Flooding in vertical tubes." A.I.Ch.E. Symp. on Multiphase Flow Systems (1974).

19. Wallis, G. B. "The influence of liquid viscosity on flooding in a vertical tube." G. E. Report GL132 (1962).

20. Brauer, H. "Flow and heat transfer at falling liquid films." VDI - Forschungsheft 457 22 (1956).

21. Belkin, H. H., Macleod, A. A., Monrad, C. C., and Rothfus, R. R. "Turbulent liquid flow down vertical walls." A.I.Ch.E. J. 5 (1959) 245-248.

22. Feind, F. "Falling liquid films with countercurrent air flow in vertical tubes." VDI - Forschungsheft 481 26 (1960).

23. Hewitt, G. F. and Wallis, G. B. "Flooding and associated phenomena in falling film flow in a vertical tube." AERE - R4022 (1963).

24. Bankoff, S. G., Tankin, R. S., Yuen, M. C., and Hsieh, C. L. "Countercurrent flow of air/water and steam/water through a horizontal perforated plate." Int. J. Heat and Mass Transfer 24 (1981) 1381-1395.

25. Hewitt, G. F. "Influence of end conditions, tube inclination and physical properties on flooding in gas-liquid flows." Harwell Report HTFS - RS 222 (1977).

26. Mishima, K. and Ishii, M. "Theoretical prediction of onset of horizontal slug flow." Trans. ASME, J. of Fluid Engineering 102 (1980) 441-445.

27. Gardner, G. C. "Onset of slugging in horizontal ducts." Int. J. Multiphase Flow 5 (1979) 201-209.

28. Wallis, G. B. and Dobson, J. E. "The onset of slugging in horizontal stratified air-water flow." Int. J. Multiphase Flow 1 (1973) 173-193.

29. Kordyban, E. S. and Ranov, T. "Mechanism of slug formation in horizontal two-phase flow." Trans. ASME, J. of Basic Engineering 92 (1970) 857-864.

30. Taitel, Y. and Dukler, A. E. "A model for predicting flow regime transitions in horizontal and near horizontal gas-liquid flow." A.I.Ch.E. J. 22 (1976) 47-55.

31. Tien, C. L. "A simple analytical model for counter-current flow limiting phenomena with condensation." Letters in Heat and Mass Transfer 4 (1977) 231-237

32. Wallis, G. B., deSieyes, D. C., Rosselli, R. J., and Lacombe, J. "Countercurrent annular flow regimes for steam and subcooled water in a vertical tube" EPRI NP - 1336 (1980).

33. Nicklin, D. J. and Davidson, J. F. "The onset of instability in two-phase slug flow." Proceedings of the Symposium on Two-Phase Fluid Flow, Inst. Mech. Engrs. (1962) 29.

34. Hewitt, G. F., Lacey, P. M. C., and Nicholls, B. "Transitions in film flow in a vertical tube." Symp. on Two-Phase Flow University of Exeter, England (1965) 401-403.

35. Jeffreys, H. "On the formation of water waves by wind." Proc. Roy. Soc. A. 110 (1926) 241-247.

36. Wallis, G. B. and Makkenchery, S. "The hanging film phenomenon in vertical annular two-phase flow." ASME Trans. J. of Fluid Engineering 96 (1974) 297-298.

37. Collier, R. P., Dworak, J. A., Flanigan, L. J., Liu, J. S. K. and Segev, A. "Steam-water mixing and system hydrodynamics program." NUREG/CR-1625 (1979).

38. Stainthorp, E. P. and Batt, R. S. W. "The effect of co-current and counter-current air flow on the wave properties of falling liquid films." Trans. Inst. Chem. Engr. 45 (1967) T372-T382.

39. Kordyban, E. "Experimental study of aerodynamic pressure at the wave surface in two-phase flow." Symp. on Basic Mechanisms in Two-Phase Flow and Heat Transfer, ASME (1980) 19-25.

40. Clift, R., Pritchard, C. L., and Nedderman, R. M. "The effect of viscosity on the flooding conditions in wetted wall columns." Chem. Eng. Sci. 21 (1966) 87-95.

41. Wallis, G. B., Crowley, C. J., and Block, J. "ECC bypass studies." A.I.Ch.E. Symp. on Light Water Reactor Safety, Boston, MA (1975).

42. Wallis, G. B., "Flooding velocities for air and water in vertical tubes." UKAEA Report AEEW - R123 (1961).

43. Pushkina, O. L. and Sorokin, Y. L. "Breakdown of liquid film motion in vertical tubes." Heat Transfer - Soviet Research 1 (1969) 56-64.

44. Dukler, A. E. and Smith, L. "Two phase interactions in countercurrent flow: studies of the flooding mechanism." NUREG/CR-0617 (1979).

45. Kamei, S., Oishi, J., and Okase, T."Flooding in a wetted wall tower." Chem. Eng.(Japan) 18 (1954) 364-368.

46. Tobilevich, N. Y., Sagan, I. I., and Porzhezihskii, Y. G. "The downward motion of a liquid film in vertical tubes in an air-vapor counter flow." J. Eng. Phys. 15 (1968) 1071-1076

47. Diehl, J. C. and Koppany, C. R. "Flooding velocity correlation for gas-liquid counter flow in vertical tubes." Chem. Eng. Prog. Symp. Series 65 (1969) 77-83.

48. Alekseev, V. P., Poberezkin, A. E., and Germsimov, P. A. "Determination of flooding rates in regular packings" Heat Transfer - Soviet Research 4 (1972) 159-163.

49. Tien, C. L., Chung, K. S., and Liu, C. P. "Flooding in two-phase countercurrent flows." EPRI NP-1283 (1979)

50. Suzuki, S. and Ueda, T. "Behavior of liquid films and flooding in countercurrent two-phase flows - Part 1, Flow in circular tubes." Int. J. Multiphase Flow 3 (1977) 517-532.

51. Block, J. A. and Crowley, C. J. "Effect of steam upflow and superheated walls on ECC delivery in a simulated multiloop PWR geometry." Creare TN - 210 (1975).

ANALYSIS OF REFLOOD PHENOMENA BY THE TWO FLUID CODE F L U T

A Hora, Ch. Michetschläger, H.G. Sonnenburg, V. Teschendorff

Gesellschaft für Reaktorsicherheit (GRS) mbH
D-8046 Garching, Germany

ABSTRACT

The computer code FLUT is being developed by GRS for the analysis of the refill and reflood phase of a Loss-of-Coolant-Accident (LOCA) in a Pressurized Water Reactor (PWR). The thermo-hydraulic model in FLUT is a two-fluid model using six field equations for the conservation of mass, momentum, and energy. The interaction between the liquid and vapour phase is modeled by constitutive equations for mass transfer rate, interfacial drag, and interfacial heat transfer. Heat transfer coefficients between rod surface and fluid are calculated by empirical correlations, based on local fluid conditions and surface temperatures, for different flow regimes. Upper and lower quench front positions are calculated explicitly by analytical correlations. At present, the FLUT code is applicable to forced feed bundle reflood experiments. Comparisons between code prediction and measured data are presented for a FLECHT and an SCTF experiment.

1 INTRODUCTION

The postulated loss-of-coolant accident of a pressurized water reactor has been the subject of intensive studies in water reactor safety analysis for more than 20 years. Large experimental and analytical effort was devoted to the investigation of the thermohydraulic behaviour of the reactor core and the effectiveness of the emergency core cooling system during all phases of a LOCA.

Recently the emphasis of investigation has shifted from blowdown to the refill and reflood phase of the LOCA and to small break analysis. The simulation of the associated physical phenomena, e.g. separation of phases and superheating of vapour, requires advanced analytical models and computer codes. Especially the large density ratio of liquid and vapour near atmospheric pressure suggests a non-homogeneous treatment of phases during the low pressure phase of a LOCA.

Considerable non-equilibrium between phases is supposed to occur in the course of a LOCA. Typical situations of thermal non-equilibrium in the refill and reflood phase are

- injection of subcooled ECC water
- superheating of vapour in the presence of liquid drops.

Predominant effects involving mechanical non-equilibrium (slip) between phases are

- bubble rise
- phase separation and water level formation
- entrainment of liquid drops
- countercurrent flow (vertical film or horizontal layer).

A promising approach for the simulation of the thermohydraulic behaviour during refill and reflood is the use of a two-fluid model. The conservation equations for mass, momentum and energy are formulated separately for vapour and liquid. Together with constitutive equations for the interaction between phases and the fluid properties, the equation system should be able to model the non-equilibrium effects mentioned above.

The computer code FLUT which is being developed by GRS for the analysis of the refill and reflood phase of a LOCA in a PWR is based on a six equation two-fluid model. The set of constitutive equations included in the present code version is preliminary and subject to further development.

For the representation of the PWR primary system a network of one-dimensional flow paths and plenum cells is developed.

For the simulation of fuel rods or electrical heaters, a one-dimensional conduction model has been linked to the code.

A package of reflood heat transfer correlations was developed and tested.

At present, the FLUT code is applicable to forced feed bundle reflood experiments. Comparisons between calculations and data will be presented for a FLECHT and an SCTF experiment.

2 HYDRAULIC MODEL

2.1 Basic Equations

The six conservation equations of the two-fluid model are shown in the appendix.

The right hand sides of the equations presented do not contain derivative terms. It is known that such a system of partial differential equations is not "well-posed", which means that not all characteristics are real. Instabilities of growing amplitude may occur in such systems.

We never observed instabilities arising from ill-posedness in our calculations, probably because enough damping is introduced by the coarse space discretisation. For the planned range of application (no critical flow), we consider these equations a reasonable approach.

2.2 Constitutive Equations

Additional relations are required to close the set of equations. Among them, the properties of liquid and vapour as functions of the independent variables pressure and enthalpy have been programmed. Spline approximations are used, and extrapolations are made for non-equilibrium conditions.

For the interaction terms between phases, this first version of FLUT uses rather simple expressions. Although it is well known that the dependence on flow regime is significant, no such dependence has been introduced in FLUT. The use of a flow regime map would require an "a priori" knowledge of the flow regime which may conflict with the system of basic equations and constitutive relations. It is planned, however, to implement more detailed constitutive relations in the near future.

The interaction terms presently used in FLUT, which are only preliminary, fulfill basic requirements, e.g. symmetry of phases and correct behaviour for a disappearing phase. Moreover, it is an advantage for the numerical solution to avoid discontinuities.

For the calculation of the mass transfer rate, the heat transport at the interface is calculated by the following expressions:

$$Q_V = c_V \alpha_V \rho_V (h_{Vs} - h_V) \cdot C_c \cdot f_{sV}$$

$$Q_L = c_L \alpha_L \rho_L (h_{Ls} - h_L) \cdot C_c \cdot f_{sL}$$

$$C_c = \sqrt{(\varepsilon_V + \alpha_V^2)(\varepsilon_L + \alpha_L^2)}$$

c_V, c_L	free parameters
C_c	accounts for interfacial area
ε_V, ε_L	small constants for start from single phase situation
f_{si}	function to switch off Q when the remaining phase is stable

The coefficient for interfacial friction is determined by the following expression:

$$K = (\alpha_V C_{KL} + \alpha_L C_{KV}) \cdot \alpha_V \alpha_L$$

$$C_{KL} = C_{KL1}(1 + C_{KL2} \cdot \rho_L \cdot |u_V - u_L|)$$

$$C_{KV} = C_{KV1}(1 + C_{KV2} \cdot \rho_V \cdot |u_V - u_L|)$$

Similar forms of the equations are discussed in the description of the COMMIX code (19).

The four parameters $C_{Ki\ 1,2}$ can be adjusted to match the drag coefficient for small bubbles or small drops respectively.

2.3 Solution Method

The system of partial differential equations is resolved for the time derivatives of the six independent variables α, p, h_V, h_L, $u_V \cdot A$, $u_L \cdot A$. This system of ordinary differential equations is then solved by a fully implicit time integration method, the time step size being controlled by user supplied accuracy criteria (1).

For space discretisation, a finite difference method is applied. Since a staggered mesh is used, additional assumptions are required to supply quantities from neighbouring locations to the actual mesh points and to approximate space derivatives. The range of these assumptions is limited by demands of accuracy and stability.

Donor cell assumptions are used for most of the terms, i.e. picking the quantities from the upstream location. This method, however, does not conserve the shape of a propagating front. In reflood calculations we observed "artificial entrainment" due to this method.

An alternate method that has been implemented in FLUT is the ASWR method (2). At present, the method's capability to calculate steep gradients is improved by inserting Lagrangian points into the Eulerian mesh. It is still under development and has not been used for the calculation of reflood experiments.

3 HEAT CONDUCTION AND HEAT TRANSFER MODEL

3.1 Heat Conduction Model

Fuel rods and heater rods can be simulated by solving the one-dimensional conduction equation for the radial mesh. An arbitrary number of axial nodes can be specified with no coupling between them. Temperature dependent properties for different materials in the radial mesh are supplied as input tables.

The time integration method is fully implicit.

3.2 Heat Transfer Model

3.2.1 Region below and around the quench front. For the nucleate boiling regime, Chen's correlation (3) is evaluated; for simpler calculations, Mc Adams' correlations (4) is supplied as an option. For the case of single phase liquid, only the macroconvective part of Chen's correlation is taken.

Zuber's correlation (5) is used to give an upper limit for the maximum heat flux. The heat flux is multiplied by $(1-\alpha)$, following a suggestion of Griffith et al. (6).

If the surface temperature is above the critical heat flux temperature, the correlation of Hsu (7) is used. The correlation is linked to Zuber's critical heat flux.

The heat is transferred entirely to the liquid phase.

3.2.2 Region above the quench front. For the inverted annular film boiling regime, the mod. Bromley correlation (8) is used. The correlation of Murao and Sugimoto (9), having the void correction term $(1-\alpha)^{1/4}$, is supplied as an option.

Above a void fraction of 0.8 or a critical vapour velocity, given by Plummer's criterion (10), a dispersed flow film boiling regime is assumed. Convection to vapour is calculated by a correlation of Chen (11).

Radiation to liquid drops is calculated by a simplified expression, derived from (12).

Additional heat transfer from the surface to the liquid drops is calculated as:

$$h_{WD} = \frac{\lambda}{\delta} (1-\alpha)^{2/3}$$

The thickness of the vapour boundary layer was chosen as δ=0.12 mm for most calculations, following Plummer's experiences (10).

For the region above the upper quench front, wettet by a liquid film, a heat transfer coefficient of $2 \cdot 10^3$ W/m^{2}K was used in most calculations, multiplied by a diminishing factor for void fractions approaching unity.

3.2.3 Determination of quench front velocity. In order to save computer time the axial discretisation of the heater rods has to be rather coarse (e.g. 6 cm). Therefore, and because axial conduction is not included, additional assumptions are necessary to determine the position of a quench front.

The analytical correlation of Yamanouchi (13) is used for the upper, and Semeria-Martinet (14) for the lower quench front velocity.

The rewetting temperature is taken from Hein (15) as

$$T_o = T_{Sat} + 160\ K + 6 \cdot \Delta T_{Sub}$$

The surface temperature on the dry side is taken from the heater rod segment above, thereby including the precooling effect.

4 PREDICTION OF A FLECHT EXPERIMENT

4.1 Experimental Conditions

One of the first applications of FLUT to a bundle reflood experiment was the blind prediction for FLECHT test No. 31701, which was part of USNRC Standard Problem No. 9, Refs. (16) and (17).

The main dimensions and boundary conditions of the experiment were:

heated length	3,66 m
no. of powered rods	159
initial power	$8,15 \cdot 10^5$ W
axial peaking factor	1,66
radial power profile	flat
system pressure	$2,76 \cdot 10^5$ Pa
cold flooding rate	0,152 m/s
liquid temperature	323 K
initial rod temp. (max.)	1144 K

4.2 Simulation

For simulation in FLUT, the heated length of the bundle was divided into 12 cells of equal length (1 ft.) and two "plenum cells" for inlet and exit.

The 159 powered rods of the bundle were modeled by one average rod, having 8 radial nodes (insulator 5, heater 1, cladding 2). Axially the rod was made up of 60 segments, 5 of them connected to each fluid cell, sharing the same fluid conditions. Since the power fraction may be input differently for each segment the axial profile was easily approximated by 60 nodes. The heat flow from the housing was neglected.

4.3 Discussion of Results

Figs. 1 and 2 show the rod temperatures at different axial positions.

As expected for high flooding rates, there was no significant temperature rise. Turnaround ocurred within 10 s for most of the axial positions. The highest temperatures observed was 1155 K at midplane elevation (1,83 m). The quench time for this position was 58 s.

Fig. 3 gives a comparison of quench front progession between experiment and calculation. For the total quench time the exact value for 115 s was predicted by the blind calculation. The lower quench front progresses too slowly, although there is sufficiently high heat transfer in the inverted annular regime.

The reason for this behaviour is revealed in fig. 4 where the collapsed liquid level is compared to some Δp data: Liquid entrainment is overpredicted during the initial phase of reflood which leads to a reduced water inventory for the rest of the transient.

5 POSTTEST CALCULATION OF AN SCTF EXPERIMENT

The Slab Core Test Facility (SCTF) of Japan Atomic Energy Research Institute (JAERI) was constructed to investigate two-dimensional thermo-hydrodynamics in the core and the communication in fluid behaviour between the core and the upper plenum during the last part of blowdown, refill and reflood phases of a LOCA in a PWR (18).

To meet these purposes, the SCTF was designed to simulate the slab shaped part of the actual PWR core with full height, full radius and one bundle width including the corresponding parts of the upper and lower plena and the downcomer. Internal structure of the pressure vessel is shown in fig. 5. The simulated core consists of eigth electrically heated simulated fuel bundles, arranged in a straight line.

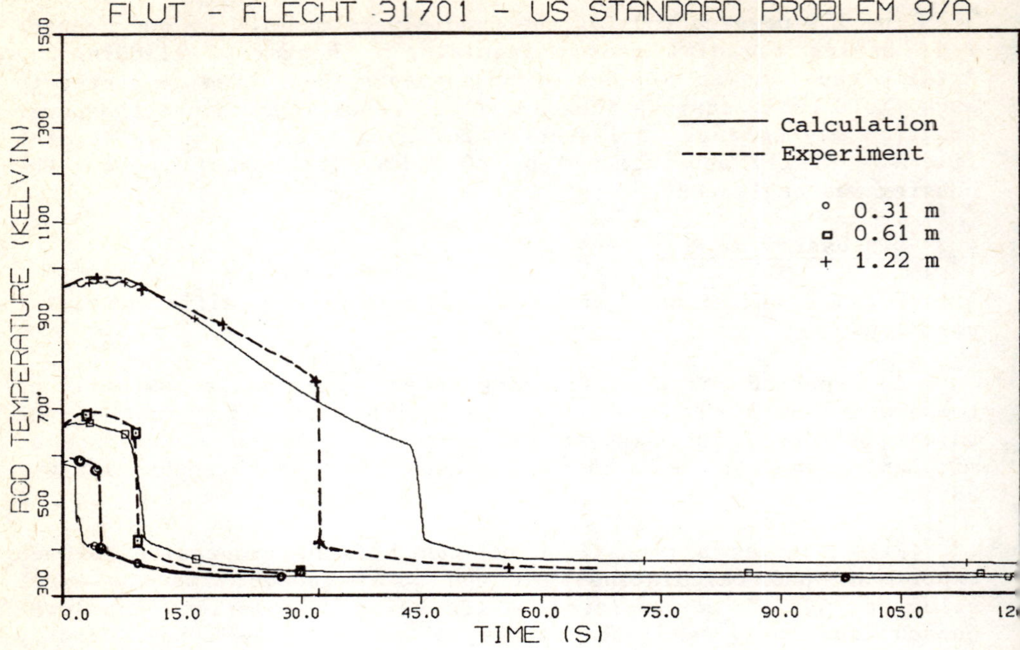

Fig.1 Heater rod temperatures for FLECHT test 31701 (bottom)

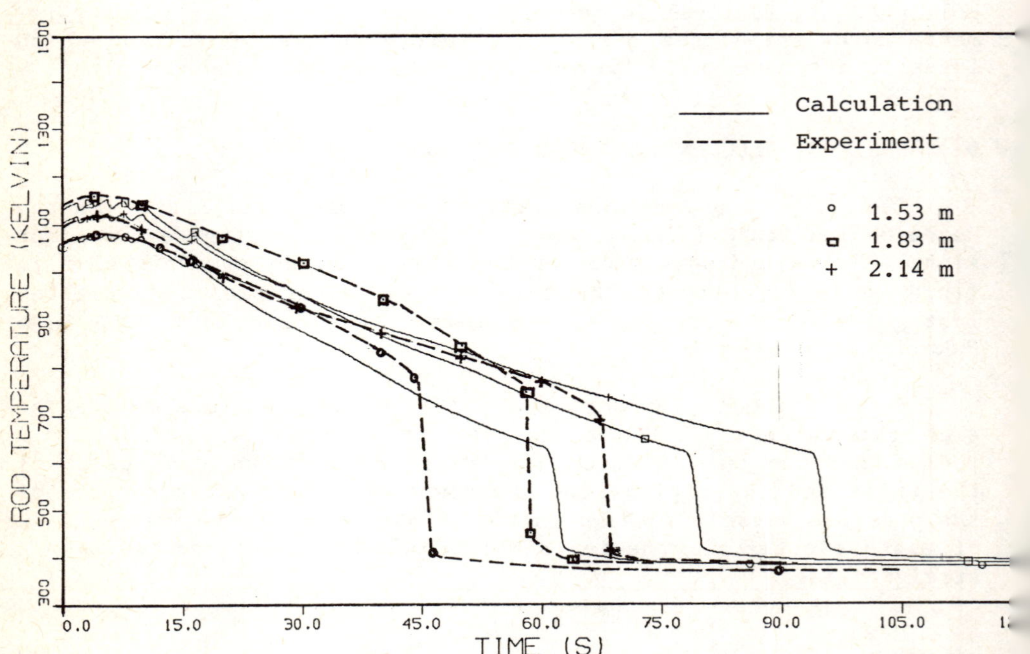

Fig.2 Heater rod temperatures for FLECHT test 31701 (center)

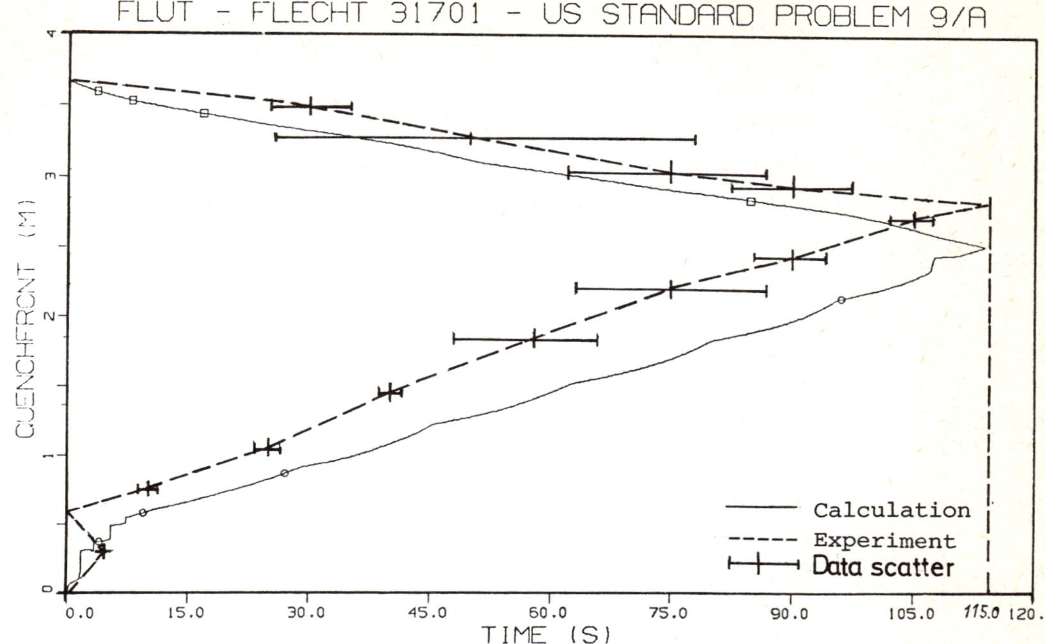

Fig.3 Upper and lower quench front for FLECHT test 31701

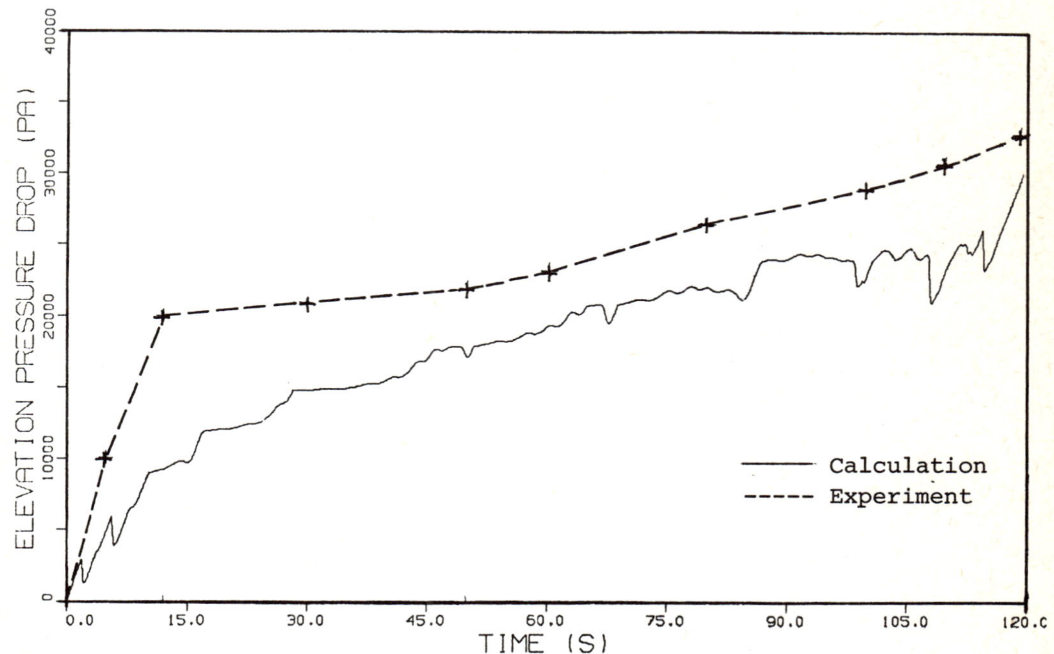

Fig.4 Collapsed liquid level for FLECHT test 31701

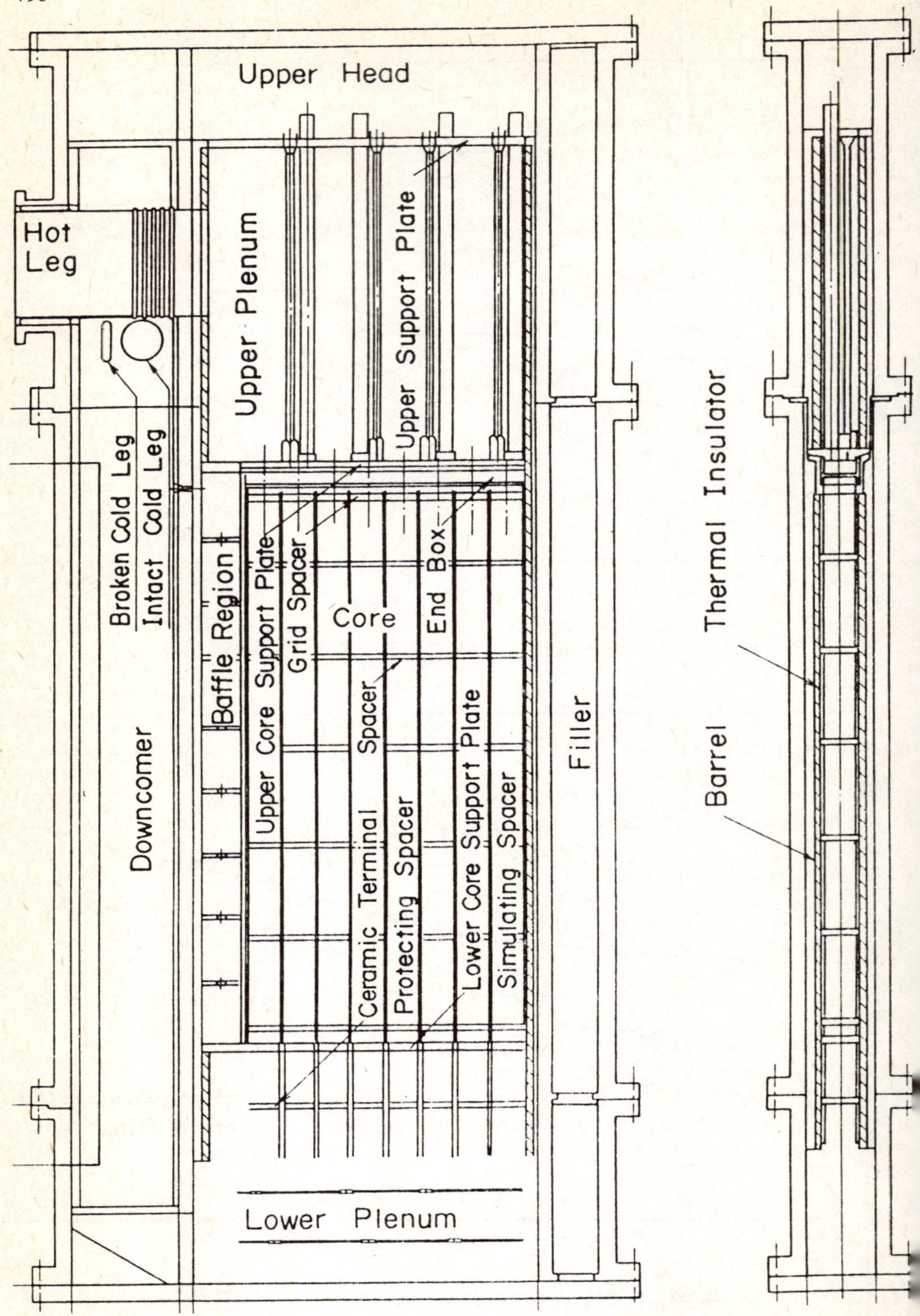

Fig.5 Vertical cross section of the SCTF pressure vessel

5.1 Experimental Conditions

SCTF test S1-01 (run 507) was chosen for analysis by FLUT because it is the base case of a test series with blocked downcomer. It can be considered nearly a forced feed test, although there was some system influence in the results.

The main dimensions and boundary conditions of the experiment were:

heated length	3,66 m
no. of powered rods	1872
initial power	$7{,}035 \cdot 10^6$ W
axial peaking factor	1,40
radial peaking factor	1,06
initial system pressure	$2{,}0 \cdot 10^5$ Pa
cold flooding rate	∿ 0,042 m/s
liquid temperature	∿ 390 K
max. rod temperature at BOCREC, bundle 2	925 K

5.2 Simulation

For the analysis of the SCTF base case test by FLUT it was first attempted to use exactly the same model and same nodalisation as for the previous FLECHT calculation.

While the test comprises a heatup period, the calculation starts at Bottom of Core Recovery (BOCREC) and ends at complete quenching of the bundle. Since FLUT is one-dimensional, the total power was averaged over the eight bundles (880 KW/bundle), which gives nearly the power supplied to bundle no. 2. Therefore the data for initial conditions and for comparison between calculation and experimental results were taken from bundle no. 2.

The nodalisation of fluid channel and heater rod was the same as for FLECHT.

The first calculation with FLUT showed a long delay in quench time. Because of the low flow rate and saturated injection, the overprediction of entrainment was even more pronounced than in FLECHT calculations. Therefore the model was changed to avoid "artificial entrainment" caused by the space discretisation method (see 2.3). Furthermore, the maximum heat flux was restricted to a smaller area around the lower quench front.

The calculation presented here were done with this modified version of FLUT.

5.3 Discussion of Results

Fig. 6 shows the rod temperatures for different axial positions. The temperature rise is calculated, but for some positions the turn around is too early, so that the maximum temperature rise is not quite reached.

Quench temperatures and quench times are in reasonable agreement.

Fig. 7 shows the quench fronts and the collapsed water level. The accumulation of liquid in the beginning is modeled well. Entrainment starts at the right time but the amount is underpredicted.

6 CONCLUSIONS AND FUTURE PLANS

The application of FLUT to forced feed bundle reflood experiments and other sample problems has shown the code's ability to simulate some of the most important reflood phenomena, e.g. entrainment, vapour superheat, and quenching.

Future model development will have to concentrate on the constitutive equations which are only preliminary at present and on their verification by separate effect experiments. Heat transfer models also need further improvement, especially in the dispersed flow film boiling regime.

At the same time, FLUT is being extended from the simulation of a one-dimensional channel to a network of flow paths and connecting cells. This should enable the next version of FLUT to calculate PKL loop experiments.

7 ACKNOWLEDGEMENTS

The development of FLUT is performed under contract of the Federal Ministry for Research and Technology (BMFT).

First knowledge of SCTF data was kindly provided by the Japan Atomic Energy Research Institute (JAERI).

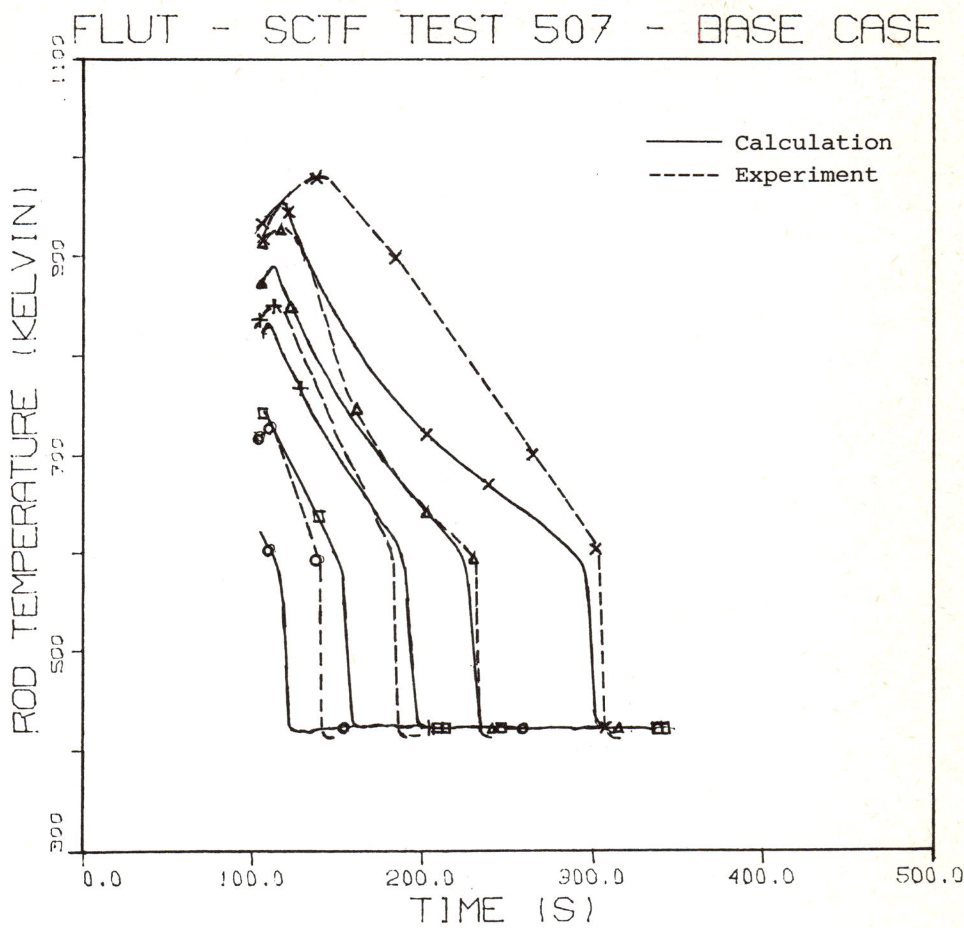

Axial positions:

Calculation	Experiment
° 0.31 m (1 ft)	
◻ 0.61 m (2 ft)	° 0.52 m
+ 0.92 m (3 ft)	+ 0.95 m
Δ 1.22 m (4 ft)	Δ 1.38 m
x 1.83 m (6 ft)	x 1.91 m

Fig.6 Heater rod temperatures for SCTF test 507

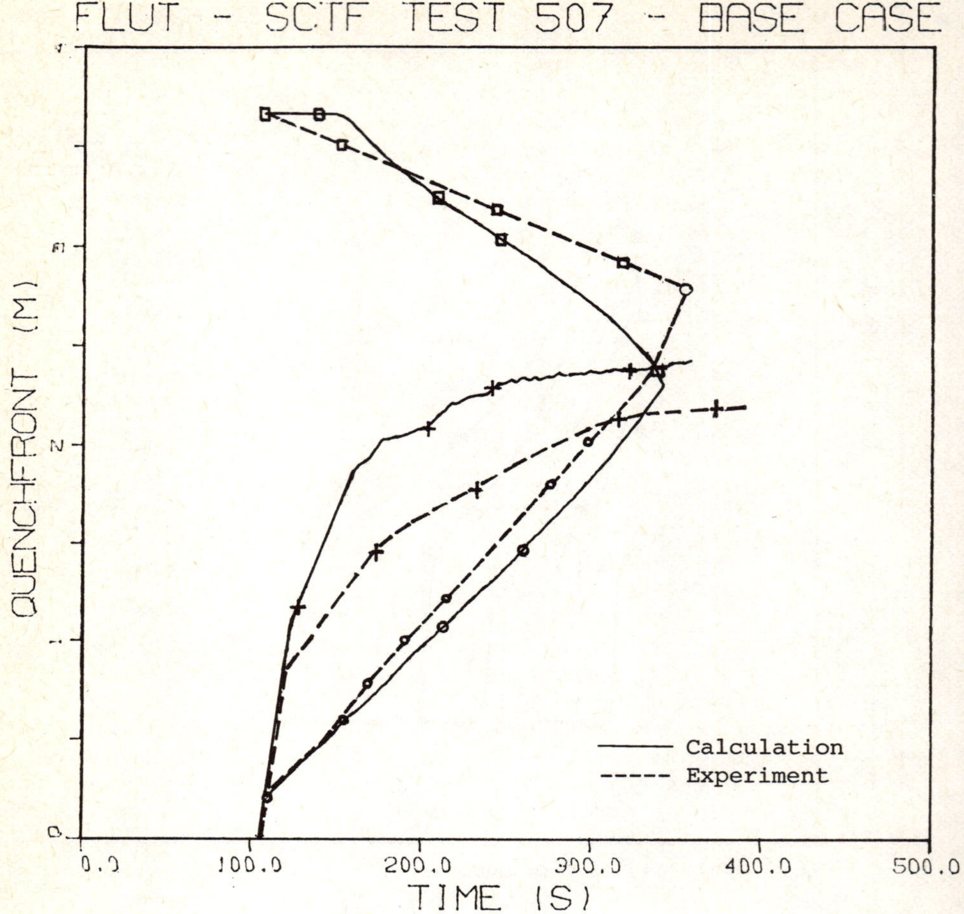

Legend:

- ° Lower Quenchfront
- ◻ Upper Quenchfront
- \+ Collapsed Liquid Level

Fig.7 Quench fronts and collapsed level for SCTF test 507

REFERENCES

1. Hofer E. A Partially Implicit Method for Large Stiff Systems of ODEs with only Few Equations Introducing Small Time-Constants, SIAM J. on Numerical Analysis, vol. 13 (1976).
2. Graf U., P. Romstedt and W. Werner. Application of the ASWR Method to Two-Phase Flow Problems, Second CSNI Specialists' Meeting on Transient Two-phase Flow, Paris (June 1978).
3. Chen C. A Correlation for Boiling Heat Transfer to Saturated Fluids in Convective Flow, ASME Paper No. 63-HT-34 (1963).
4. Mc Adams W.H. et al. Ind. Eng. Chem. 41 (1949) 1945.
5. Zuber N. et al. The Hydrodynamic Crisis in Pool Boiling of Saturated and Subcooled liquids, International Development in Heat Transfer, Part II, No. 27 (1961).
6. Avedisian C.F. and P. Griffith. Report 80620-84, MIT (1974).
7. Hsu Y.Y. Boiling Heat Transfer, Von Karman Institute for Fluid Dynamics Lecture Series 1978-5, Two Phase Flow in Nuclear Reactors, Brussels (1978).
8. Bromley L.A. Heat Transfer in Stable Film Boiling, Chemical Engineering Progress, vol. 46 (1956) 221-226.
9. Murao Y. and J. Sugimoto. Correlation of Heat Transfer Coefficient for Saturated Film Boiling during Reflood Phase Prior to Quenching, Journ. Nucl. Sci. Techn. 18(4) (1981).
10. Plummer D.N. Post Critical Heat Transfer to Flowing Liquid in a Vertical Tube, Ph.D. Thesis, MIT (1974).
11. Chen J.C., F.T. Ozkaynak and R.K. Sundaram. Vapour Heat Transfer in Post-CHF Region Including the Effect of Thermodynamic Non-Equilibrium, Nucl. Eng. Des. 51 (1979).
12. Deruaz R. and B. Petitpain, Modeling of heat transfer by radiation during the reflooding phase of a LWR, Specialists' Meeting on the behaviour of water reactor fuel elements under accident conditions, Spatind, Norway (1976).
13. Yamanouchi A. Effect of Core Spray Cooling in Transient State after Loss of Coolant Accident, Nucl. Sci. Techn. 5 (11) (1968).
14. Semeria R. and B. Martinet. Calefaction Spots on a Heating Wall: Temperature Distribution and Resorption, Proc. Inst. Mech. Engr. 180 (1966).
15. Hein D. Hydraulics during Reflood, Europ. Two-Phase Flow Group Meeting, Glasgow (1980).
16. Hochreiter L.E. et al. PWR FLECHT-SEASET Unblocked Bundle, Forced and Gravity Reflood Task: Task Plan Report NRC/EPRI/Westinghouse Report No. 3 (March 1978).
17. Hochreiter L.E. Westinghouse Electric Corporation Proposal for Standard Problem 9, Enclosure to letter to L.E. Phillips (USNRC) (Febr. 14, 1978).
18. Adachi A. et al. SCTF-Core I Test Results, 9th LWR Safety Research Inf. Meeting, Gaithersburg (1981).
19. Domanus H.M. et al. COMMIX-2, NUREG/CR-1807, ANL-81-10 (1980).

APPENDIX

Conservation Equations for FLUT

1. Vapour Mass Balance

$$\frac{\partial}{\partial t}(\alpha_V \rho_V) + \frac{\partial}{\partial z}(\alpha_V \rho_V u_V) = \Gamma$$

2. Liquid Mass Balance

$$\frac{\partial}{\partial t}(\alpha_L \rho_L) + \frac{\partial}{\partial z}(\alpha_L \rho_L u_L) = -\Gamma$$

Mass Transfer Rate

$$\Gamma = -\left(\frac{Q_V + Q_L}{\Delta h}\right)$$

$- (Q_V + Q_L) > 0$ (Net-evaporation)

$$\Delta h = h_{VS} - h_L, \ \Gamma > 0$$

$- (Q_V + Q_L) < 0$ (Net-condensation)

$$\Delta h = h_V - h_{LS}, \ \Gamma < 0$$

3. Vapour Momentum Balance

$$\frac{\partial}{\partial t}(\alpha_V \rho_V u_V) + \frac{\partial}{\partial z}(\alpha_V \rho_V u_V^2) = -\alpha_V \frac{\partial p}{\partial z}$$

$+ \alpha_V \cdot f_{WV}$ Wall friction

$+ K \cdot (u_L - u_V)$ Interfacial friction

$- \alpha_V \rho_V g \cos \theta$ Gravitation

$+ \Gamma \cdot u_L$ (for $\Gamma > 0$)
or
$+ \Gamma \cdot U_V$ (for $\Gamma < 0$) Phase change

4. Liquid Momentum Balance

(analogous form)

APPENDIX (cont.)

5. Vapour Energy Balance

$$\frac{\partial}{\partial t}\left[\alpha_V \rho_V \left(h_V + \frac{1}{2}u_V^2\right)\right] + \frac{\partial}{\partial z}\left[\alpha_V \rho_V u_V \left(h_V + \frac{1}{2}u_V^2\right)\right] - \alpha_V \frac{\partial p}{\partial t} =$$

$$K \cdot (u_L - u_V) \cdot u_V$$

$$- \alpha_V \rho_V g \cos\theta \cdot u_V$$

$+ \alpha_V K \cdot (u_V - u_L)^2$ Dissipation by interfacial friction

$+ W_{aV}$ Wall heat transfer

$+ Q_V$ (see mass transfer rate)

$+ \Gamma \cdot (h_{VS} + \frac{1}{2}u_L^2)$ (for $\Gamma > 0$)

or

$+ \Gamma \cdot (h_V + \frac{1}{2}u_V^2)$ (for $\Gamma < 0$)

6. Liquid Energy Balance (analogous form)

NOMENCLATURE

A	flow area
f	friction factor
g	gravitational acceleration
h	enthalpy
h_{WD}	heat transfer coefficient wall to drop
k	interfacial friction coefficient
p	pressure
Q	interfacial heat flux
t	time
T	temperature
u	velocity
W_a	wall heat flux
z	space coordinate
α	void fraction
δ	thickness of vapour layer
Γ	mass transfer rate
λ	thermal conductivity
ρ	density
θ	angle of inclination

Subscripts

L	liquid
S, SAT	saturation
Sub	subcooling
V	vapour
W	wall

RESEARCH IN TWO PHASE FLOW AND HEAT TRANSFER TO ASSIST THE MANAGEMENT OF NUCLEAR REACTOR ACCIDENTS

R. B. Duffey and M. Merilo

Electric Power Research Institute
3412 Hillview Avenue
Palo Alto, California 94304

ABSTRACT

As a result of the accident at Three Mile Island, a number of new requirements have been imposed on nuclear power plants and their operations. Many of these imply an increased awareness of the role of two-phase flow and heat transfer in operating procedures for accident and transient conditions. As a result, safety research has changed its focus from hypothetical large break accidents to the evaluation of more probable accidents where operator action is of crucial importance. This paper emphasizes the research aspects of these topics, giving a brief indication of current industry status.

1. INTRODUCTION

There is great motivation, particularly during an accident, in preserving a utility's investment in a nuclear plant. The equivalent economic loss from an accident like at TMI-2 (\$$10^9$) considerably outweighs the public exposure to radiation (15 CI* and no immediate fatalities).

The major concern is loss of adequate water inventory in the primary circuit, leading to core uncovery, heat-up, and potentially extensive core damage. Thus, many new requirements have been proposed by U.S. licensing authorities which have the same basic thrust, and imply an increased awareness of two-phase flow and heat

----

*One death is variously valued in the legal and insurance arena as between 10^5-10^6\$, based on lifetime earning power, etc.

transfer in operating procedures for accident and transient conditions. These include:

1. reactor water level and inadequate core cooling instrumentation;
2. installation of saturation "meters" and cool-down rate limitations to avoid additional boiling or vapor formation;
3. resetting (in BWRs) of vessel level instrumentation and their verification by testing;
4. safety and relief valve position indicators, and performance tests;
5. provision of vents in the vessel head for bubble removal;
6. monitoring of Critical Safety Functions in the control room;
7. procedures that emphasize water inventory in small break accident conditions (e.g., pump operation and tripping, mandated safety and emergency injection);
8. training of operators for natural circulation conditions;
9. improvements in small break modeling;
10. recomputation of small break trasients for procedure writing and operator response guidelines.

These requirements all need some degree of validation, and hence research and development to demonstrate feasibility, degree of improvement, ranking of importance, and development needs, if any, in these "fixes" or desirable features.

As a result, safety research has changed its primary thrust also, from being biased towards hypothetical large accidents, that might bound large events, to the evaluation of more probable accidents and sequences where operator action is of crucial importance. Helpful further studies, e.g., on probabilistic risk analysis (PRA), define which paths or sequences have greatest probability of occurrence, and safety (but not economic) significance. Discussions on so-called Class IX events (core melt and debris bed cooling) assist in determining sheltering, evacuation, and emergency response policy and planning, whilst providing guidance for some important additional mitigating features on the plant, such as for hydrogen control and management.

We can therefore trace a common thread in current reactor safety research and development. This consists of providing adequate operator guidance to determine the true plant safety and functional status, particularly for determining the causes, rates, and consequences of coolant inventory loss. Consequently the requirements for adequate heat sink capability can be determined.

In this paper we emphasize the research aspects of these topics, giving a brief indication of current industry status, and the extent to which these so-called "high risk" or more probable accidents are being studied.

2. PROVIDING ADEQUATE OPERATOR GUIDANCE

It has been established that one can consider operator response as of three complementary types:

(1) skill based, from the experience and learning derived by handling similar situations;

(2) rule based, wherein the actions are based on given written procedures;

(3) knowledge based, whereby the fundamental understanding of the physical principles of the plant and its operation are paramount.

All three have drawbacks. The first is derived from prior experience and the adequacy of training and simulation. The second requires all circumstances be covered by writing, which actually then means only simplified guidelines ("paths") can be given. The third requires an appreciation in real time of all complex responses of the plant even under degraded instrumentation and/or plant conditions. Nevertheless, the current trend of providing more computerized operational support systems must derive from a firm basis of knowledge and understanding plant behavioral principles.

Plainly, as far as the details of heat transfer and two-phase flow are concerned, the operator sees little. The plant process parameters are all on dials, gauges, and recorders. Thus, to overcome this immediate skill barrier, we can supply an engineering sense of the plant by defining the plant status in relation to its critical Availability and Safety Functions. The plant systems may then be grouped according to such functional requirements, and the plant operated to ensure that plant limits and flow requirements, whether licensing or operationally derived, are met.

The present requirements of NUREG 0696 for Safety Parameter Display Systems are being met by the nuclear industry, and can be regarded as a simplified subset of a complete plant-wide parameter monitoring system (DASS). The verification testing and development of such systems is being carried out on reactor simulators. This ensures that the displays, speed of response and information guidance are adequate for the purpose desired--that of improved safety via improved operation.

It should be pointed out, however, that no matter how the information is "packaged" for the operators, the quality of this information is critically dependent on the adequacy of the plant instrumentation. This point is crucial in an accident where the instruments are required to perform under conditions which can be entirely different than normal operation.

Operator training, plant engineering studies, and procedure development rely on the use of simulators with a high fidelity vis-a-vis plant response. It is now necessary to take simulators into regions well beyond the design basis of the system. This taxes the fidelity of the current generation of simulators. Upgrading of simulators is underway in the industry to permit two-phase saturated conditions, complex interactive failures, inadequate core cooling, and natural circulation to be studied. Real-time simulators must have more simplified plant representation than nodal accident and transient analysis codes embodied by RELAP [1], RETRAN [2] and TRAC [3, 4] which were designed to analyze large break LOCAs. The so-called modular approach, MMS [5], to plant wide simulation then appears very attractive, both for speed of computation and ease of modeling by building a library of modules.

For small breaks and complex transients, several additional simulation requirements are evident. The physics demand that one models long duration transients with stratified and separated flows, steam generator secondary side transients, two-phase natural circulation, and the circulation and accumulation of permanent gases in the primary loop. Most of these phenomena are within the capability of present technology to handle, provided one is seeking trends in plant behavior (e.g., heating, cooling, pressurization, inventory, systems availability and effect) and not requiring excessive and unnecessary accuracy.

Plainly an actual plant will respond differently to a simulator not only due to modeling uncertainties, but also the uncertainties in the as-built condition (e.g., in flow resistances, and in instrumentation failures and accuracy).

3. MAINTAINING ADEQUATE CORE COOLING

During normal operation, heat generated in the reactor core is transferred out of the primary system through the steam generators to the secondary side. The thermal-hydraulics on both sides of the steam generator are strongly coupled, and determine the heat removal ability under both normal and off-normal operation. Consequently transients both on the primary and secondary side affect the overall operation. As an example, for a once through steam generator the heat removal capability is directly related to the secondary side water level even during normal operation.

The initial perturbation at TMI-2 was a loss of both the main and auxiliary feedwater on the secondary side. When the secondary side water had boiled off, heat removal capability was lost from the core by this means and had to be reestablished by the relief valve operations, safety injection, and makeup and letdown. Studies of the primary and secondary side interactions due to transients on either side have been and continue to be performed both through scaled experiments [6, 7] and analysis [8]. A schematic diagram of the scaled U-Tube Steam Generator Simulation Facility is shown in Figure 1.

When there is water on the secondary side of the steam generator, heat can be removed from the primary side by forced convection when the pumps are running, and by single phase and two phase natural circulation when they are off. The elevation difference between the core and the steam generators is all that is needed to exploit the buoyancy forces acting on the heated fluid in the core. Indeed, TMI-2 has been cooled by single phase natural circulation for several years. This mode of heat removal is a preferred, stable, and passive mode.

If there is a substantial loss of inventory from the primary coolant system then single phase natural circulation is replaced by two phase circulation. Experiments have been conducted on both single phase and two phase natural circulation in small scale models of both once through and U-Tube steam generators [9, 10]. Larger scale experiments have been performed in LOFT and, during start up tests, in actual nuclear plants. A diagram of a small scale two loop simulation facility with U-Tube steam generators is shown in Figure 2.

The results of experiments performed to date show that there are two fundamental stable modes of cooling under two-phase conditions. If the hot legs are covered there is a flow of two-phase mixture through the primary system. Conversely, if the inventory in the primary system is low enough to uncover the hot legs, then the heat rejection to the steam generators switches into the reflux condensation mode. In this mode, the local decrease in pressure created by steam condensation continuously draws steam from the vessel, and the condensate runs back to the vessel in countercurrent flow to the steam. The flow in the hot legs under these conditions is stratified. It is clear that homogeneous codes, or simple models, cannot represent the details of this type of counter current flow. Either drift flux or a two fluid formulation is necessary.

It should be noted that boiling water reactors (BWR) operate stably and routinely to power levels of the order of 30% of full power. When the downcomer is uncovered as the two-phase level falls below the separators, then internal circulation within the vessel is possible. In this latter mode, water in the core bypass feeds the boiling flow in the core bundles.

One of the major results to date from natural circulation experiments is that both two-phase natural circulation and reflux condensation are very tolerant of non condensible gases. Thus, stable cooling with the core covered is possible in a PWR even with boiling and two-phase flow on the primary side. The system generally attains an equilibrium condition in pressure and temperature when the secondary heat removal simply matches the primary heat generation.

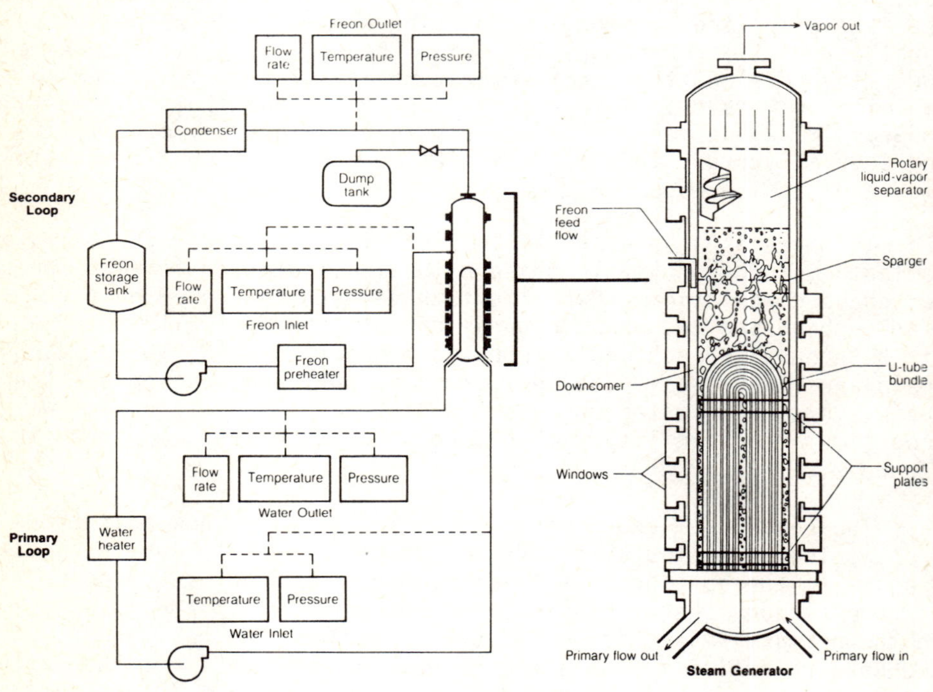

Figure 1. Steam Generator Simulation Facility.

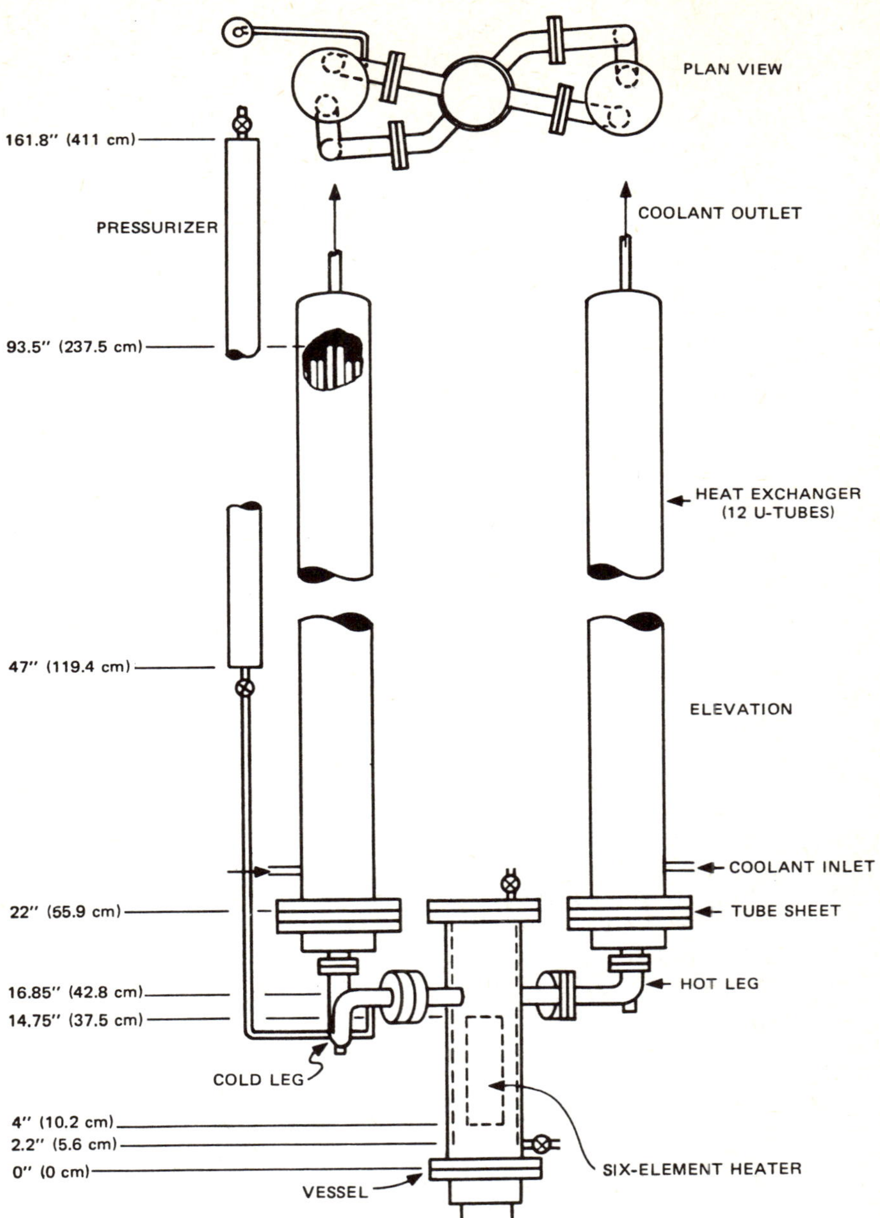

Figure 2. Two-Loop Natural Circulation Test Facility

4. DETERMINING PLANT CONDITIONS DURING OFF-NORMAL OPERATION

As pointed out previously, the information available to the operators on the nuclear plant conditions must be obtained from the plant instrumentation. Since the existing instruments were generally selected, designed, and calibrated for normal operating conditions they are frequently out of range or even nonexistent for parameters which are important during off-normal transients. For example, the existing flow instrumentation is not sensitive enough to quantify, or even detect natural circulation flows. Also, until very recently at least, PWRs had no means of detecting water level in the vessel or inadequate core cooling in the core.

To understand the primary system heat transfer conditions an assessment of both the coolant inventory and the flow is required[11].

This is illustrated in Figure 3, where a conceptual map of adequate and inadequate core cooling, (ICC) is shown for a PWR on a plot of coolant inventory versus flow. From this we see that ICC occupies a small region where approximately 80 percent or more of the primary system coolant inventory has been lost for boiloff under stagnant conditions, and even more for the situation where the pumps are running.

Pump characteristics also affect the system operation and can be shown on the inventory flow map. These indicate conceptually the degradation of the volumetric flow through the pumps as a two-phase mixture with increasing void is pumped through the primary system. We see that the volumetric flow does not decrease until steam enters the coolant loops and starts to affect the pump performance. This means that water can be lost from the pressurizer, or even the upper plenum of the reactor pressure vessel until the hot legs uncover, without any degradation of coolant flow.

At zero flow, even though the coolant inventory may be less than that required to cover the core with single phase liquid, it can still provide adequate cooling if the steam generated in the liquid raises the two-phase level sufficiently to cover the core.

Therefore, it is clear that ICC conditions can be reached through loss of both inventory and flow, and that these variables can define the "approach" to or "early warning" of ICC.

A large number of measurement concepts and instrumentation system designs have been suggested for the measurement of parameters that might indicate an approach to or the existence of inadequate core cooling conditions in PWRs. Only a few of the many proposed systems have actually been tested on reactors or under reactor conditions. These include level measurement by external neu-

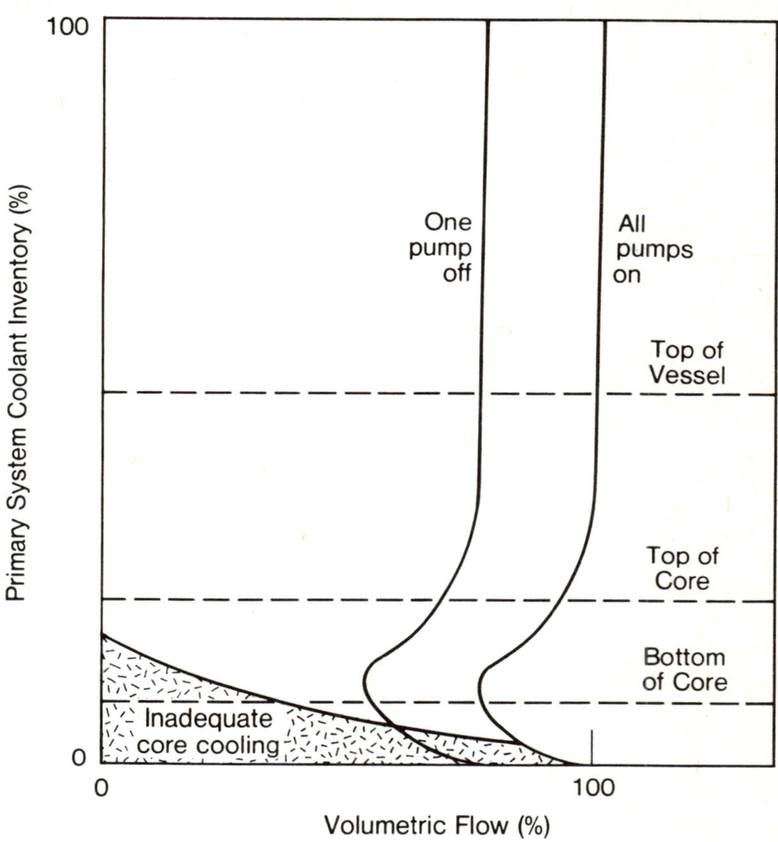

Figure 3. Conceptual Inventory-Flow Map

tron detectors, differential pressure transducers and heated junction thermocouples. All of these suffer from one or more problems [11] including lack of sensitivity, interpretation of results when the pumps are running, or incomplete coverage of the vessel.

Other work on upgrading plant instrumentation is also in progress [12]. For flow measurement and void detection ultrasonic methods are quite promising.

They are particularly attractive because they are totally non intrusive, and no penetration of the pressure boundary is required. The problems associated with these devices involve survivability and operation at high temperatures, and attenuation and interpretation of the signals for two phase flow.

5. OPERATOR ACTIONS

The operator is given indications in a PWR control room of the onset of saturation conditions on the primary side (from pressurizer pressure, hot leg temperature and saturation "meters" or curves). The operating procedures call for maintenance of an effective heat sink, via

1) steam generators in PWRs;
2) safety injection;
3) make-up and let-down;
4) primary system relief valve operation.

In conjunction with core outlet thermocouple and reactor vessel level indication, the onset of inadequate core cooling by core uncovery due to excessive inventory loss can be determined and remedied. Experiments and plant simulations are aimed at aiding operator guidelines, action, and procedures. These particularly determine the timescales and various core-cooling options which are available. As discussed previously, water level or inventory measurement systems are being installed in PWRs, and differential pressure based level measurement systems already exist in BWRs.

6. CORE UNCOVERY

If the fluid inventory falls sufficiently, as predicted by various BWR and PWR small break analyses, the two phase level will drop and uncover the reactor core. As at TMI-2, this can occur only if the ECC injection systems fail to maintain the liquid inventory.

The portion of the core below the two-phase mixture is well cooled by convection and nucleate boiling in water. The uncovered upper region is only cooled by the steam flow from the boiling and flashing in the lower water-covered region.

The analysis of the uncovery proceeds from the estimation of the two-phase mixture height, utilizing a void fraction correlation. This is necessary because the vapor separates from the two phase region due to gravitational forces. Many analyses have been conducted for the TMI-2 case, this being the most widely publicized and documented.

The results of such an analysis [13] for a single tube is shown with the experimental data on Figure 4.

For BWRs with internal circulation, the loss coefficient of the bundle side-entry orifice is important determining the bundle inlet flow for cases where the high pressure core spray is inactive. Given a proper representation of the system, good agreement between drift flux theory and full-height bundle experiments can be obtained. The sensitivity to the exact form of the void fraction correlation (i.e., drift-flux versus empirical) is reduced at the low powers and flows, and the high pressures (∿1000 psia) of interest to small break transients in BWRs and PWRs. A direct analytical solution is then possible for the level trajectory.

This ability to accurately track the level leads to the intriguing possibility of advising the operator--in real time--of the core level status. This is only possible if all the systems providing water to the reactor, including pump seal leakage and PORV flows, are known.

For BWRs and some PWRs, core uncovery can be mitigated by the action of (a) core sprays, (b) upper head, (c) upper plenum or (d) hot leg injection above the core. To study the spray distribution, full-scale experiments are performed since the spray is strongly affected by geometric and environmental conditions. These must be coupled with multidimensional analyses of the spray and water subcooling distribution, since these determine the available cooling of the core.

For some accident scenarios, fluid may be held up in the upper plenum by the steam flow from the core. This is a counter-current flow condition common to partially-flooded tubes. The reactor level indicators would give an ambiguous reading for this case. It has been found, however, that injection of cold water, e.g., from the HPCS, can cause the steam to condense and hence cause sudden entry of the water into the channels. The processes of turbulent mixing and interfacial condensation are important for detailed modeling of these phenomena. Typical results of a two-fluid analysis [14] of such an experiment [15] are shown in terms of subcooling contours in Figure 5.

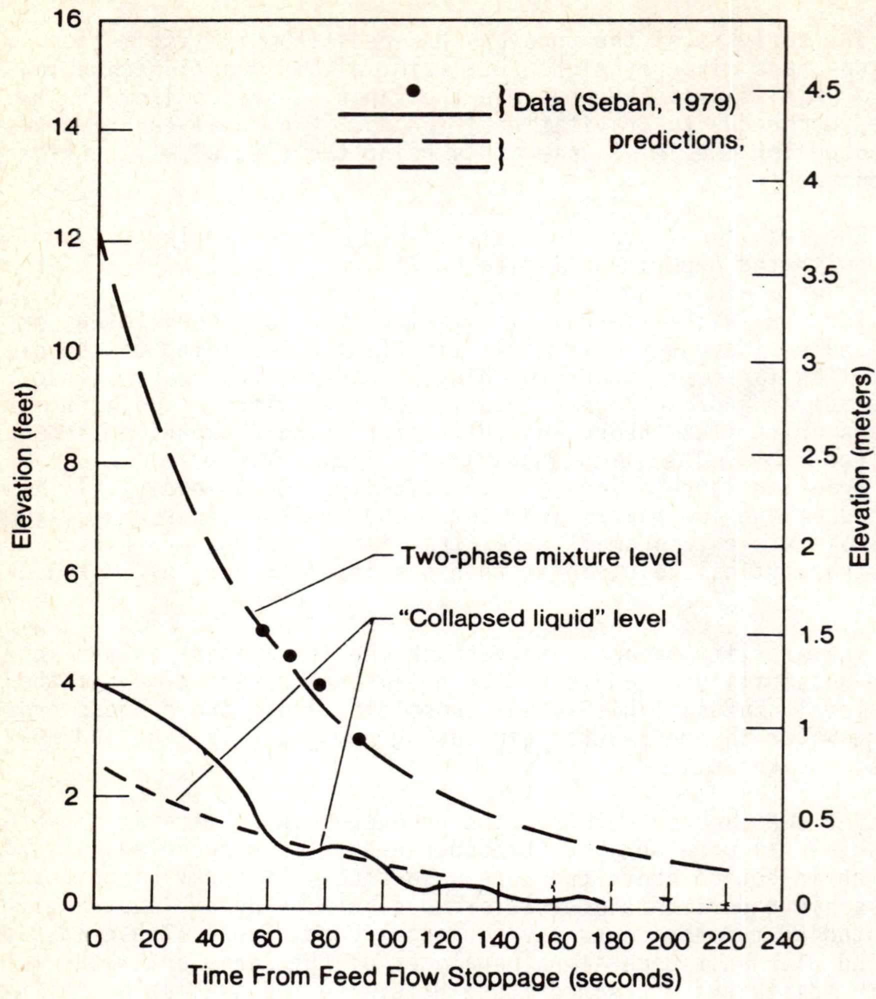

Figure 4. Single Tube Boiloff Experiments and Prediction

7. COOLING DISTORTED CORES

The TMI-2 reactor is thought to have a highly degraded core geometry, but is still coolable. It has been known for some time that Zircaloy deformation causes extensive subchannel blockage. In bundle tests of Zircaloy tubes, up to 90% local blockage has been obtained. It is also known that there is a dynamic interaction between the applied stress on the clad, its swelling rate, and the external heat removal.

The analysis of cooling distorted regions stems from considering the flow diversion between parallel blocked and unblocked regions. The local pressure drop balance and flow resistance in the blocked and unblocked regions determines the flow split. The reduction in flow is at least partially offset by increases in the undistorted heat transfer due to a combination of (a) crossflow both upstream and downstream, (b) downstream recirculation, and turbulence in the wake of the blockage, and (c) effects on the thermal boundary layer of the flow contraction and expansion. This enhancement is of order one to three times, is highly localized to within a few diameters of the blocked region and is a function of flowrates and regime. Most experiments to date have been performed on relatively small bundles and interconnected subchannels. Two exceptions are the large scale studies using 2000 pins in Japan and the 12 ft. long fuel bundle, in-reactor, reflood experiments in the NRU reactor at Chalk River [16]. Both these experiments show good coolability. Progressing to more severe fuel damage extensive experimental and analytical programs have been started on the coolability of rubble beds.

8. AEROSOL TRANSPORT, RETENTION, AND SCRUBBING

If a nuclear accident continues without any of the emergency core cooling systems functioning,fuel could start melting releasing core materials into the primary system in the forms of gases and vapors. The former can be expected to experience essentially no primary system retention. Of the latter, refractory material (such as metal oxides) vapors can be expected to condense in the immediate region above the melt both onto surfaces at colder temperatures than the hottest portions of the core and unto particles, by homogeneous and heterogeneous nucleation. Vapors of more volatile materials may exist as vapors or aerosol particles throughout the primary system and may, in fact, change phase during transport. The aerosol transport mechanisms expected to play a role are well understood in a microscopic sense, that is, as expressions for transport velocities of individual particles. Much less well understood, however, is their superposition onto the complex gas flow patterns that must be expected almost everywhere in the primary system.

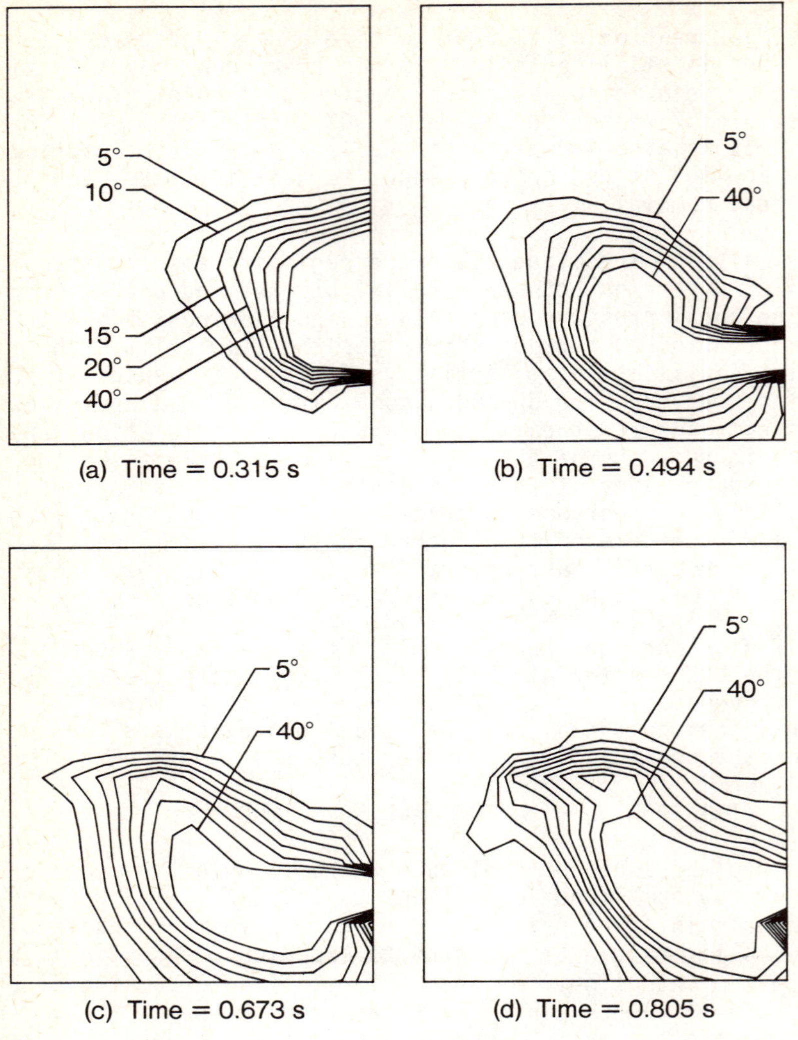

Figure 5. Transient Liquid Subcooling Contours

Consequently, experiments have been started on aerosol transport and retention in reactor primary system components. The microscopic transport mechanisms which are expected to play a role are sedimentation, Brownian diffusion, thermophoresis, inertial and diffusive deposition from turbulent flow, inertial impaction on obstructions, and resuspension of wall deposits.

Aerosols which are released from a reactor primary system often must bubble through pools of water, such as BWR suppression pools, PWR pressurizers, and quench tanks. Substantial attenuation of the radionuclide transport is expected from scrubbing in these pools. Therefore experiments to determine realistic scrubbing decontamination factors have been started [17].

CONCLUSIONS

The various cooling mechanisms for ensuring adequate core cooling for high risk accident sequences have been discussed. In addition, the importance of operator actions and alternatives available to him are presented. It is evident that significant research and development is necessary to support this understanding of the evaluation of reactor accidents in order to minimize excessive economic loss and undue radiation exposure.

REFERENCES

1. Ransom, V. H. et al "RELAP 5/MOD 1 Code Manual" NUREG/CR-1826, 1980.

2. Moore, K. V. et al "RETRAN-A Program for One-Dimensional Transient Thermal-Hydraulic Analysis of Complex Fluid Flow Systems", EPRI CCM-5, 1978.

3. Safety Code Development Group, "TRAC-P1A, An Advanced Best-Estimate Computer Program for PWR LOCA Analysis", Los Alamos Scientific Laboratory.

4. Spore, J. W. et al "TRAC-BD1: An Advanced Best Estimate Computer Program for Boiling Water Reactor Loss-of-Coolant Accident Analysis", NUREG/CR-2178.

5. DiDomenico et al "Power Plant Performance Modeling: Dynamic Model Evaluation", EPRI CS/NP 2086, 1981.

6. Kalra, S. P., Duffey, R. B., and Adams, G., "Loss-of-Feed Water Transients in PWR U-Tube Steam Generators: Simulation Experiments and Analysis", AIChE Symposium Series, Vol. 199, pp. 36-40, 1980.

7. Kalra, S. P., and Adams, G., "Thermal-Hydraulics of Steam Line Break Transients in Thermal Reactors--Simulation Experiments", ANS Trans., Vol. 35, pp. 297, 1980.

8. Kalra, S. P., "Dynamic Thermal-Hydraulic Behavior of PWR U-Tube Steam Generators--Simulations Experiments and Analysis", Electric Power Research Institute Report EPRI NP1837-SR, 1981.

9. Kiang, R. L., "Two-Phase Natural Circulation Experiments in a Test Facility Modeled After Three Mile Island Unit-2" EPRI NP-2069, 1981.

10. Kiang, R. L. and Marks, J. S., "Two-Phase Natural Circulation Experiments on Small-Break Accident Heat Removal", EPRI NP-2007, 1981.

11. Merilo, M. and Bailey, P. G., "Some Coolant Inventory Measurement Methods for Pressurizer Water Reactors", ASME 82-NE-38, 1982.

12. Merilo, M., "Electric, Ultrasonic, and Microwave Measurements in Two-Phase Flow", Trans. of the ANS, pp. 1001, 39, 1981.

13. Sun, K. H., Duffey, R. B. and Peng, C. M., "The Prediction of Two-Phase Mixture Level and Hydrodynamically Controlled Dryout Under Low Flow Conditions", Int.J. Multiphase Flow, pp.521,7,1981.

14. Sargis, D. A., Khadem, M., and Merilo, M., "Correlation and Multidimensional Analysis of Scaled CCFL Breakdown Tests", ASME 82-WA/HT-9, 1982.

15. Merilo, M., Colah, M., and Duffey, R. B., 1979, "Condensation Induced Transition From Bubbling to Liquid Downflow in a Turbulent Two-Phase Pool", 18th National Heat Transfer Conference, San Diego.

16. Mohr, C. L. et al, "Loss of Coolant Accident Simulations in the National Research Universal Reactor", NUREG CR-1208 (1981).

17. Flanigan, L. J. et al, "Radionuclide Scrubbing in Water Pools: Bubble Hydrodynamics" Second International Topical Meeting on Nuclear Reactor Thermal-Hydraulics, January 11-14, 1983, Santa Barbara, California.

PRACTICAL ASPECTS OF CONDENSER AND REBOILER DESIGN

Jerry Taborek
Heat Transfer Research, Inc.
Alhambra, California

ABSTRACT

The selection of the most suitable type of condensers and reboilers within given restraints of operating conditions (usually a compromise between conflicting requirements) is a crucial problem in proper design. Even with the use of sophisticated computer programs, this shall remain an engineering-experience decision. The main elements of selection are shell types, inside versus outside tube flow, vertical versus horizontal orientation, plain or finned tubes, ease and frequency of cleaning, start-up conditions, etc. Guidelines are presented for the application of the latest research in condensation and boiling to obtain the most effective design.

INTRODUCTION

Correlational equations for condensation and boiling are as a rule developed on laboratory type apparatus and under often ideal, well-controlled conditions. However, if applied to industrial heat exchanger design, proper interpretation of such methods and sometimes substantial correction may be required. Typical examples are (a) the classical Nusselt analysis of condensation which assumes conditions rarely encountered in practice and (b) nucleate pool boiling data based on single tube, which need substantial corrections for complex flow in reboiler geometries. The science and the art of heat exchanger design are to adjust these basic equations to a variety of heat exchanger construction types, while at the same time respecting a number of provisions for operational practices, controls, start-up conditions, etc.

The increasing availability of complex computer programs will make the often very tedious calculational processes easier for the design engineer and thus permit investigation of alternate solutions. However, the ultimate design for a well-operating heat exchanger requires considerations which are impractical or impossible to express in the format of computer logic. Engineering judgment remains the final criterion.

This paper attempts to focus on some of the practical aspects connected with the operational characteristics as they influence the selection and design of condensers and reboilers. The importance of this subject is supported by the observation that the *really serious* problems with two-phase flow exchangers in industry are mostly not due to lack of knowledge of the basic phenomena, but rather poor attention (and experience) to the overall design features. While there is still much to be learned about two-phase flow from rather fundamental experimental work, such results will improve the accuracy of exchanger design but will not eliminate the basic errors of judgment.

For presentation at the NATO Advanced Research Workshop on Two-Phase Flow Heat Transfer, Munich, 1982.

PROCESS CONDENSERS

Under this term we understand condensers used for all types of duties commonly encountered in the broad spectrum of process industrial applications, such as refineries, petrochemical, chemicals, refrigeration, as well as some power plant applications, especially systems using organic fluid (geothermal, solar), waste heat recovery, and similar. The design of process condensers is difficult because of the multiple choices of condenser types with respect to operational and construction parameters. There is virtually no one single selection, and the best choice in most cases is a compromise between some conflicting demands. The main interactive parameters are illustrated in the following:

Vapor:
- a) Single component (isothermal). Typically steam, refrigerants, or pure distillation products.
- b) Multicomponent mixtures, all condensable (condensing temperature range).
- c) Vapors with noncondensable gases (temperature range and large change of coefficients between inlet and outlet).

Flow Regimes: Gravity- or shear-controlled, annular, bubbly, slug flow, etc., requiring flow regime maps.

Geometry: Inside or outside of tubes in vertical or horizontal orientation.

Tube type: Plain: external crossfin (horizontal); internal longitudinal fin; grooved, fluted, or rope tubes.

Heat Duty: Vapor can enter either superheated, at dew point, or as two-phase flow mixture. Vapor can exit as fully condensed, partially condensed, or with condensate subcooling.

Coolant: Most frequently no-phase-change fluid (water). If boiling coolant is used, this results in more restrictive considerations.

A. Operational Variables.

The following is a summary of the considerations for condenser type selection:

1) Allocation of Vapor and Coolant. Several considerations and their combination dictate this most important selection criterion.

a) If vapors are corrosive and/or at high pressure, tubeside condensation is highly preferable.

b) If coolant is water, fouling and consequently ease of cleaning indicate tubeside coolant as best choice.

If the combination of (a) and (b) conditions is present, shellside cleaning will usually be accepted as the necessary evil, unless very frequent cleaning is required.

c) If coolant is a boiling liquid, considerations must be given equally to advantages and disadvantages of both processes. As a general rule it is preferable to respect the laws of gravity and use a vertical unit with boiling in upflow and condensate in downflow.

2) Temperature Profile. In some cases of close temperature approaches between the vapor and the coolant, a strict counterflow may be unconditionally required, as otherwise (multitubepass) a temperature "pinch" may be present. In such cases this requirement will override any other considerations.

3) Desuperheating. Tube wall temperature should be calculated to decide if "wet" or "dry" wall condition exists. Small desuperheating duties are often neglected as the additional calculational procedures are not warranted. In most cases the "wet" wall condition will exist and essentially the condensing coefficient will apply. If "dry" wall conditions exist, the coefficient may be very low, unless special provisions are made, such as in feedwater heaters.

4) Subcooling of Condensate. Small subcooling duties can be accommodated economically (one exchanger) in some types of units (as described later), but the subcooling heat transfer coefficients are usually very low unless special provisions are taken, such as in feedwater heaters. However, if more substantial subcooling duties are required, consideration should be given to use a separate exchanger properly designed for this service.

5) Noncondensable Gas-vapor Mixtures. If noncondensable gases are present, the heat transfer coefficient may vary by an order of magnitude from vapor entry to vapor exit. Stepwise calculations are unconditionally required.

Furthermore, in many vapors small traces of noncondensables are present, especially under vacuum conditions. These must be removed as they otherwise would accumulate and deteriorate the performance (steam condensers). Unit designs with effective provisions for the gas removal must be selected, requiring venting outlets at locations where the gases are easily separated from the condensate. Failure to accommodate such provisions constitutes one of the most frequent reasons for poor condenser performance.

6) Flow Regime Determination. The classical Nusselt analytical solution based on laminar film buildup is unfortunately valid only under limited conditions of very low vapor velocity. Much more typically, the vapor enters at high velocity, creating a highly turbulent rather than laminar film. As the vapor velocity decreases, the condensate layer is already turbulent; thus never reaching Nusselt's assumptions.

Basically different condensation mechanisms exist depending on the flow regime, usually characterized as "gravity-controlled" or "shear-controlled." Other subdivisions exist depending on the exchanger type and tube orientation (horizontal, vertical, tubeside, shellside). The flow regimes, of course, change from vapor entry to vapor exit and thus require a sectional calculation along the path of the condensate flow, applying the appropriate equation to each segment. Flow regime "maps" have recently been developed to identify the probable conditions of the condensate-vapor interaction and made it possible to improve substantially the accuracy of calculations.

7) Enhanced Surfaces. The ever increasing cost of energy and therefore the need for most efficient heat exchangers led to the development of various heat transfer enhancement techniques, usually tube surface modifications. Unfortunately, none of them is universally applicable. The most typical ones are

a) External low fins

These have been successfully used for many years for shellside condensation in the process and refrigeration industry. The limitations for effective usage for condensers are the liquid surface tension and the fin density to high-ratio, as illustrated in Fig. 1.

Under the effects of the surface tension and high fin density, the grooves of the fins fill up with condensate which does not readily drain out, thus making the surface ineffective. This phenomenon known as "retention" has been in open literature recently identified by Rudy and Webb (1) and is further documented by photographic techniques in Ref. 2. For all practical purposes it eliminates the use of finned tubes for condensation of water vapors due to the high surface tension.

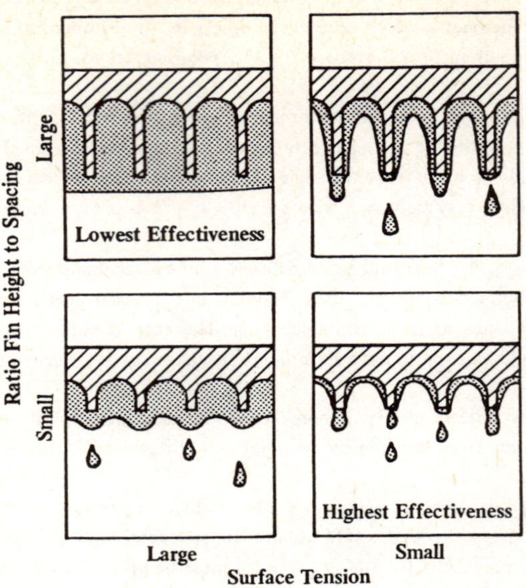

Fig. 1 **Condensate retention effect as a function of fin geometry and fluid.**

b) Internal longitudinal fins

Available commercially just recently is Cu-alloy materials. These are very effective, especially in vertical tube condensers.

c) Grooved and spirally fluted tubes

Distortion of the tubular surface by twisting or grooving has been commercially developed to utilize the Gregorig effect of condensate drainage (Fig. 2).

Recent surveys by Bergles (3) and Carnavos (4) summarize well the state of the art. While various degrees of improvement were observed, it is invariably connected with substantial increase of the tubeside pressure drop and difficulty of cleaning, thus eliminating the use of such tubes from many otherwise potential applications.

8) Effects of Pressure Drop. In single-phase flow the coefficient, and hence the heat duty, increases for higher pressure drops (under otherwise identical conditions). For condensing vapors, however, higher pressure drop usually will also increase the coefficient but at the same time decrease the effective temperature difference, due to the lower vapor pressure. This results in the phenomenon that, at least in some cases, an optimum pressure drop will exist which will produce the highest heat duty. This optimum is usually close to using approximately 10 percent of the vapor pressure as pressure drop.

In terms of exchanger design elements, this means that baffle spacing should not be decreased below a certain level and the tube diameter-to-tube length ratio must be tested to produce the best overall result.

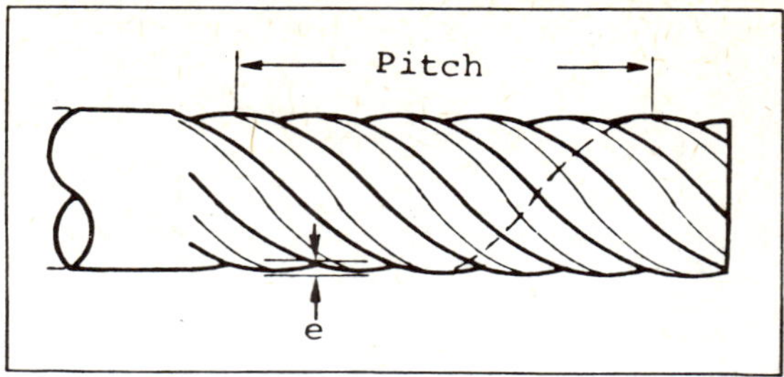

Fig. 2 Tube geometry illustration.

B. Shell and Tube Condenser Types and Their Characteristics.

The most frequently used tubular condensers and their typical applications, advantages, and limitations are described in the following.

1) Vertical Tubeside Downflow Condenser. In many aspects, this is a "classical" condenser in the sense of the condensation mechanism simplicity resulting in the predictive methods being probably most reliable. Other characteristics include

- Can handle both sensible heat or boiling coolant on the shell side.
- Suitable for corrosive and high pressure vapors.
- Low pressure drop and gravity-aided condensate drainage.
- Effective removal of noncondensables.
- Good for subcooling.
- Difficult cleaning of the shellside coolant (if applicable).
- Internally finned tubes are applicable.

2) Vertical Outside Tube Condenser. Main usage is confined to vertical thermosiphon reboilers with tubeside boiling where a condensing vapor is used as the heating medium. Baffles act as a condensate stripping device, possibly enhancing performance. However, noncondensable venting is difficult.

3) Horizontal Tubeside Condensers. In some process applications horizontal intube condensation may be preferable to vertical design, mainly for construction reasons. Single tubepass is preferred because of sensitivity to inclination. TEMA E, J, or X shell types are acceptable, depending on coolant. Condensation is less effective and calculations are more uncertain than for vertical orientation, requiring good flow regime maps.

4) Horizontal Shellside Condensers. This type is extremely popular as it offers many advantages and flexibilities of design. TEMA E, J, or X shell can be used, often depending on pressure drop requirements. The main features favoring this design are

a) coolant is tubeside and therefore easy to clean;

b) external low fin tubes are effective and can substantially decrease the cost of exchangers and;

c) pressure drop can be easily controlled by design elements to obtain the maximum effectiveness, contrary to tubeside condensation where this is rather difficult. However, subcooling is ineffective and inert gas removal is difficult unless good design provisions are made. Proper attention to condensate drainage must be given to prevent accumulation.

C. Concluding Comments.

Substantial improvements in the understanding of condensation processes have been made in the past decade, mainly in the application of flow regime maps and correspondingly the ability to use appropriate correlations for specific types of condensate film behavior. For horizontal tubeside condensation the typical flow patterns are shown in Fig. 3 below.

Several recently developed flow regime maps replaced the obsolete Baker map, as summarized by Breber et al. (Ref. 6). It should be noticed that the flow regimes are separated by rather broad transitional areas. Furthermore, identification of flow regimes makes practical sense only where corresponding correlating equations are available. For these reasons a very simple but quite effective flow regime map was proposed in Ref. 6 as shown in Fig. 4 below, where $j_g{}^*$ is the Wallis parameter

$$j_g{}^* = \frac{G_t\, y}{\sqrt{D_i\, g\, \rho_v\, (\rho_\ell - \rho_v)}}$$

For shellside flow the best published flow regime map is that of Grant (7), while for condensation again the main important distinction is that between shear- and gravity-controlled flow.

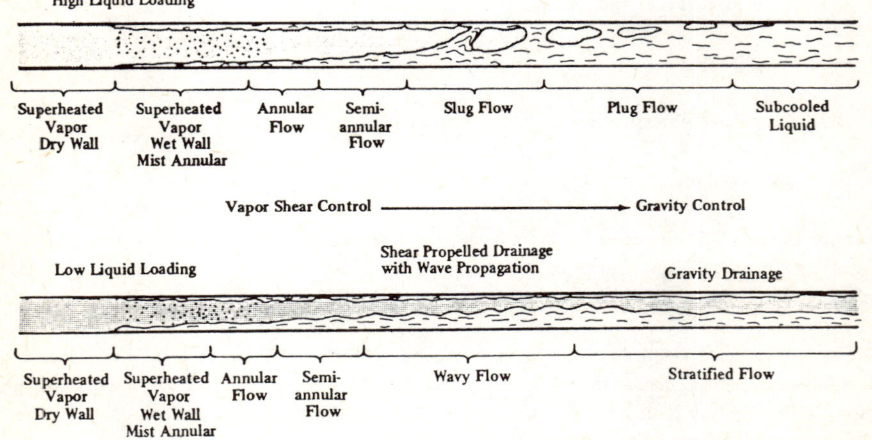

Fig. 3 Schematic illustrations of flow regimes in tubeside condensation (from Ref. 5 by permission).

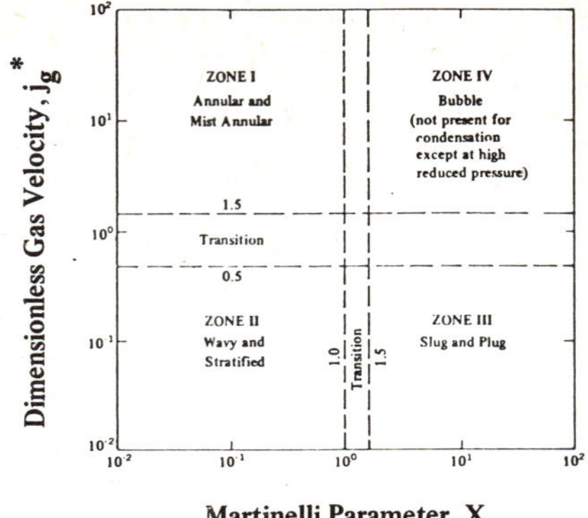

Fig. 4 Simplified criteria for horizontal tubeside condensation flow regimes, based on available data.

The main problems encountered in poor condenser design are usually due to a relatively few items.

- Poor venting of noncondensable gases.
- Inadequate drainage of condensate.
- Faulty analysis of the temperature profile, especially in complex mixtures and boiling coolants.
- Lack of understanding of the flow regime effects and the associated hydrodynamics (rather than heat transfer aspects of the process).

REBOILERS

A. Basic Principles.

The comments offered here relate strictly to process type reboilers as classified below. Unlike for condensers there are not that many combinations of process variables and equipment types, but a careful selection process is equally important (8) for proper operation.

The basic process is nucleate boiling, as obtained from fundamental data from laboratory experiments. However, there is a strong effect of pressure, boiling range (mixtures) and surface type. These effects are not well understood and predictive equations produce rather widely scattered results. There are two types of predictive equations in the literature

a) Based on critical pressure and correlated in the form (9)

$$\alpha_{nb} = f(\dot{q}, P_c, F_p, F_c, F_s)$$

where F_p pressure correction factor

F_c boiling range correction factor

F_s surface type adjustment factor, applicable to enhanced surfaces

b) Based on numerous combinations of physical properties groups.

Neither of the two types predicts α_{nb} with good accuracy but type (a) is less sensitive to physical properties estimation. Future research will hopefully clarify these basic problems, but the interaction of the multiple variables poses a formidable chore.

However, of great importance is the fact that pure nucleate boiling rarely exists in industrial exchangers. Usually, nucleate boiling acts in interaction with convective (flow-velocity-induced) boiling. In many instances convective boiling can predominate, thus making the premises of nucleate boiling less important. For the combination of nucleate and convective boiling process, the additive formula is an accepted form of a reasonable correlating equation

$$\alpha_b = s\,\alpha_{nb} + \alpha_{tp}$$

where α_b the combined boiling coefficient

s nuclear boiling suppression factor (see Ref. 9) due to flow boiling effects on α_{nb}

α_{tp} the two-phase flow or convective boiling coefficient, as documented in the various references.

B. Reboiler Types and Their Selection.

Most industrial reboilers will operate not in pure nucleate boiling but, as flow velocity almost instantly develops, effects of flow boiling will become important. Neglecting these effects accounted for severe underestimation of boiling coefficients in the past (but often compensated by poor understanding of other effects, especially pressure and boiling range).

The following reboiler types are most frequently used in industry and are characterized as follows (8).

1) Kettle Reboilers. The basic performance characteristics of tube bundle boiling were documented in a pioneering paper (13) which related single tube to bundle boiling (Fig. 5). Large enhancement from single tube to bundle is obvious. Other features of the basic kettle reboiler design are as follows:

- Least sensitive to hydrodynamics and process fluctuations.
- Efficient at low ΔT where other reboiler types are difficult to operate.
- Enhanced surfaces are effective.
- Most reliable in vacuum and near critical pressure where other reboilers are difficult to design.
- Dirt collectors as they operate close to complete evaporation.
- Ineffective for large boiling range mixtures.
- The operation is not by pure nucleate boiling mechanisms, but includes the previously mentioned flow-induced effects, which increase the performance.

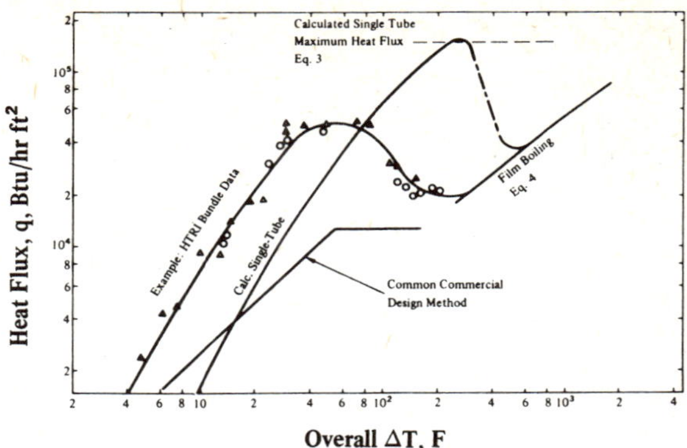

Fig. 5 Typical tube bundle boiling data compared to single tube.

2) Tubeside Thermosiphon Reboilers. Again, while nucleate boiling is present and dominating at entry, flow boiling effects increase and prediction of flow regimes becomes imperative to good design. Common features include:

- Suppression of fouling due to flow velocity.
- Best design in medium pressure range and low boiling range mixtures.
- Sensitive design in vacuum and near critical pressure unless dependable data are available.

3) Horizontal Thermosiphon Reboilers. Characteristic items include:

- Sensitive to proper hydrodynamic calculations, which are less precise than for tubeside (possible vapor separation).
- Preferable for wide boiling mixtures as design is more "permissive" than vertical tube.
- Large surfaces can be accommodated.
- Lower static head than for vertical units is possible.

4) Forced Flow Reboilers. This type of unit is used mainly for cases with very viscous fluids and where heavy fouling would be expected in any other design type. Low vapor fraction is necessary for good design. Cost factor is high but unavoidable for given conditions.

5) Other Type Reboilers. Many other special constructions of reboilers are used in the process industry for particular purposes, but these are outside of the limited objective of this paper.

C. Concluding Comments.

In general, process reboilers are designed in the basic nucleate boiling region, avoiding transition or film boiling which can cause numerous problems, especially fouling and poor control characteristics. The uncertainty of the nucleate boiling prediction calls for caution, but the demonstrated tube bundle enhancement effect (Fig. 5) compensates for the previously unwarranted safety factors.

Enhanced surfaces are becoming an increasingly important factor in shellside boiling, as documented in Ref. 10 and by HTRI work on tube bundles (11). However, the performance of enhanced surfaces in tube bundle boiling produces an anomaly as shown in Fig. 6 (from Ref. 11).

While plain tube bundle performance is strongly shifted toward lower ΔT with respect to single tube performance, for the enhanced surfaces (typically the Wieland GEWA-T or Hitachi Thermoexcel-E) the tube bundle performs about the same as the single enhanced tube (12). This effect is further elaborated in Ref. 9, but basically the following features are of basic importance:

- At a given ΔT a plain tube bundle would not produce boiling while the enhanced surface one would.
- This phenomenon is especially important for heat recovery applications where often very low ΔT is dictated.
- Cost of the enhanced surfaces may be a factor in other than very low ΔT applications.

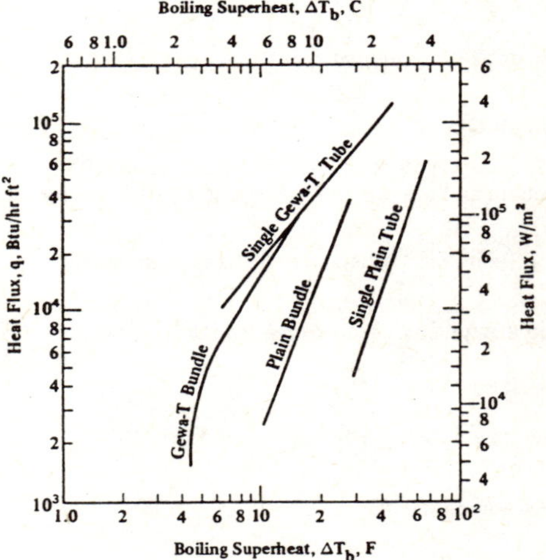

Fig. 6 Comparison of performance for single enhanced and plain tubes to that of enhanced and plain tube bundles with p-xylene at 15 psia (103 kPa).

REFERENCES

1. Rudy, T. M. and Webb, R. L., Advances in Enhanced Heat Transfer – 1981, ASME-HTD-Vol. 18, 1981, pp. 35-42.

2. HTRI Movie, "Visual Study of Shellside Condensation of Pure Fluids on Horizontal Low Integral Finned Tube Bundles," 7th Intl. Heat Transfer Conf., Munich, 1982.

3. Bergles, A. E., Advances in Enhanced Heat Transfer – 1981, ASME-HTD-Vol. 18, 1981, pp. 21-34.

4. Carnavos, T. C., "An Experimental Study: Condensing R-11 on Augmented Tubes," ASME Paper 80-HT-54, 1980.

5. Palen, J. W., Breber, G., and Taborek, J., "Prediction of Flow Regimes in Horizontal Tube-Side Condensation," Heat Transfer Engineering, Vol. 1, No. 2, Oct.-Dec. 1979, pp. 47-57.

6. Breber, G., Palen, J. W., and Taborek, J., "Prediction of Horizontal Tubeside Condensation Components Using Flow Regime Criteria," ASME J. Heat Transfer, Vol. 102, No. 3, 1980.

7. Grant, I. D. R., "Flow and Pressure Drop with Single-Phase and Two-Phase Flow on the Shellside of Segmentally Baffled Shell-and-Tube Heat Exchangers. Advances in Thermal and Mechanical Design of Shell-and-Tube Heat Exchangers," Natl. Eng. Lab. Report No. 590, 1975.

8. Palen, J. W., Heat Exchanger Design Handbook, Section 3.6.1, Hemisphere Publishing Corporation, Washington, 1982.

9. Palen, J. W., Taborek, J., and Yilmaz, S., "Comments to the Application of Enhanced Boiling Surfaces in Tube Bundles," ICHMT Symposium, Dubrovnik, 1981, published in Heat Exchangers: Theory and Practice, Hemisphere Publishing Corporation, 1981.

10. Brisbane, T. W. C., Grant, I. D. R., and Whalley, P. B., "A Predictive Method for Kettle Reboiler Performance," ASME Paper 80-HT-42, 19th Natl. Heat Transfer Conf., Orlando, 1980.

11. Yilmaz, S., Palen, J. W., and Taborek, J., "Enhanced Boiling Surfaces as Single Tubes and Tube Bundles," Advances in Enhanced Heat Transfer – 1981, ASME-HTD-Vol. 18, 1981, pp. 123-130.

12. Yilmaz, S. and Westwater, J. W., "Effect of Commercial Enhanced Surfaces on The Boiling Heat Transfer Curve," Advances in Enhanced Heat Transfer – 1981, ASME-HTD-Vol. 18, 1981, pp. 73-91.

13. Palen, J. and Taborek, J., "Characteristics of Boiling Outside Large Scale Horizontal Multitube Bundles," AIChE Symp. Series, Vol. 68, No. 118.

BOILING HEAT TRANSFER IN COMPACT AND FINNED HEAT EXCHANGERS

J. W. Westwater

Department of Chemical Engineering
University of Illinois at Urbana-Champaign
Urbana, Illinois 61801 U.S.A.

ABSTRACT

The object of this paper is to describe the past and present state of knowledge, and suggest the future, for flow boiling and flow condensation in compact heat exchangers. Already boiling and condensation heat transfer services are carried out in some compact, finned heat exchangers, even though no standard, reliable design procedure has been available. The earliest calculation schemes were simple modifications of expressions used for one-phase flow. For example, one could calculate the heat transfer and pressure drop for all-liquid flow and for all-vapor flow and use either the worst-case calculations or some simple averaging procedure to obtain conservative predictions for two-phase flow. Subsequent improvement for flow boiling involved the assumption that it is composed of two processes which are purely additive: pool boiling and single-phase forced convection. A recent design method by Robertson and coworkers involves the assumption that boiling heat transfer can be described in terms of the mass velocity and the vapor quality, whereas the ΔT is ignored. Westwater and coworkers propose the local assumption for flow boiling, namely the local heat transfer coefficient is fixed by the local velocity and the local ΔT.

As the design methods are improved, the competitive edge will shift from tube-and-shell exchangers to compact heat exchangers. In principal, almost any non-fouling fluid should be used in compact heat exchangers. Many processes for which heat transfer in narrow passages with fins is unknown today will see such use in the future.

1. INTRODUCTION

Compact heat exchangers have grown steadily in popularity

during the last several decades. Their low weight-to-area ratio makes them ideal for aerospace use. Their low ratio of outside shell area to internal heat transfer area makes them ideal where heat leakage causes serious harm. Thus they are common now in liquid-air separation plants, liquified natural gas plants, and in various cryogenic applications. The exchanger cost per unit area is modest, because the manufacturing technique is well developed. Usually the entire exchanger core, including all fins is brazed as a one-shot operation in a molten salt bath or controlled environment.

A compact heat exchanger may be defined as one having more than about 700 m^2 of heat transfer area per 1 m^3 volume. This is roughly 10 times the value for ordinary tube and shell exchangers. Compact exchangers can be large; Mikus [1] reports one manifold assembly with a surface area of over 37,000 m^2. The large area is obtained by the incorporation of a huge number of closely-spaced fins. Compact exchangers may be used over a wide range of service conditions, including pressures up to 700 kPa and temperatures from 4 to 1070 K, depending on the metal used. Good design methods exist for heating or cooling gases and liquids, provided no phase change occurs.

The one significant disadvantage of compact exchangers is that the small flow passages (say 2 or 3mm in width) may clog if a fouling fluid is used. A lesser disadvantage is the lack of accepted design techniques for boiling or condensation heat transfer service. The object of this writing is to discuss possible solutions to the problem of phase change in compact exchangers.

Compact heat exchangers are usually made of aluminum alloy. The dip brazing process for joining parts was invented at Alcoa in 1939. Heat exchangers made by this process became commercially important about 10 years later. Today there are nearly a dozen manufacturers of such exchangers in the USA. Other manufacturers are in Great Britain, Belgium, France, and Japan. Available materials of construction, in addition to aluminum alloy, include stainless steel, copper, and cupro nickel. Considering the availability, desirable properties, and economic advantages of compact heat exchangers, it is inevitable that their numbers will increase in the process industries, wherever non-fouling (or slow-fouling) streams are encountered.

2. BOILING IN COMPACT HEAT EXCHANGERS - PRIOR WORK

A standard procedure for designing a compact heat exchanger for a gas or a liquid, with no change in phase, was described by Lenfestey [2] in 1961. A number of later writers attempted to adapt this procedure for boiling or condensation. Lenfestey's method is described briefly below.

Lenfestey adopted as a guide the Colburn [3] analogy between heat transfer and friction. She developed Eq. (1) as the key relationship which must be satisfied.

$$\frac{\Delta P}{\eta h A} = \frac{f}{C' j}\left(\frac{G^2}{2\rho\, g_c}\right) \qquad (1)$$

(Lenfestey - 1961)

where

$$C' = \frac{W\, C_p\, (A_1 + A_2\, \eta)}{(Pr)^{2/3}\, (A_1 + A_2)} \qquad (2)$$

f is defined by: $\Delta P = 4f \left(\frac{G^2}{2\rho\, g_c}\right)\left(\frac{L}{d}\right)$ (3)

j is defined by: $\frac{h}{C_p G} = j\, (Pr)^{-2/3}$ (4)

Eq. (1) shows that the heat transfer coefficient and the pressure drop are interrelated. The factors f and j cannot be predicted for compact exchangers from theory. They must be determined experimentally. Manufacturers furnish graphs of f and j vs. Re for air flowing in their equipment. A number of such graphs have been published [4,5]. It is seen that for each geometry, the factors f and j depend on the Reynolds number in very similar ways. Consider the example illustrated in Fig. 1. The Reynolds number covers a range of nearly two orders of magnitude, however the ratio of f to j is almost constant at a value of about 4.3. If one assumes that the ratio is truly constant, one can proceed in a systematic manner to select the dimensions of a proposed heat exchanger. Thus the Colburn idea, originally proposed for turbulent flow inside straight, round tubes, was adapted by Lenfestey for flow in very small non-round channels, even including Reynolds numbers as low as 200.

The fin efficiency in Eq. (2) is calculated by the usual fin formula derived for constant heat transfer coefficient h. This may be justified when no phase change occurs. When phase change occurs, this procedure is incorrect; h is not constant, and it is not fixed by Eq. (4).

In her 1961 paper, Lenfestey suggested that for a condenser-reboiler, one should estimate h = 8400 to 10,000 W/m^2K on the boiling side and 21,000 to 25,000 W/m^2K on the condensing side for typical air separation applications. These numbers were stated to be reasonable if the flow passages were 1.5 to 1.8m long, contained 472 to 590 fins/m of width, and the fin height was 5.1 to 6.4mm.

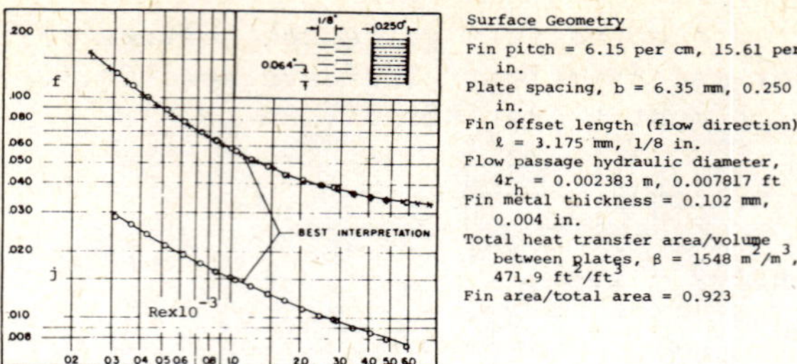

Fig. 1. Sample f-factors and j-factors for a particular plate-type, offset fins, compact heat exchanger. From Shah [5].

Ten years later, Lenfestey [6] wrote that for a condenser-reboiler one should use empirical data and that the expected heat transfer coefficient is 800 to 1800 W/m^2 K. It is not clear whether these newer numbers are for the overall or individual coefficients. For a subcooled liquid feed which exits as superheated vapor, he wrote that the exchanger should be considered as three discrete steps, with assumed constant heat transfer coefficients in both streams for each step. One step is to bring the feed liquid to its boiling point, one step is to vaporize, and one is to superheat. To estimate the pressure drop, he wrote that a stepwise procedure is also needed. A different mean quality is used for each step, and the method is that developed by Martinelli and Lockhart [7] for isothermal flow.

In the same year, Clark and Thorogood [8] endorsed the Martinelli-Lockhart scheme as an approximate way to predict ΔP for boiling flow in a compact heat exchanger. They suggested that a good value to use for the boiling side heat transfer coefficient is 3000 to 4000 W/m^2 K. Timmerhaus and Schoenhals [9] also endorsed the Martinelli-Lockhart scheme, or as an alternate, the modifications by Chenoweth and Martin [10] to predict the flow boiling ΔP. In addition, they suggested that a conservative way to design for the heat transfer coefficients and ΔP for flow boiling or condensation is to design for all-liquid or all-vapor flow and then select the more conservative of the answers.

Among the early equations produced specifically for flow boiling in small channels was that of Sydoriak and Roberts [11], Eq. (5).

$$Q = A \lambda (\rho_L \rho_V g x \Delta P)^{1/2} \tag{5}$$

(Sydoriak & Roberts - 1957)

This equation includes the flow rate indirectly by the pressure drop and fluid quality. The quality x is at the exit of the flow channels. Data for flow boiling of nitrogen and hydrogen in annular channels of 0.18 to 0.56mm were obtained by Richards et al. [12] to test the expression. They concluded that Eq. (5) is at best an expression for the order of magnitude.

Flow boiling in ordinary geometries was considered by Rohsenow [13] with some success as being an additive process. Nucleate pool boiling heat transfer is added to forced convection (non-boiling) heat transfer, Eq. (6).

$$q = q_{pool} + q_c \tag{6}$$

(Rohsenow - 1953)

J. C. Chen modified this into Eq. (7) which he presented at the 1963 National Heat Transfer Conference and published elsewhere [14].

$$h = h_{pool} + h_c \tag{7}$$

(J. C. Chen - 1963)

Collier [15] described the Chen method as the "best available". The value of h_{pool} is obtained from the Forster-Zuber equation for nucleate boiling [16] modified as Eq. (8).

$$h_{pool} = 0.00122 \left[\frac{k_L^{0.79} C_L^{0.45} \rho_L^{0.49}}{\sigma^{0.5} \mu_L^{0.29} \lambda^{0.24} \rho_v^{0.24}} \right] \Delta T^{0.24} \Delta p^{0.75} S \tag{8}$$

(J. C. Chen - 1963)

The factor S was not present in the original Forster-Zuber equation. It was introduced by Chen to account for the fact that the thermal boundary layer during flow boiling is not identical with the thermal boundary layer during pool boiling. The factor S is predicted from a graphed correlation of S vs. the Reynolds number. For the convective coefficient in Eq. (7), Chen presents Eq. (9).

$$\frac{h_c d}{k_L} = 0.023 \left(\frac{d\,G\,(1-x)}{\mu_L} \right)^{0.8} \left(\frac{C_L \mu_L}{k_L} \right)^{0.4} F \tag{9}$$

(J. C. Chen)

This is a modified form of the Dittus-Boelter equation. The factor F is obtained from a graphed correlation of F vs. the Martinelli factor X_{tt}.

Johannes and Mollard [17] conducted tests with liquid helium-I in small channels with heated walls. The smallest channel gap was 0.35mm, which is comparable to that in compact exchangers. The maximum flow length was about 20cm. The size of the gap was found to be a strong variable if it was less than 2mm. For example, the maximum heat flux varied with the gap dimension as shown in Fig. 2. Also the maximum heat flux varied with the channel length according to

$$q_{max} = \text{const.}\ (L)^{-0.21} \tag{10}$$

(Johannes & Mollard - 1972)

Fig. 3 shows the Johannes-Mollard boiling data (curves 2 and 4) compared to pool boiling, flat plate results (curves 1 and 3). It is clear that ΔT is a strong variable. The equipment with channels operated as a thermosiphon reboiler, and the fluid velocity for every data point was different. Fig. 4 shows that the data replotted to demonstrate the velocity effect. Here the ΔT for every point is different. Figs. 3 and 4 illustrate the puzzle posed by data for two-phase flow with phase change. There exists a thermal and hydrodynamic coupling of the flow phenomenon and the heat transfer phenomenon. The flow rate, flow pattern, channel dimensions, fluid quality, pressure, ΔT driving force, and heat flux are interrelated. Fig. 3 provides little information concerning hydrodynamic factors, and Fig. 4 provides little concerning thermal factors. One way of trying to correlate most factors is Fig. 5. The ΔT is not included in this graph. This means that it is assumed that any change in ΔT will affect some other variables such as the quality x and the Nusselt number just right to stay on the correlating lines. Fig. 5 is equivalent to Eq. (11).

$$\frac{h\,d}{k_L} = 5.5 \left(\frac{d\,G}{\mu_L}\right)^{0.8} x^{0.8} \left(\frac{L}{d}\right)^{-0.8} \tag{11}$$

(Johannes & Mollard - 1972)

Eq. (11) and Fig. 5 are for (L/d) < 50. When (L/d) > 50, Eq. (12) is recommended.

$$\frac{h\,d}{k_L} = 0.24 \left(\frac{d\,G}{\mu_L}\right)^{0.8} x^{0.8} \tag{12}$$

(Johannes & Mollard)

These two equations are for helium - I at atmospheric pressure and are untested for other systems.

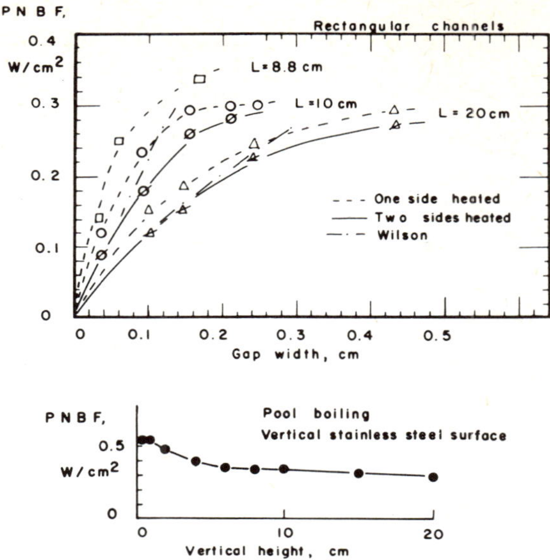

Fig. 2. Effect of gap width on flow-boiling maximum heat transfer to helium - I in parallel plate channels. From Johannes and Mollard [17].

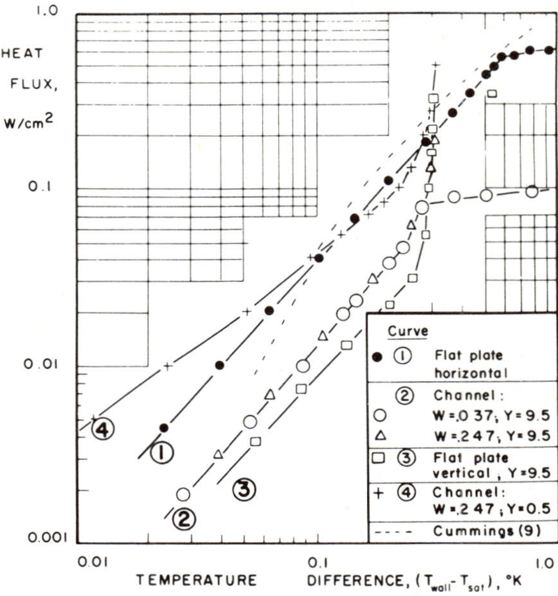

Fig. 3. Effect of ΔT on heat transfer to helium - I. Curves 2 and 4 are for flow boiling. From Johannes and Mollard [17].

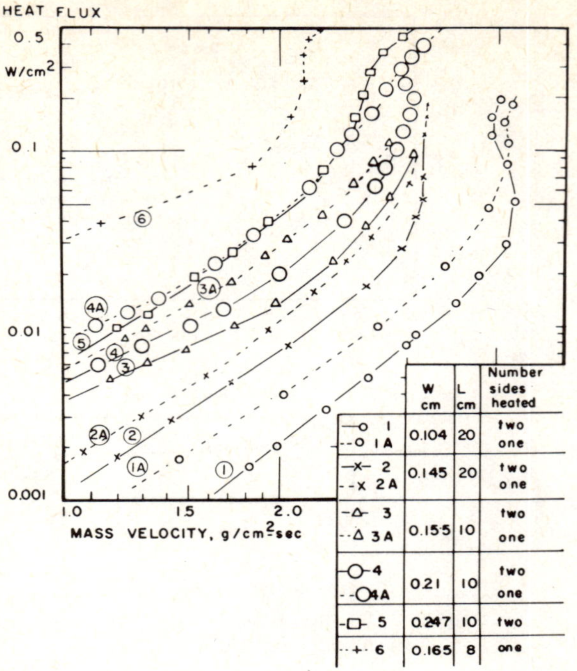

Fig. 4. Effect of velocity on heat transfer to helium - I during flow boiling in parallel-plate channels of width indicated by W. From Johannes and Mollard [17].

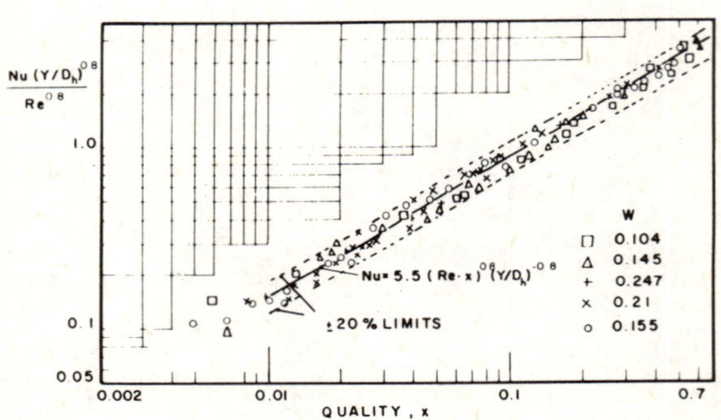

Fig. 5. Heat transfer correlation for flow boiling of helium - I in narrow channels. The symbols Y and D_h are the channel length and hydraulic diameter. From Johannes and Mollard [17].

Collier and coworkers [18] ran tests with a vertical plate-fin heat exchanger 3.5m long x 75mm wide. The gap between the plates was 6.35mm and corrugated fins spaced 590 fins/m. Liquid nitrogen was the test fluid. The authors presented no heat transfer data for this test section, but they stated that the heat transfer could be predicted satisfactorily with the use of the equations of J. C. Chen [14]. Collier stated that he used a step-by-step integration method involving trial and error. A demonstration of their calculation results is Fig. 6. This reproduces the data used for one of the runs in Fig. 3 and includes a new smooth curve by Collier, with the vapor quality indicated in parentheses. Several assumptions were made, the principal one being that the inlet subcooling was 0.15 K (which is substantial for liquid helium). The agreement in Fig. 6 is good.

Collier points out that the design methods then in use (1974) had definite weaknesses. One was that the formula for fin efficiency, Eq. (13), is valid only when the heat transfer coefficient is uniform over the fin surface.

$$\eta = \frac{\tanh m}{m} \tag{13}$$

However when boiling occurs on a fin, the coefficient is not constant.

For calculating the pressure drop in a plate-fin exchanger, Collier adopted Eq. (14)

$$-\frac{dp}{dz} = \left(\frac{dp}{dz} f\right)_{LO} \phi^2_{LO} + G^2 \frac{d}{dz}\left[\frac{(1-x)^2}{\rho_L(1-\alpha)} + \frac{x^2}{\rho_V \alpha}\right] + g\,[(1-\alpha)\rho_L + \alpha\rho_V] \tag{14}$$

(Collier - 1974)

In Eq. (14), the three terms on the right side represent the frictional, momentum, and static head losses. The two-phase frictional multiplier ϕ_{LO} and the void fraction α must be selected from one of several published correlations. Eq. (14) is integrated step by step. Fig. 7 shows the result of this procedure with boiling nitrogen in the 3.5m long exchanger.

Sydoriak [19] considered the problem of the interaction among the thermal and hydrodynamic properties and proposed the following argument. Surface friction is assumed to be negligible, and assume that all available potential energy is converted to kinetic energy. The two-phase mixture is assumed to be homogeneous. The heating surfaces are assumed to be covered with a thin layer of tiny bubbles which reduce the cross sectional area for flow. The procedure thus becomes one of curve fitting to select the best thickness to use for the bubble layer. Typical values used were 0.11mm for liquid nitrogen and 0.13mm for liquid hydrogen.

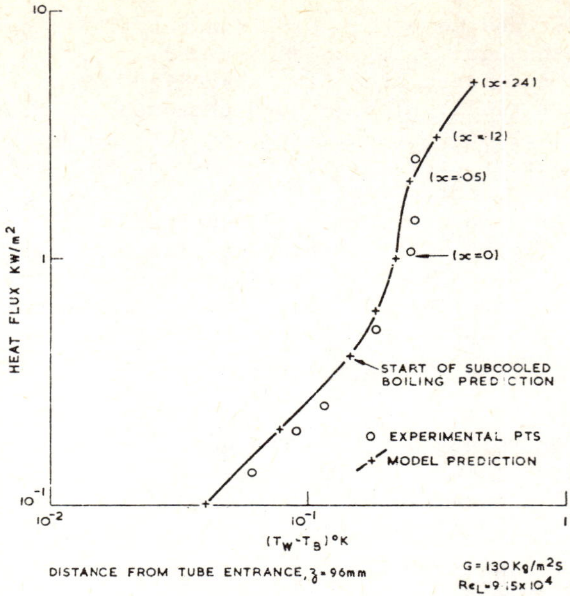

Fig. 6. Local conditions for one run from Fig. 3 computed by Collier, Kennedy, and Ward [18].

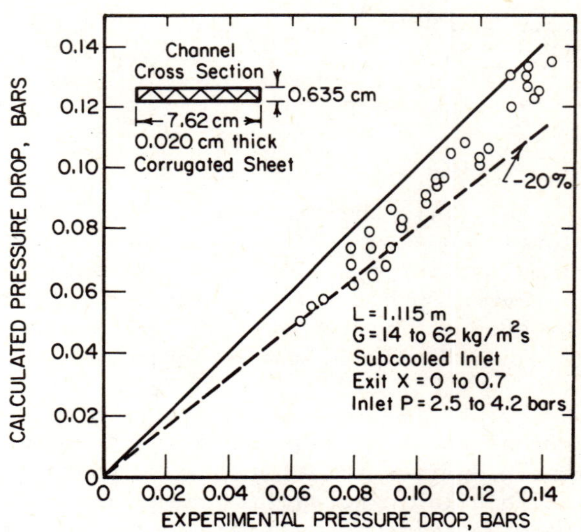

Fig. 7. Comparison of measured and calculated pressure drop for flow boiling nitrogen in plate-fin channels. From Collier, Kennedy, and Ward [18].

Shorin and coworkers [20] used a 1.625m long heat exchanger with boiling oxygen and condensing nitrogen. The gap between plates was fitted with fins such that each flow passage was 6 x 4.22mm. The oxygen velocity was 0.03 to 0.35m/sec (presumably at the inlet) and the heat flux was 600 to 4500 W/m^2. The pressure was 1.23 to 1.55 bars. They succeeded in correlating the results by Eq. (15).

$$\frac{W \Delta H}{Q} = 1 + 0.15 \left(\frac{d\,G}{\mu_L} \right)^{1.1} \left(\frac{\rho_L\ q}{\lambda\ \rho_v\ \mu_L} \right)^{-0.75} \left(\frac{L}{d} \right)^{-1.1} \tag{15}$$

(Shorin, et al. - 1974)

Figure 8 shows a graph of Eq. (15) compared to their data and also data for a 12mm I.D. tube. The ΔT was not known; it was estimated using a method of Labuntsov. Note that the flow rate occurs on both sides of Eq. (15), in W and in G. The heat transfer rate is also on both sides, in Q and q. This means that the agreement apparent in Fig. 8 is closer than it would be if the variables were truly separated.

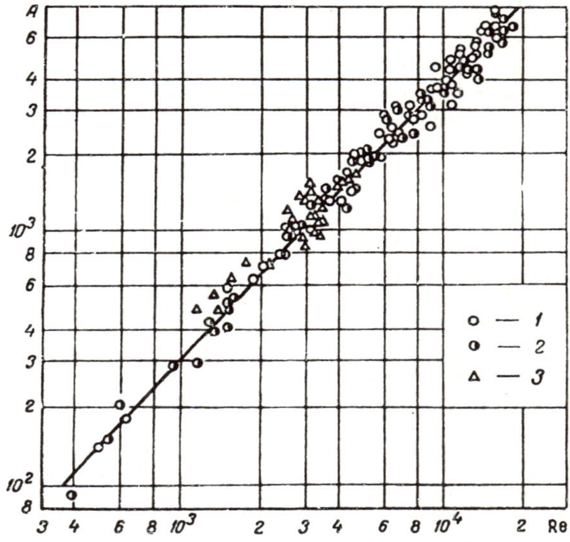

Fig. 8. Graph of Eq. (15) for flow boiling oxygen in a plate-fin exchanger (data-3) compared with two tubular models (data-1 and data-2). The abscissa is the Reynolds number and the ordinate is all else in Eq. (15). From Shorin, Sukhov, Shevyakova, and Orlov [20].

Galezha et al. [21] made tests with Freon-12 and Freon-22 flowing upward in five different compact heat exchangers, all having discontinuous fins. The core size was 22 x 20 x 4cm and the plates were heated electrically. The hydraulic diameter was from 2.5 to 4.6mm. The flow rate was low, less than $16 kg/m^2$ sec, corresponding to a maximum inlet velocity of 0.012m/sec. Apparently this flow rate was too small to be significant. They obtained the ordinary pool-boiling behavior for each exchanger, expressed as $h = const. (Q/A)^{1/3}$. The constant was dependent on the geometry, the pressure, and the molecular weight of the fluid, but not the flow rate.

M. M. Shah [22] worked out a correlation for flow boiling without bubble nucleation (evaporation). His correlation may be rearranged to Eq. (16).

$$h = 1.8\, h_L\, x^{0.64} (1-x)^{0.16} \left(\frac{\rho_L}{\rho_v}\right)^{0.4} \tag{16}$$

(M. M. Shah - 1976)

Here h is the local heat transfer coefficient for flow boiling, and h_L is the heat transfer coefficient for all the mass flowing as a liquid, as calculated by the Dittus-Boelter expression.

$$\frac{h_L d}{k_L} = 0.023 \left(\frac{dG}{\mu_L}\right)^{0.8} \left(\frac{C_L \mu_L}{k_L}\right)^{0.4} \tag{17}$$

Eq. (16) has the weakness that it predicts h = zero as the quality x goes either to zero or one. Shah applied his equation to 10 liquids with a wide range of test conditions and concluded that the fit was within 17 percent, on the average.

Robertson [23,24] conducted research with a much longer compact heat exchanger than tested previously. The core size was 3.4m x 76.2mm x 6.35mm; serrated fins were spaced 591 per m. The test fluid was boiling nitrogen, the flow was up to $110 kg/m^2 sec$, and the ΔT was 0.4 to 1.5K. Fig. 9 shows the results. The author stated that the graph shows a laminar flow regime, where h is independent of G, and a turbulent flow regime, where h increases with G. For the turbulent regime ($G < 30 kg/m^2 sec$) he gives Eq. (18).

$$h = 375 \times G^{0.8} \tag{18}$$

(Robertson - 1979)

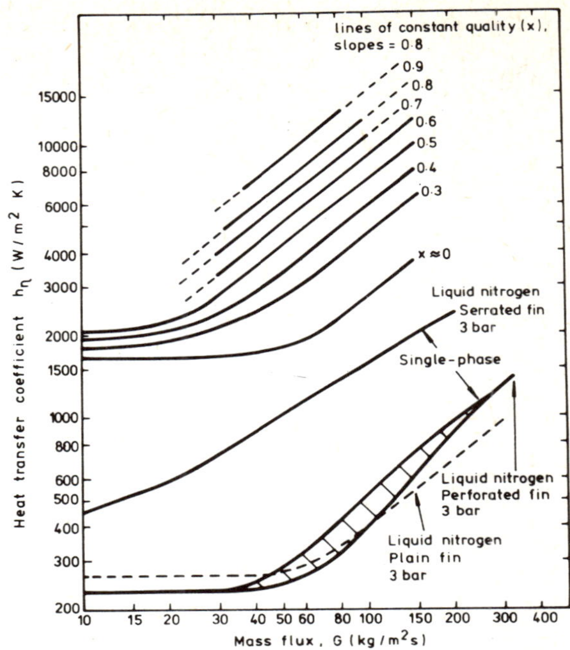

Fig. 9. Smoothed data for flow boiling of nitrogen at 3 bars in a 3.4m long, plate-fin heat exchanger. From Robertson [23].

The exponent on the flow rate G is the same as used in Eq. (11), but the exponent on the quality x is slightly higher. The coefficient 375 is dimensional and holds only for nitrogen at 3 bars in the geometry used. A demonstration of Eq. (18) is Fig. 10. Robertson stated that the correlation of J. C. Chen is not satisfactory for the Robertson tests because the pool boiling equation employed as part of the Chen correlation is not adequate for nitrogen. In fact, Robertson concluded that the boiling mechanism is unimportant, compared to forced convection, for heat transfer to boiling nitrogen--at least for his test conditions.

The exchanger used by Robertson was more sophisticated than that used by others. A brief description is worthwhile. The core included two long, vertical, parallel plates, electrically heated. They measured 3.4m long, 76.2mm wide, and 6.35mm thick. Sandwiched in a 6.35mm gap between the plates were the fins. Along the flow path, at about 12 positions, were located thermocouples in holes which had been drilled in from the edge. A similar number of thermocouples was located in the flow path. Thus an excellent record of the fluid and wall temperature was achieved.

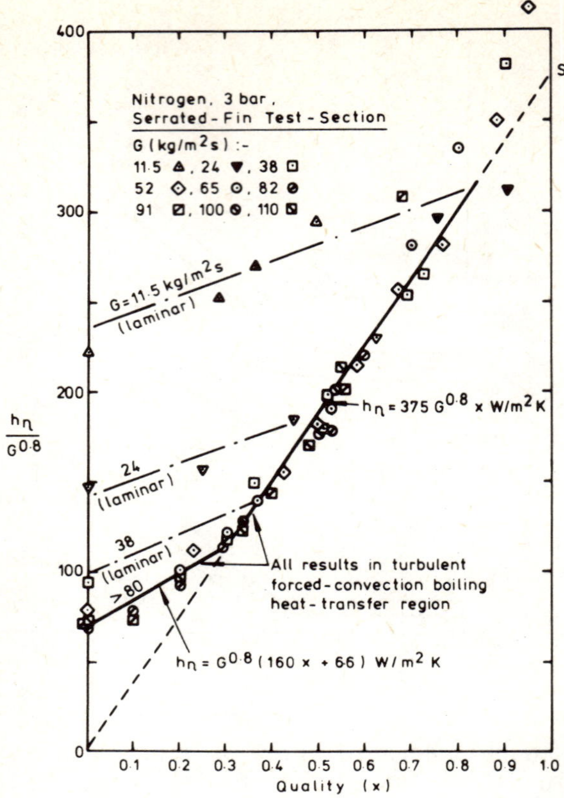

Fig. 10. Graph of Eq. (18) and data of Fig. 9. From Robertson [23].

The same heat exchanger was used to boil Freon-11 [25]. The results are shown in Fig. 11. Many of the data points are well off to the left of the straight line, and the authors state that those represent laminar flow conditions. Some of the data fit the straight line and thus may be described by $h = (\text{const.}) \times Re^{0.8}$ which uses the same exponents on flow rate and quality as used in Eq. (18). The authors stated that a Reynolds number of about 1100 corresponds to the division between viscous and turbulent flow. However graphs of friction factors f and Colburn factors j versus the Reynolds number do not show two lines (viscous and turbulent) for serrated heat exchangers with numerous flow disturbances. Rather such graphs show continuous curves. Thus the Robertson explanation should be considered tentative.

It is clear from the prior works discussed above that a state of uncertainty exists. No accepted, standard method is yet available for predicting the heat duty of a compact heat exchanger used for flow boiling.

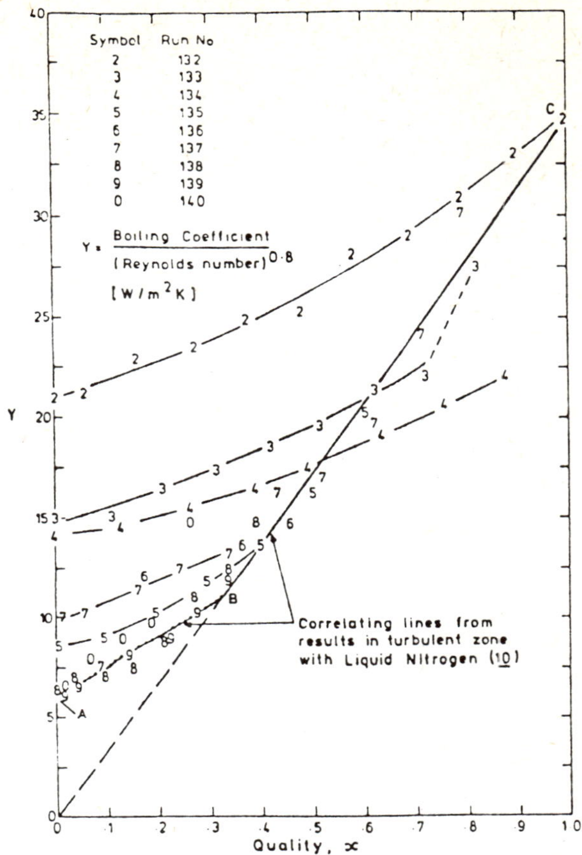

Fig. 11. Correlation of data for flow boiling of Freon-11 in same exchanger used for Figs. 9 and 10. From Robertson and Lovegrove [25].

3. BOILING IN COMPACT HEAT EXCHANGERS--WORK AT UNIVERSITY OF ILLINOIS

Work on compact heat exchangers for boiling heat transfer began at the University of Illinois in 1971. The starting principle was the "local assumption". This is the assumption that the heat transfer coefficient at any point on a fin is determined by the local conditions, such as fluid velocity and metal-to-fluid ΔT, and is independent of the conditions anyplace else. This guiding rule was developed through prior research starting in 1961 with boiling on single fins of widely varying shapes; with arrays of 2 to 10 spines having spacings from 0.8 to 27.4mm; and on single tubes having external transverse fins with spacings from 0.8 to 19.3mm and having varying thicknesses and heights [26-34]. Other laboratories also have confirmed the local assumption for single fins and small arrays [35-39].

Only two papers have been published [40,41] wherein the local assumption was applied to a finned, compact heat exchanger. Work continues using two other compact exchangers. The published work includes tests made with a brazed aluminum exchanger having a core 8.25 x 8.25 x 7.87cm. It contained 22 parallel plates separated by fins. It was used as a thermosyphon reboiler with Freon-113 and isopropanol flowing upward in channels containing 11,200 fins while steam flowed horizontally in channels containing 23,660 fins. Fig. 12 indicates the orientation of the exchanger and shows the double-stack, single-pass array on the boiling side and the single-stack, double-pass array on the condensation side. The fins on the boiling side had the shape shown in Fig. 13. A standard nomenclature for compact heat exchanger has not been agreed on. Fins as in Fig. 13 are called multi-entry fins as well as interrupted, serrated, step, lanced, and offset fins. Each fin was 0.101mm thick, 3.88mm high, and 4.67mm long in the flow direction. The boiling liquid flowed in a set of 672 parallel flow channels each 3.78 x 1.46mm in cross section. The flow was split 4.67mm downstream by the next row of staggered fins. The test results are shown in Fig. 14 as individual data points.

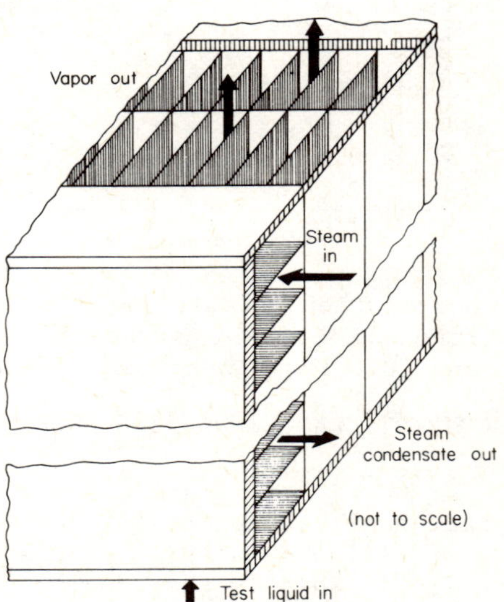

Fig. 12 Overall specifications and cutaway view of compact heat exchanger used at University of Illinois. From Panitsidis, Gresham and Westwater [40].

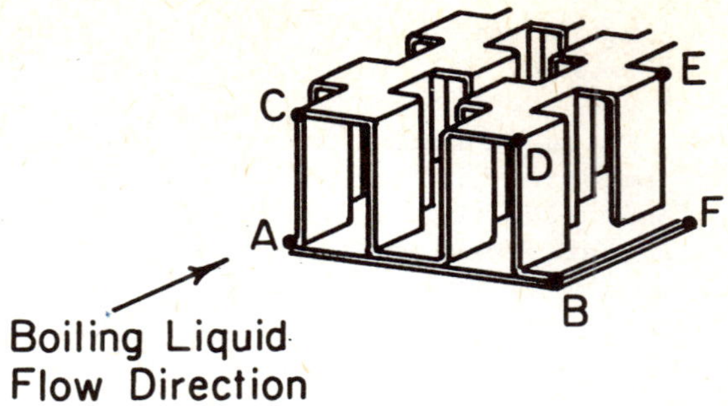

Fig. 13. Sketch of one plate and arrangement of the fin rows for the exchanger of Fig. 12. From Chen, Loh, and Westwater [41].

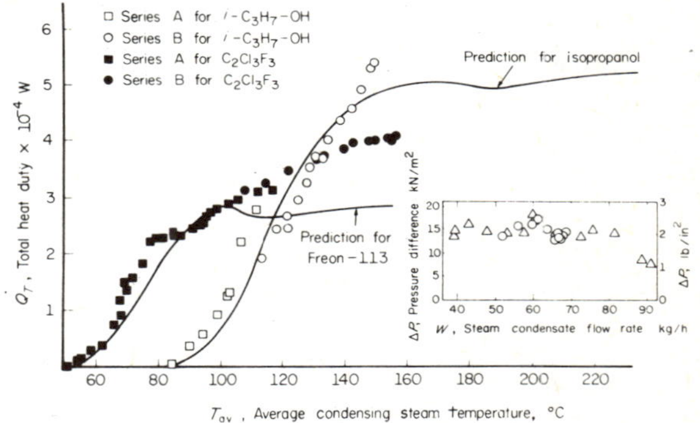

Fig. 14 Flow boiling data compared to the one-fin, pool-boiling model using the local assumption. From Panitsidis, Gresham, and Westwater [40].

3.1 Exclusion of Velocity Effect

Pool boiling curves from Freon-113 and isopropanol boiling on a 6.35mm copper tube were already available and these were used to generate the predicted performance of the exchanger shown by the curves in Fig. 14. This Model-I design may be called a one-fin pool-boiling model. It was based on one typical fin, composed of 26 nodes, with the heat transfer coefficient at each node being exactly equal to that at a point on the pool boiling curve at the same ΔT. The total heat duty was predicted by multiplying the one-fin duty by the total number of fins on the boiling side. The agreement between the predictions and laboratory measurements in Fig. 14 are surprisingly good, and it was decided that further refinement of the design procedure was warranted. In particular, the velocity effect could no longer be ignored.

One can imagine applying the local assumption with maximum precision to a compact heat exchanger having cross flow with fins on both the hot and cold sides. It would require a massive computer program. The procedure would involve predicting the local temperature at every position on every fin. Also it would require prediction of the local fluid conditions (velocity, quality, pressure, and temperature) at every position in every flow passage. Then the heat duty would become known. This procedure is too ambitious at this time. Simplifications are called for.

3.2 Inclusion of the Velocity Effect

Computational Model-II [41] had two significant improvements over the earlier one-fin, pool-boiling model. First, the pool-boiling data were replaced with forced-flow boiling data, when selecting local heat transfer coefficients. Second, instead of assuming that one fin in the center of the exchanger is representative of all the fins, it was assumed that one fin must be selected in every transverse row of fins. The procedure is summarized briefly below.

Only one paper exists [42] showing the effect of velocity on the entire boiling curve q vs. ΔT (or h vs. ΔT). The measurements had been made using boiling Freon-113 at atmospheric pressure flowing upward across a horizontal, 6.4mm diameter, steam-heated, copper tube. The data covered nucleate, transition, and film boiling, and included velocities up to 6.8m/s. Fig. 15 shows the flow boiling results for this cylinder. Note that at any ΔT, an increase in velocity causes an increase in the heat flux. The effect is very pronounced in transition and film boiling. It is not so great in the region of nucleate boiling.

For the improved Model II, the data in Fig. 15 were used with the local assumption. Interpolations were used to get h at velocities not tested. The data for the tube were assumed to be applicable to the fins in the compact exchanger, even though the two geometries were different. The calculations proceed in the following manner, with the inlet fluid assumed to be a saturated liquid.

First the heat duty is calculated for a single fin in the first row of fins using the boiling heat transfer coefficient corresponding to the saturated liquid inlet velocity and to the local ΔT. This is a computer calculation with the fin divided into many nodes (39 in our practice). The velocity is the same at every node, but the ΔT is not. In other words, the h varies over the fin. The heat duty for the one fin is then multiplied by the number of fins in the row to get the heat duty for the row. This is used to compute the amount of vapor formed at the first row of fins.

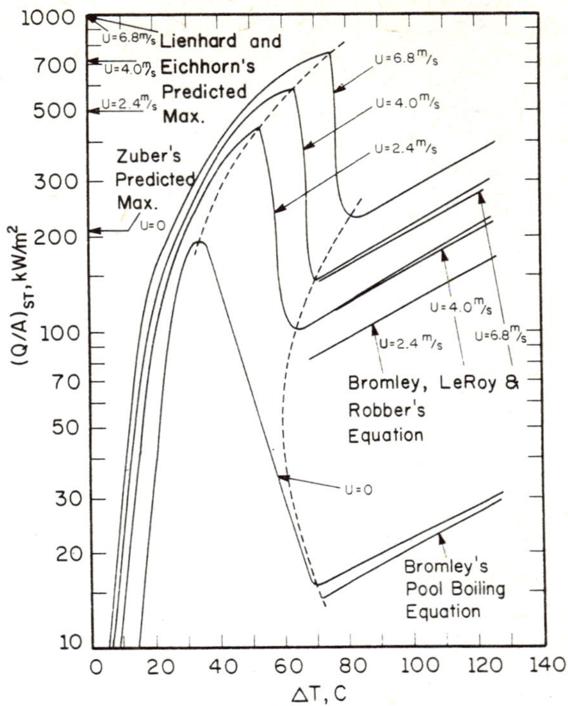

Fig. 15. Effect of velocity on the boiling curve for Freon-113 flowing upward normal to a 6.4mm, steam-heated copper tube. This plot was used to get values of the heat transfer coefficient for the local assumption in the finned exchanger of Figs. 12 and 13. From Yilmaz and Westwater [42].

The fluid, of quality somewhat higher than zero, now passes to the second row of fins. The liquid-vapor mixture is assumed to be homogeneous, and its velocity is greater in accordance with the known volume increase due to phase change at the first row.

Is it reasonable to use a homogeneous model for two-phase flow in a compact heat exchanger? Perhaps not for an exchanger having straight, smooth, flow channels. But for interrupted fins as depicted in Fig. 13, the assumption seems mandatory. Here the flow path is not straight nor smooth. The flow is split every 4.67mm downstream. Well-developed boundary layers never build up. Well-developed flow profiles cannot form. The many published two-phase flow correlations developed for long, smooth straight flow paths appear hopeless for this case.

The computational procedure used for the first row of fins is repeated for the second row. For a single fin in the second row of fins, the heat duty is calculated by the same procedure. This fin also has 39 nodes and the same new velocity exists at every node. The ΔT is different at each node. Again a computer is used to select the proper distribution of h over the fin corresponding to the local ΔT and the new velocity. The heat duty for the second fin is then used to compute the amount of vapor formed at the second row of fins. The increase in fluid volume will result in a new, higher fluid velocity at the next row of fins.

The procedure is repeated from fin-row to fin-row until all rows have been transversed (16 rows from our tests). It is assumed that all fins which are at one particular distance from the inlet are exactly alike (672 fins/row for our tests). These are here referred to as a single fin row. The sum of the heat duties for all the rows is the predicted duty for the entire exchanger.

Figure 16 illustrates the change in fluid quality from row to row as the 16 rows are transversed. This graph is for saturated liquid entering the first row of fins at 0.03m/s. Three cases are shown with different ΔT_w, the temperature difference from the base plate to the boiling liquid. Case 1 is for ΔT_w = 20K, and the exit fluid is partially vaporized. Case 2 for ΔT_w = 24.8K results in the exit fluid being saturated vapor. Case 3 for ΔT_w = 30K results in the fluid becoming completely vaporized at the 12th row and slightly superheated thereafter. The velocity of the exit fluid for these three cases is 3.7, 6.1, and 6.2m/s. The heat duty for the exchanger for the three cases is 15.3, 25.1, and 25.6 kW.

The effect of a change in velocity is very different for the three illustrative cases. For Case 3, a significant increase in inlet velocity causes a significant increase in heat duty. This occurs because liquid becomes available in fin-rows 12-16 which were dry previously. For Cases 1 and 2 the increase in heat duty is much smaller, because all fin rows were in contract with liquid already at the first velocity.

The inlet velocity is a weak variable provided all fins in the heat exchanger have a supply of liquid. The inlet velocity is a strong variable if some of the fins are starved for liquid and the exit fluid is all superheated vapor. If the latter case occurred in industry, a pump might prove of advantage.

The design procedure discussed here indicates that it is possible to obtain the optimum dimensions for a compact heat exchanger for a prescribed heat duty. Such optimization studies are now underway at the University of Illinois. But let us continue with the discussion of the actual heat exchanger tested during the present investigation.

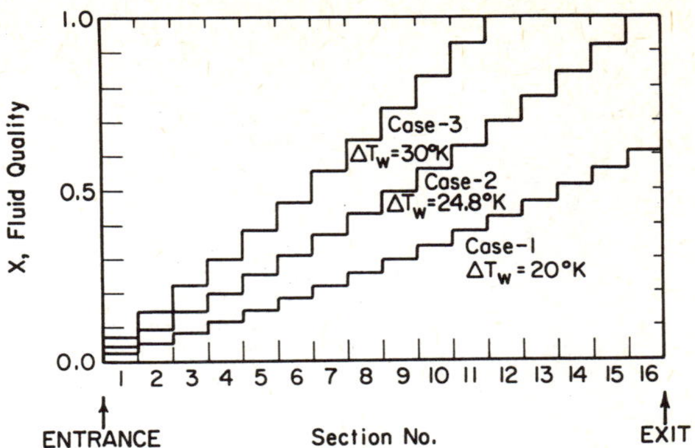

Fig. 16. Illustrative change in fluid quality from fin-row to fin-row for the finned exchanger of Figs. 12 and 13. The inlet fluid is saturated liquid Freon-113 at 0.03m/s. From Chen, Loh, and Westwater [41].

Figure 17 illustrates the predicted velocity effect according to the Model II calculations described herein. The total heat duty is shown as a function of ΔT_w for three liquid inlet velocities: 0.015, 0.030 and 0.045m/s. Recall that with a thermosiphon reboiler the inlet velocity is uncontrolled in the sense that it is a dependent variable fixed by the geometry, the physical properties, and the temperature difference. In particular, V_{in} changes as ΔT_w changes.

Figure 17 shows that for ΔT_w in the region of 10 K, good agreement exists between the measured and the predicted heat duties for liquid inlet velocities of roughly 0.015-0.045m/s. For ΔT_w around 65 K, the inlet velocity would need to be about 0.045m/s to produce the correct heat duty. This last corresponds to an exit velocity of 9.5m/s and a quality of 1.0. The other cases correspond to a quality of 1.0 or less and to exit velocities smaller than 9.5m/s. All of these inlet and outlet velocities are practical for and typical of reboilers.

The Model II predictions in Fig. 17 show that the experimental heat duties can be explained better than before. Panitsidis et al. [40] used a design method which was too conservative in that the local boiling heat transfer coefficients were assumed to be independent of the fluid velocity. When the velocity effect is included, the highest observed heat duties become quite rational.

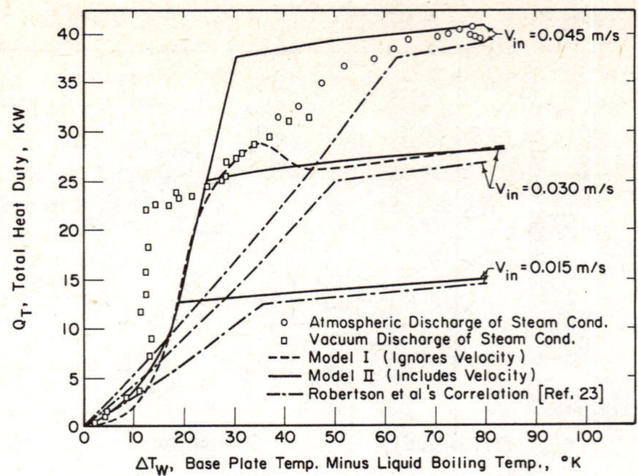

Fig. 17. Experimental data for Freon-113 compared to two predictions based on the local assumption. The Robertson correlation [25] is included for comparison. From Chen, Loh, and Westwater [41].

Robertson [23,25] reported experimental results with boiling liquid nitrogen and Freon-11. The flow was vertically upward through an electrically heated rectangular channel with an internal finning similar to that in Fig. 13. The boiling heat transfer coefficient was assumed to be independent of ΔT but to be a function of the local mass quality and the Reynolds number at the inlet. The wall-to-bulk temperature difference was quite small, less than 2 K. Robertson's correlation [25] was used with Freon-113 to estimate the performance of the compact heat exchanger used at the University of Illinois. The computation procedures were the same as those of Model II except that the heat transfer coefficient was assumed to be uniform over a fin, and the concept of fin efficiency was used. Extrapolations of Robertson's data were necessary when Re < 200; (V_{in} < 0.03m/s). Three predictions by this method with V_{in} = 0.015, 0.030, and 0.045m/s are shown in Fig. 17 labelled 'Robertson'.

At low ΔT_w(< 10 K), the predictions by the two methods are close. At intermediate ΔT (~ 25 K), Robertson's predictions are only about half the measured values. For Robertson's correlation to give a good fit in this region, the inlet velocity would have to be about 1.0m/s, which is far in excess of the velocities actually used. This value also is far out of the velocity range of the thermosyphon reboiler described in the present study. Furthermore, this V_{in} causes large discrepancies in the heat duty at both lower and higher

ΔT's. The superiority of Model II appears in this intermediate ΔT region. At some high ΔT the predictions by Robertson's method and Method II come close again as expected. The inlet liquid flow will be all evaporated ($x = 1.0$) eventually as ΔT increases. The difference in heat duty will arise only because of the different small amount of sensible heat received by the superheated vapor. Since Model II predicts an earlier dryout with the same V_{in}, the total heat duty of the compact heat exchanger is a little higher.

It is not possible to test Robertson's data with the new method described in this paper until such time as complete flow boiling curves for Freon-11 or liquid nitrogen become available.

4. CONDENSATION IN COMPACT HEAT EXCHANGERS

The work of Colburn and Carpenter [43] was originally developed for condensation inside vertical, straight, round tubes. It was modified later by others for flow condensation in very small, non-round channels.

$$\frac{h\,\mu_L}{k_L\,\rho_L^{0.5}} = 0.065\left(\frac{C_L\,\mu_L}{k_L}\right)^{0.5}\tau^{0.5} \tag{19}$$

(Carpenter & Colburn - 1951)

Here τ is the interfacial shear and is related to the interfacial friction factor by $\tau = 1/2\, f\, G_v^2/\rho_v$. The mean mass velocity of the vapor is given by $G_v^2 = 1/3\,(G_1^2 + G_1G_2 + G_2^2)$, where G_1 and G_2 refer to the vapor at the inlet and outlet. The friction factor f is obtained from the usual graphs for one-component flow. A key feature of Eq. (19) is that the heat transfer coefficient is predicted to be independent of the ΔT but is fixed by hydrodynamic conditions and the geometry.

A large number of subsequent workers have been concerned with refining and improving the relationship. One example is by Ananiev et al. [44]. They had flow condensation of steam in a horizontal 13mm diameter tube. They suggested that one could calculate the heat transfer coefficient h_L for all-liquid flow and then correct it for two-phase flow by use of $h = h_L\,(\rho_L/\rho)^{0.5}$. Here, h and ρ are two-phase, local values. For an entire tube, the average heat transfer coefficient is given by Eq. (20).

$$\text{ave. } h = 1/2\, h_L\,\left(1 + \sqrt{\rho_L/\rho_v}\right) \tag{20}$$

(Ananiev, et al. - 1961)

The Ananiev approach is really very similar to that of Carpenter and Colburn as is that of various others [45-48].

Gopin et al. [49] condensed Freon-12, Freon-22, and Freon-142 flowing downward in a compact heat exchanger. The core size was 4 x 22 x 0.8cm, and was provided with four different sets of fins: one plain, and three offset. The channel equivalent diameter was 2.50 to 4.64mm. The maximum heat transfer was 12 KW/ m^2, and the maximum vapor velocity was 0.3 m/s. The data are shown in Fig. 18. As the heat flux varies the heat transfer coefficient (for any fin array) also varies, thus it is clear that the ΔT is a significant variable. All the fin arrays were correlated as shown in Fig. 19. The best fit line is Eq. (21).

$$\frac{h\,d}{k_L} = 1.06 \left[\left(\frac{\lambda}{C_L \Delta T}\right)\left(\frac{C_L \mu_L}{k_L}\right)\left(\frac{d^3 \rho_L^2 g}{\mu_L^2}\right)\right]^{1/4} \left(\frac{L}{d}\right)^{-0.125} \tag{21}$$

This is easily rearranged into Eq. (22).

$$\text{ave. } h = 1.06 \left[\frac{k_L^3 \rho_L^2 g \lambda}{L \mu_L \Delta T}\right]^{1/4} \left(\frac{L}{d}\right)^{1/8} \tag{22}$$

(Gopin et al. - 1976)

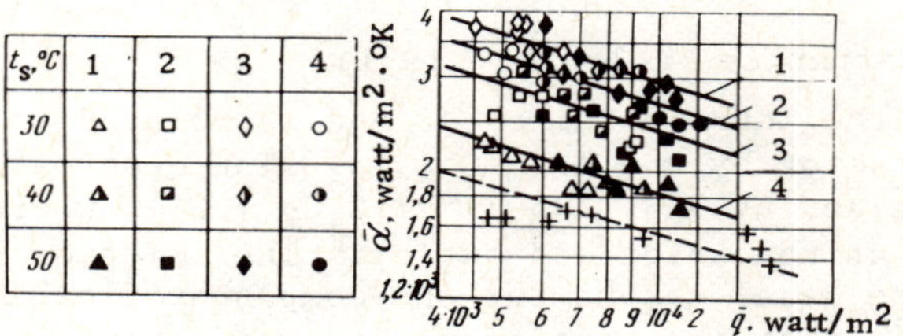

Fig. 18. Data for flow-condensation of Freon-22. Lines 1-4 indicate different plate-fin exchangers. The dashed line is for a flat vertical slot. From Gopin, Usyukin, and Averyanov [49].

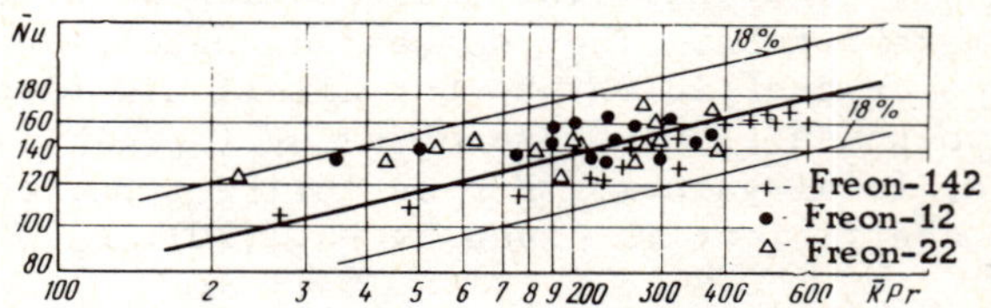

Fig. 19. Correlation of flow-condensation data for three Freons. The ordinate is the Nusselt number, and the abscissa is the first two groups on the right side of Eq. (21). The heat exchanger was No. 3 of Fig. 18. From Gopin, Usyukin, and Averyanov [49].

This is obviously a modification of Nusselt's equation for condensation on a vertical wall. The new group L/d accounts for the different geometries. The coefficient 1.06 is 12% greater than the value from Nusselt. Although Eq. (22) worked well for the described exchangers, it will probably fail for exchangers using large vapor velocities. Note that Carpenter and Colburn used vapor velocities about 100 times those used by Gopin.

M. M. Shah [50] worked out a correlation for flow condensation assuming that the wall everywhere is wet. His correlation is Eq. (23)

$$h = h_L \left[(1-x)^{0.8} + \frac{3.8\,x^{0.76}\,(1-x)^{0.04}}{P_R^{0.38}} \right] \tag{23}$$

(M. M. Shah - 1979)

Here h is the local heat transfer coefficient for flow condensation in a channel with any orientation, and h_L is the heat transfer coefficient for all the mass flowing as liquid as given by the Dittus-Boelter expression, Eq. (17). Eq. (23) is similar to Shah's correlation for flow boiling, Eq. (16). Both predict $h = 0$ at $x = 1$, however Eq. (23) sensibly predicts $h = h_L$ at $x = 0$. Shah suggests that the mean h for condensation may be obtained from Eq. (23) by using the mean quality for x. Figure 20 illustrates the Shah correlation for Freon-11 and Freon-22 in vertical and horizontal pipes of 6.8 to 20mm I.D. at mass velocities of 83 to 660 kg/m^{2}s. The mean deviation of the data from the curve is 17%. Equally close fits were obtained for 8 other liquids.

Another correlation which omits ΔT was presented by Tovazhnyanskiy and Kapustenko [51]. This is Eq. (24).

$$\frac{h\,d}{k_L} = \text{const.} \left(\frac{d\,G}{\mu_L}\right)^{0.73} \left(\frac{C_L\,\mu_L}{k_L}\right)^{0.43} \frac{1}{2} \left\{ \left[1 + x_1\left(\frac{\rho_L}{\rho_v} - 1\right)\right]^{1/2} + \left[1 + x_2\left(\frac{\rho_L}{\rho_v} - 1\right)\right]^{1/2} \right\} \tag{24}$$

(Tovazhnyanskiy & Kapustenko-1980)

Here x_1 and x_2 refer to the quality at the inlet and the exit. The authors made tests with steam at 1 to 3 bars condensing in a plate-type heat exchanger having corrugations 5mm high with a 18mm spacing, at a 60 degree angle with the long axis. No data were published, but the authors said the equation was satisfactory with the constant chosen as 0.15 for this geometry. The inlet velocity varied from 1 to 60 m/sec, the heat flux from 22 to 400 kW/m^2, and the exit quality from 0 to 0.6.

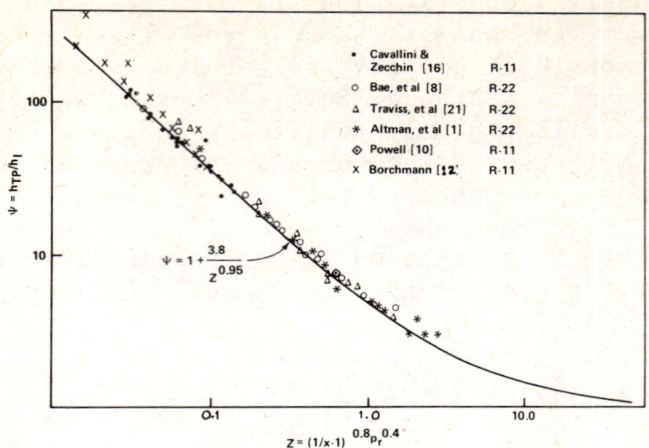

Fig. 20. Correlation by Eq. (23) for flow condensation of Freons in pipes as small as 6.8 mm diameter. From M. M. Shah [50].

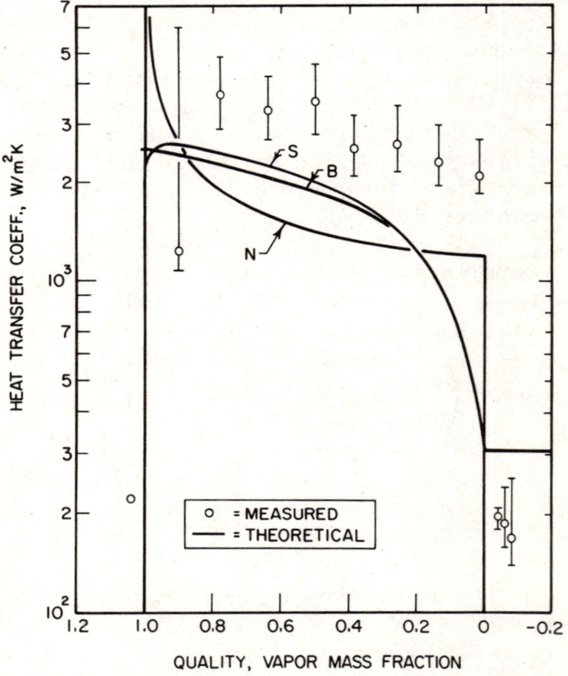

Fig. 21. Local values of quality and heat transfer coefficient for flow condensation of nitrogen in a compact exchanger. Comparison curves refer to Nusselt (N), Boyko and Kruzhilin (B), and M. M. Shah(S). From Haseler [52].

Condensation measurements for nitrogen in a compact heat exchanger were reported by Haseler [52]. The hydraulic diameter was 2.1mm. The channel was vertical, 3.2m high, 76mm wide, and provided with 709 fins/m. The fin area was 5.1 times the primary area. The heat transfer coefficient was assumed to be constant over any fin and therefore given by Eq. (13). Figure 21 shows Haseler's data for one run. The local heat transfer coefficients were larger than the Nusselt prediction by about 100% or so. They were also larger than predicted by Shah [50] or by Ananiev, Boyko and Kruzhilin [44], as shown in the graph.

It is clear from the work discussed in this section that a state of uncertainty exists. No accepted, standard method is yet available for predicting the heat duty of a compact heat exchanger used for flow condensation.

5. ACKNOWLEDGMENT

The National Science Foundation provided support for the preparation of this writing and for the described research at the University of Illinois. Important contributions by Mr. Cheng-Chung Chen are appreciated.

6. NOMENCLATURE

A	total surface area, m^2
A_1, A_2	primary and secondary surface area, m^2
C_p	specific heat, J/kg K
C_L	specific heat of liquid, J/kg K
C'	defined by Eq. (2)
d	equivalent diameter of flow passage, m
f	friction factor, dimensionless
F	correction factor in Eq. (9)
g	gravitational acceleration, m/s^2
g_c	conversion factor for English units, 32 lb m ft/lb f s^2
G	mass velocity, kg/sm^2
G_v	mean mass velocity of vapor, kg/sm^2
G_1, G_2	mass velocity at inlet, outlet, kg/sm^2
h	heat transfer coefficient, $W/m^2 K$
h_c	heat transfer coefficient for forced convection, $W/m^2 K$
h_L	heat transfer coefficient for all-liquid flow, $W/m^2 K$
h_{pool}	heat transfer coefficient for nucleate pool boiling, $W/m^2 K$
ΔH	enthalpy for 2-phase mixture, $C_L \Delta T + x\lambda$, J/kg
j	Colburn heat transfer factor, dimensionless
k_L	thermal conductivity of liquid, W/m K
L	length of flow channel, m
m	fin parameter, $y(2h/kt)^{1/2}$, dimensionless
q	heat flux, W/m^2

q_c	heat flux for forced convection with no phase change, W/m^2
q_{pool}	heat flux for nucleate pool boiling, W/m^2
Q	heat transfer rate, W
Δp	vapor pressure difference corresponding to ΔT, Pa
P	pressure, Pa
P_R	reduced pressure, dimensionless
Pr	Prandtl number, $C_p\mu/k$, dimensionless
ΔP	pressure drop, Pa
Re	Reynolds number, dG/μ, dimensionless
S	correction factor in Eq. (8), dimensionless
t	fin thickness, m
ΔT	wall temperature minus fluid temperature, K
W	flow rate, kg/s
x	vapor quality, dimensionless
y	fin height, m
z	distance from channel entrance, m
α	void fraction, dimensionless
η	fin efficiency, dimensionless
λ	latent heat of vaporization, J/kg
μ_L	viscosity of liquid, Pa s
ρ, ρ_L, ρ_v	density of fluid, of liquid, of vapor, kg/m^3
σ	surface tension, N/m
τ	shear stress, Pa
ϕ_{LO}	multiplier for 2-phase friction, dimensionless

7. REFERENCES

1. Mikus, H. J. "Cryogenics and the plate-fin exchanger." Cryogenic Engineering News (Sept. 1967) 20-23.
2. Lenfestey, A. G. "Low temperature heat exchangers." Progress in Cryogenics, Vol. 3, edited by K. Mendelssohn, Academic Press, New York (1961) 25-47.
3. Colburn, A. P. "A method of correlating forced convection heat transfer data and a comparison with fluid friction." Trans. AIChE 29 (1933) 174-210.
4. Kays, W. M. and A. L. London, Compact Heat Exchangers, 2nd edition, McGraw-Hill Book Co., New York (1964).
5. Shah, Ramesh K. "Compact heat exchangers." Heat Exchangers, edited by S. Kakac, A. E. Bergles and F. Mayinger, Hemisphere Publishing Co., Washington, D.C. (1981) 111-151.
6. Lenfestey, A. G. "Heat transfer and fluid flow." Advanced Cryogenics, edited by C. A. Bailey (1971) 155-181.
7. Martinelli, R. C. and R. W. Lockhart. "Proposed correlation of data for isothermal, two-phase, two-component flow in pipes." Chem. Eng. Progress 45, No. 1 (1949) 39-48.

8. Clark, J. A. and R. M. Thorogood. "Selection and design of heat exchangers." Cryogenic Fundamentals, ed. by G. G. Haselden, Academic Press, New York, Chapter 3 (1971) 150-166.

9. Timmerhaus, K. D. and R. J. Schoenhals. "Design and selection of cryogenic heat exchangers." Advances in Cryogenic Engineering 19 (1974) 445-462.

10. Chenoweth, J. M. and M. W. Martin. "Turbulent two-phase flow." Petroleum Refiner 34, No. 10 (1955) 151-155.

11. Sydoriak, S. G. and T. R. Roberts. "Study of boiling in short narrow channels and its application to design of magnets cooled by liquid H_2 and N_2." J. Applied Physics 28, No. 2 (1957) 143-148.

12. Richards, R. J., R. F. Robbins, R. B. Jacobs and D. C. Holten. "Heat transfer to boiling liquid nitrogen and hydrogen flowing axially through narrow annular passages." Advances in Cryogenic Engineering 3 (1960) 375-389.

13. Rohsenow, W. M. "Heat transfer associated with nucleate boiling." Proceedings of Heat Transfer and Fluid Mechanics Institute, Stanford University Press, (1953) 123-141.

14. Chen, J. C. "Correlation for boiling heat transfer to saturated fluids in convective flow." Ind. Eng. Chem. Process Des. Dev. 5 (1966) 322-329.

15. Collier, John G. Convective Boiling and Condensation, McGraw-Hill International Book Co., New York, 214-219 (1981).

16. Forster, H. K. and N. Zuber. "Dynamics of vapor bubbles and boiling heat transfer" AIChE J. 1 (1955) 531-535.

17. Johannes, C. and J. Mollard. "Nucleate boiling of Helium I in channels simulating the cooling channels of superconducting magnets." Advances in Cryogenic Engineering 17 (1972) 332-341.

18. Collier, J. G., T.D.A. Kennedy, and J. A. Ward. "Thermal design of plate-fin reboilers." Cryotech 73 Prod. Use Ind. Gases Proc. Conf., IPC Technol Press, Guildford, England (1974) 95-100.

19. Sydoriak, S. G. "Interpretation and correlation of deviations from a hydrodynamic channel boiling theory of data on nitrogen and hydrogen in forced and free flow boiling. Advances in Cryogenic Engineering 18 (1973) 73-80.

20. Shorin, C. N., V. I. Sukhov, S. A. Shevyakova, and V. K. Orlov. "Experimental investigation of heat transfer with boiling oxygen in vertical channels during condensation heating." Intern. Chem. Eng., 14, (1974) 517-521.

21. Galezha, V. B., I. P. Usyukin and K. D. Kan. "Boiling heat transfer with freons in finned-plate heat exchangers." Heat Transfer-Soviet Research 8, No. 3 (1976) 103-110.

22. Shah, M. M. "A new correlation for heat transfer during boiling flow through pipes." ASHRAE Transactions 82, No. 2 (1976) 66-86.

23. Robertson, J. M. "Boiling heat transfer with liquid nitrogen in brazed aluminum plate-fin, heat exchangers." AIChE Symposium Series 75, No. 189 (1979) 151-164.

24. Robertson, J. M. "Review of boiling, condensation and other aspects of two-phase flow in plate-fin, heat exchangers." Compact Heat Exchangers HTD-Vol. 10, American Society of Mechanical Engineers, New York (1980) 17-27.

25. Robertson, J. M. and P. C. Lovegrove, "Boiling heat transfer with freon-11 in brazed aluminum plate-fin, heat exchangers." ASME Preprint 80-HT-58, 19th National Heat Transfer Conference, Orlando (1980).

26. Haley, K. W. and J. W. Westwater. "Heat transfer from a fin to a boiling liquid." Chem. Eng. Science 20 (1965) 711.

27. Haley, K. W. and J. W. Westwater. "Boiling heat transfer from single fins." Proc. 3rd Int. Heat Transfer Conf., Chicago 3, New York (1966) 245-253.

28. Cash, D. R., G. J. Klein and J. W. Westwater. "Approximate optimum fin design for boiling heat transfer." J. Heat Transfer 93 (1971) 19-24.

29. Klein, G. J. and J. W. Westwater. "Heat transfer from multiple spines to boiling liquids." AIChE J. 17 (1971) 1050-1056.

30. Bondurant, D. L. and J. W. Westwater. "Performance of transverse fins for boiling heat transfer." Chem. Eng. Prog. Symp. Ser. 67, No. 113 (1971) 30-37.

31. Shih, C. C. and J. W. Westwater. "Use of coatings of low thermal conductivity to improve fins used in boiling liquids." Int. J. Heat Mass Transfer 15 (1972) 1965-1968.

32. Westwater, J. W. "Development of extended surfaces for use in boiling liquids." Chem. Eng. Prog. Symp. Ser. 69, No. 131 (1973) 1-9.

33. Shih, C. C. and J. W. Westwater. "Spheres, hemispheres and discs as high-performance fins for boiling heat transfer." Int. J. Heat Mass Transfer 17 (1974) 125-133.

34. Shih, C. C. and J. W. Westwater. "Effect of geometry and wall thickness on heat transfer from longitudinal fins to boiling liquids." Proc. 5th Int. Heat Trans. Conf., Tokyo, 4 (1974) 15-19.

35. Petukhov, B. S., S. A. Kovalev, V. M. Zhukov and G. M. Kazakov. "Study of heat transfer during the boiling of a liquid on the surface of a single fin." Heat Transfer-Soviet Research 4 (1972) 148-156.

36. Dul′kin, I. N., N. I. Rakushina, L. I. Royzen, and V. G. Fastovskiy. "Heat transfer with water and Freon-113 boiling on a non-isothermal surface." Heat Transfer-Soviet Research 3 (1971) 61-69.

37. Cumo, M., S. Lopez and G. C. Pinchera. "Numerical calculations of extended surface efficiency." Chem. Eng. Prog. Symp Ser. 61, No. 59 (1965) 225-233.

38. Lai, F. S. and Y. Y. Hsu. "Temperature distribution in a fin partially cooled by nucleate boiling." AIChE J. 13 (1967) 817-821.

39. Siman-Tov. "Analysis and design of extended surfaces in boiling liquids." Chem. Eng. Prog. Symp. Ser. 66, No. 102 (1970) 174-184.

40. Panitsidis, H., R. D. Gresham, and J. W. Westwater. "Boiling of liquids in a compact plate-fin, heat exchanger." Int. J. Heat Mass Transfer 18, (1975) 37-42.

41. Chen, C. C., J. V. Loh and J. W. Westwater. "Prediction of boiling heat transfer duty in a compact plate-fin heat exchanger using the improved local assumption." Int. J. Heat Mass Transfer 24, (1981) 1907-1912.

42. Yilmaz, D. and J. W. Westwater. "Effect of velocity on heat transfer to boiling Freon-113." J. Heat Transfer 102 (1980) 26-31.

43. Carpenter, F. G. and A. P. Colburn. "The effect of vapor velocity on condensation inside tubes." Proc. of General Discussion on Heat Transfer, Institute of Mechanical Engineers/American Society of Mechanical Engineers (1951) 20-26.

44. Ananiev, E. P., L. D. Boyko and G. N. Kruzhilin. "Heat transfer in the presence of steam condensation in a horizontal tube." International Developments in Heat Transfer, Proc. 2nd Int. Heat Transfer Conference 2 (1961) 290-295.

45. Akers, W. W., H. A. Deans and O. K. Crosser. "Condensing heat transfer within horizontal tubes." Chem. Eng. Progress 54, No. 10 (1958) 87-90.

46. Akers, W. W., O. K. Crosser and H. A. Deans. "Condensing heat transfer within horizontal tubes." Chem. Eng. Progr. Symp. Ser. 55, No. 29 (1959) 171-176.

47. Akers, W. W. and H. F. Rosson. "Condensation inside a horizontal tube." Chem. Eng. Progress Symp. Ser. 56, No. 30 (1960) 145-149.

48. Altman, M., F. W. Staub and R. H. Norris. "Local heat transfer and pressure drop for Refrigerant-22 condensing in horizontal tubes." Chem. Eng. Progress Symp. Ser. 56, No. 30 (1960) 151-159.

49. Gopin, S. R., I. P. Usyukin, and I. G. Aver′yanov. "Heat transfer in condensation of freons on finned surfaces." Heat Transfer-Soviet Research 8, No. 6 (1976) 114-119.

50. Shah, M. M. "A general correlation for heat transfer during film condensation inside pipes." Int. J. Heat Mass Transfer 22 (1979) 547-556.

51. Tovazhnyanskiy, L. L. and P. A. Kapustenko. "Heat transfer from steam condensing in an extended-surface slot channel." Heat Transfer-Soviet Research 12, No. 4 (1980) 34-36.

52. Haseler, L. E. "Condensation of Nitrogen in brazed aluminum plate-fin, heat exchangers." ASME Paper 80-HT-57, 19th National Heat Transfer Conference, Orlando (1980).

INVESTIGATIONS ON TWO-PHASE FIOW IN STEAM ACCUMULATORS

S. ISHIGAI*, S.NAKANISHI* and H. SUGITA**

* Faculty of Engineering
Osaka university
Osaka, Japan
** Kobe University of Merchantile Marine
Kobe, Japan

ABSTRACT

The characteristics of the steam accumulator system in the electricity utility system in comparison with the pumped storage hydraulic system demands that the steam accumulator system be used for shaving off peak loads of shorter duration, say one or two hours, than the pumped storage. Responce of the bulk water and the steam space to rapid discharging of steam was investigated up to 2 m/s average steam velocity out of the water surface. Bulk water flow in the charging period was found dependent on the baffle configuration and a method of prediction was presentd.

1. INTRODUCTION.

The number of installed steam accumulators in indutrial plants has steadily been increasing in Japan since the energy crisis in 1973, and steam accumulator application to the electricity utility system is under investigation to supplement the pumped storage hydraulic plants whose site finding is becoming increasigly difficult in Japan while the hourly or daily load fluctuation is getting worse.

In spite of the long history of the steam accumulator since the 1929 installation in Berlin-Charlottenburg Staion, no significant progress has been realized up to the present time. Our preliminary study has indicated that the steam accumulator system integrated in the electricity generation system is nearly equal in the exergy efficiency to the pumped hydraulic storage system, but is much inferior in the storage capacity. It is therefore reasonable to use the accumulator system for shaving off the peak load of shorter duration than the pumped storage system. In other words, not only the amount of storable energy but also the time rate of energy transfer, and consequently the mass flow rate of steam to or from the accumulator is important. Responce of the two-phase flow in the bulk of water and in the steam space of the accumulater is treated in this paper.

2. VISUAL STUDY OF LIQUID SPACE BEHAVIOR IN DISCHARGING PERIOD.

Freon-113 was charged to the predetermined level in two cylindrical test sections with 100 mm I.D. and about 1000 mm height. One is made of pyrex glass and heated by a hot water bath. The other is made of acrylite resin and is heated by an external heating loop. The former is referred to as the homogeneous nucleation experiment, and the latter, the abundant nuclei experiment. After Freon-113 was heated to the specfied temperature, the blow valve at the top of the cylinder was quickly opened. Observations were made untill the pressure of the test section dropped to the atmospheric pressure.

Results with the homogeneous nucleation experiments: Fig. 1 is a typical example of the pressure trace. The pressure curve is accompanied by a number of large pressure surges. Each surge corresponds to an intermittent bulk boiling. At the beginning of each surge, a tiny bubble appears in the central part of the liquid space, and grows very rapidly to a large vapor slug inducing very sharp pressure rise. As the slug rises, a bulk boiling zone is formed just below the liquid level and grows to a foaming layer. Once the vapor slug escapes into the vapor space, the bulk boiling calms down, the foaming layer shrinks, and the pressure sharply falls down. Incidentally, a new tiny bubble appears in the bulk and the same process as above repeats again. When the pressure comes down to the atmospheric pressure, the liquid superheat still is there. Such case with no or very few nucleation site is unlikely to be there in practical applications, especially in the steam accumulator.

Results with the abundant nuclei experiments: In this case, there are abundant nuclei for boiling particularly around the bottom connection to the heating loop. Fig.2 is a typical example of serial photographs of the discharging process. The experimental parameters are as follows: the initial superheat of liquid above the normal saturation temperature, ΔT_o, is 13.4 C, the initial liquid level from the bottom, H_o, 480 mm and the size of the steam blow line is 1/2 B. The origin of time is taken at the moment when the blow line was opened. At 0.4 sec, a small bubble enters from the heating pipe and goes up growing rapidly to a large vapor slug. In the wake of the slug, a two-phase bulk boiling regeon is formed and expands over the whole liquid volume. In contrast with the glass tube experiment,

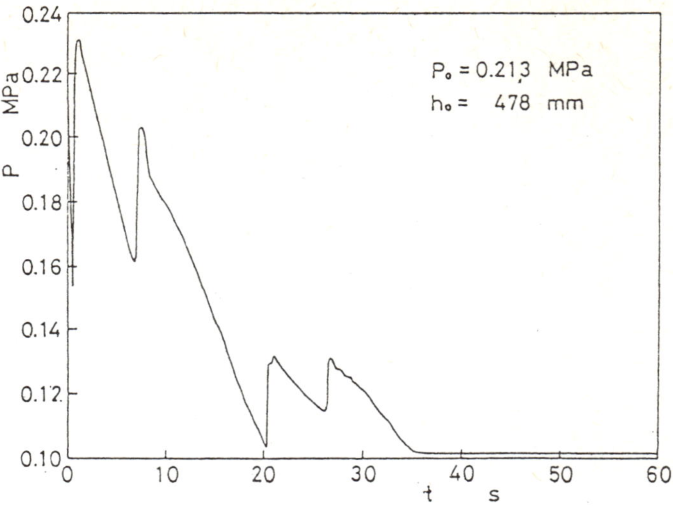

Fig. 1 Pressure curve for a homogeneous nucleation experiment with glass test section.

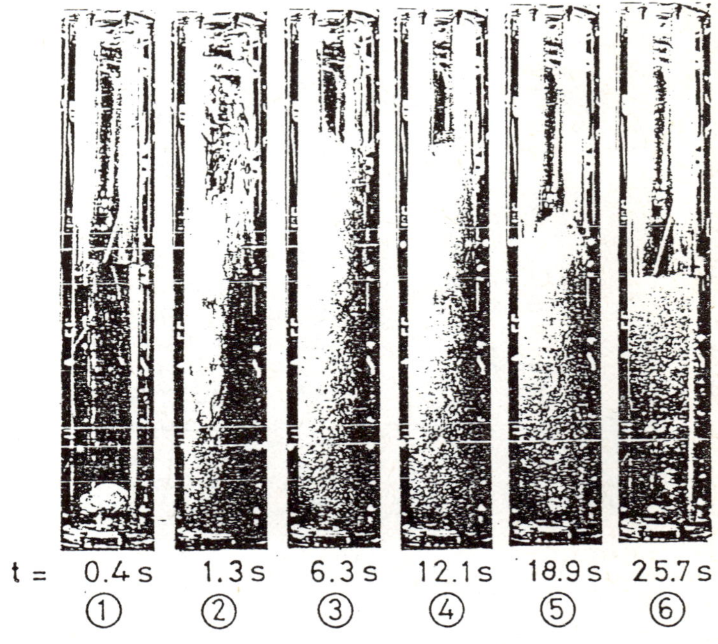

Fig. 2 Serial photographs of depressurization in an abundant nuclei experiment with acrylite resin test section.

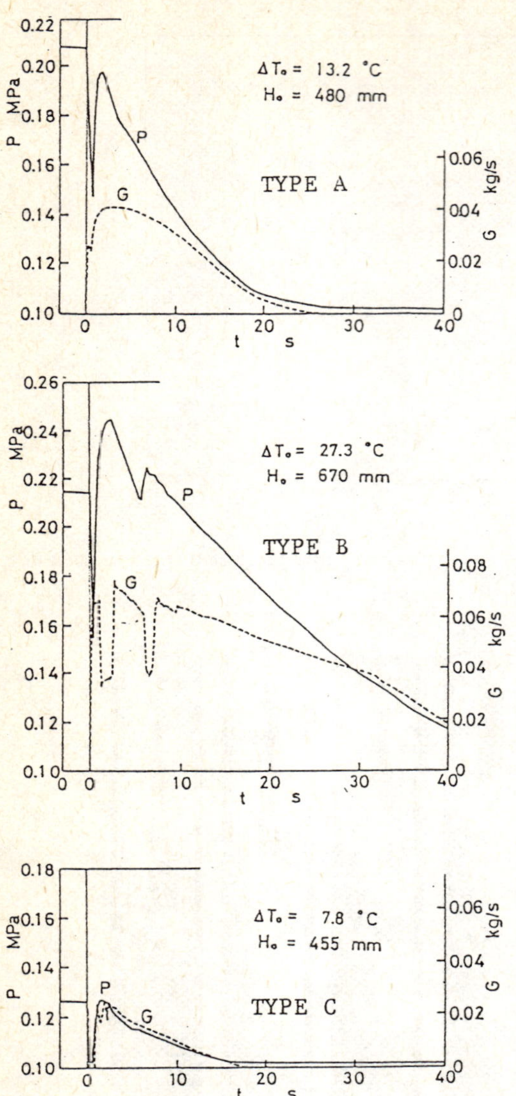

Fig. 3 Patterns of pressure curves.

the bulk boiling region does not begin to shrink immediately after the slug departure. At about 15 sec, it starts to shrink. The pressure curves during the runs can be classified into three types as shown in Fig. 3. The pattern is dependent both on the initial superheat and on the liquid level. In type A, the pressure sinks gradually to the atmospheric one after initial undershoot and restoration. Fig. 2 corresponds to this pattern. The pattern B is the pattern A superposed with additional small spikes and is seen when the liquid is larger in quantity. The pattern C is seen when the initial superheat is rather small and is similar to the pattern with glass tube test section.

Fig. 4 is the pattern map for this experiment. The patterns A and B correspond to lower and higher liquid levels, respectively. This fact indicates that the bulk boiling region can not extend to whole liquid level at once when the liquid level is high. The rate of depressurization was found to have some effect on the pattern.

3. EXPERIMENTAL RESULTS WITH WATER.

The experiment with Freon-113 gives us an insight into the mechanism of steam discharge by depressurization, but the result is not directly applicable to the apparatus with water as its working fluid because of the lack of the suitable similarity law. Experiments with water were carried out by using the test rig as shown

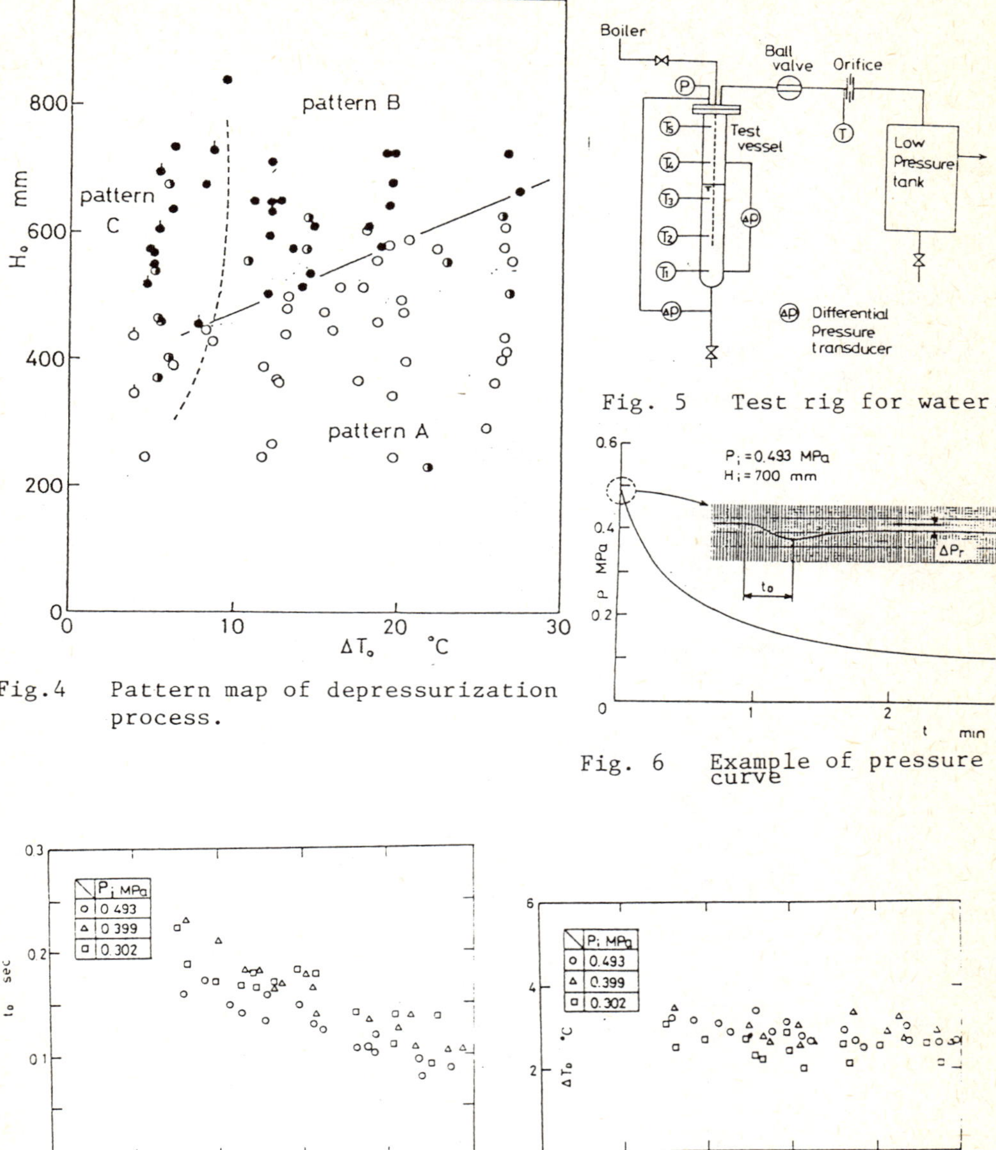

Fig.4 Pattern map of depressurization process.

Fig. 5 Test rig for water.

Fig. 6 Example of pressure curve

Fig.7 Delay time of incipient boiling.

Fig. 8 Superheat at incipient boiling.

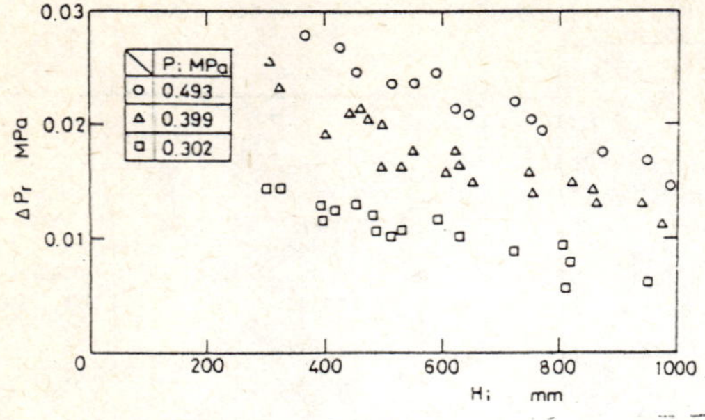

Fig. 9 Pressure recovery.

in Fig. 5. The test vessel is made of a carbon steel pipe with 141 mm I.D. 1500 mm height and the maximum allowable pressure is 7 MPa. After repeated charge and discharge of steam for degassing and temperature equalization of the test rig, the pressure and the water level in the vessel are set as specified. Then the steam blow off valve is quickly opened to start the depressurization. The profile of the pressure curve as indicated in Fig. 6 can be classified as the pattern A of Fig. 4, though the pressure undershoot at the initial stage is much smaller than that of Freon. In the enlarged record in Fig. 6, t_D is the time delay of the incipience of vapor generation, and the corresponding liquid superheat is ΔT_D which is reproduced in Fig. 8. Both t_D and ΔT_D are much smaller than those for Freon. However, their effect on the performance of steam accumulator is not neglisible, and some method of accelarating nucleation will certtainly help to improve the accumulator performance.

Lienhard and his coworkers proposed semiempirical formulae for ΔP_D and ΔT_r based on data of very rapid depressurization tests of hot water [4,5];

$$\Delta P_D = 1.33 \frac{\sigma^{1.5}\, T_r^{13.73}(1 + 14\Sigma^{0.8})^{0.5}}{\sqrt{G_b}\sqrt{kT_c}\,(1 - v'/v'')} \quad ; \qquad (1)$$

$$Ja \equiv \frac{\rho' c_p' \Delta T_r}{\rho''(h'' - h')} \simeq 1.26 \quad , \qquad (2)$$

where ΔP_D: pressure difference corresponding to ΔT_D, atm;

ΔT_r: temperature difference corresponding to ΔP_r;

T_c: critical temperature;

T_r: reduced initial temperature;

c_p': isobaric specific heat of boiling water;

k : Boltzmann's constant;

Gb: Gibbs number, taken 28.6 in [5];

Ja: Jakob number;

ρ' ρ'': specific densities of boiling water and dry saturated steam, respectively;

σ: surface tension;

Σ: rate of depressurization, Matm/s.

Comparison of our experimental data with the above equations are presented in Figs. 10 and 11. Equation (2) fits our ΔT_r data well as seen from the figure, while Eq. (1) estimates about two times larger values. This discrepancy is attributable to the lower rate of depres-

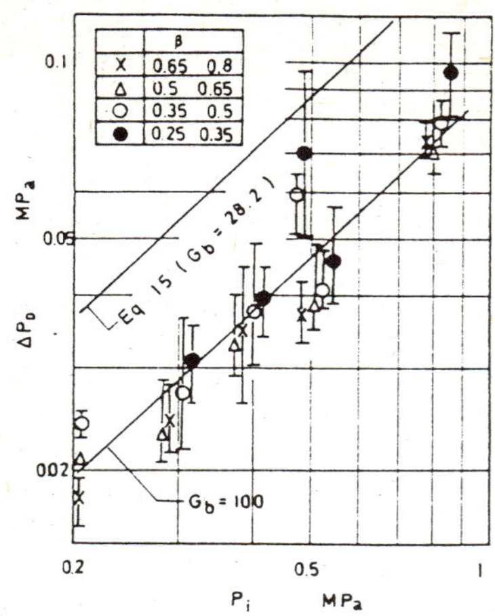

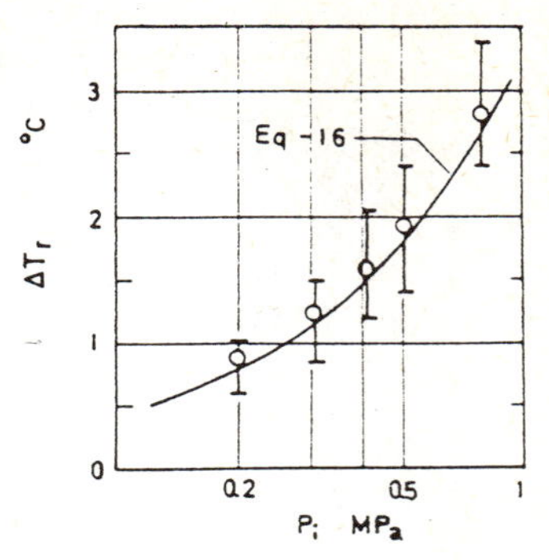

Fig. 11 Comparison of our data with Eq.(2)

Fig. 10. Comparison of our ΔP_D data with Eq.(1).

surization and the more abundant nucleation sites than the experiments by Lienhard et al [4,5]. If the Gibbs Number is replaced to 100 instead of the original 28.6, estimation by Eq.(1) is improved. Though the assumptions for Eq. (1) are not satisfied in our experiment, The equation is still valid for prediction of pressure undershoot.

4. THERMOHYDRAULICS OF CHARGING PROCESS.

Nonuniformity of temperature distribution in the accumulator induces some discrepancy between the pressure in the steam space and the saturation pressure corresponding to the mean water temperature, and results in irreversible loss of work at steam generation. The mixing motion in the water space is induced by the momentum of charged steam and the density difference due to the temperature distribution. As it is known that the latter driving force is considerably weaker than the former, the mixing by the momentum of the condensing steam jet at charging period is crucial for realization of uniform temperature distribution. In the following, the mixing process is investigated for the two-dimensional models.

Two test rigs are used; one is for Freon-113 and another is for air-water simulation. The test section for Freon-113 is made from acrylite resin plate, and its inside size is 200 mm in width, 250 mm in height and 30 mm in depth. The charging pipe is located on the vertical center line of the section and feeds vapor upward. After setting the liquid level, vapor charging starts; the pressure in the vapor space and the temperature at several points in the liquid space are recorded, and visual observation is conducted. Fig. 11 shows a typical example of temperature recording; the uppermost curve indicates the saturation temperature corresponding to the pressure in the vapor space. It is seen from the figure that there exists significant

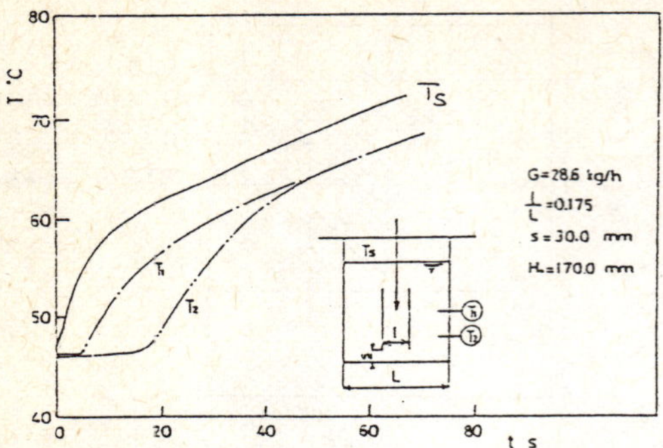

Fig. 11. Temperature curves for charging process.

nonuniformity of the temperature distribution. It is found that the degree of nonuniformity depends on the geometrical arrangement of two baffle plates which simulate the circulation tube of the steam accumulator.

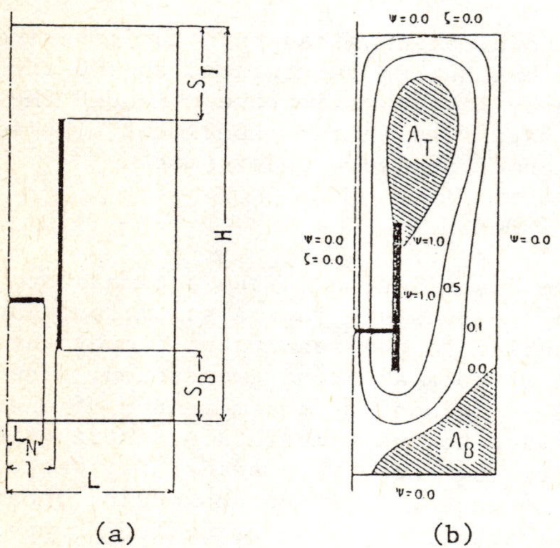

(a) (b)

Fig. 12. Nomenclature.

The test section for air-water simulation is also made from acrylite resin plate; its inside size is 470 mm in width, 1000 mm in height and 50 mm in depth. In this rig, the flow pattern induced by an air jet is visually observed. In this simulation, large isolated circulating flow zones are observed at the bottom corner, the backside of the baffle plate, and just below the liquid level; the appearance of such zones is responsible to the malmixing of liquid space. It is found that the area of the circulating zones largely depends on the size and arrangement of the baffle plates.

As it is too tedious to determine the area of circulating zones and to optimize the arrangement of the baffles by experimental works, we attempt to do them by integrating numerically the two-dimensional Navier-Stokes equations. The model for calculation is shown in Fig. 12,a; the gas-liquid interface is assumed straight, and the incoming vapor flow is replaced with single-phase liquid flow. Numerical integration is conducted by using the vorticity, ζ, and the stream function, ψ [6]. The Reynolds number, Re, is fixed to 100 for all calculation, because it was shown in the preliminary calculations that the results using this value well agreed with the pattern observed at the air-water simulation and the rate of convergence was reasonable. The Reynolds number is defined

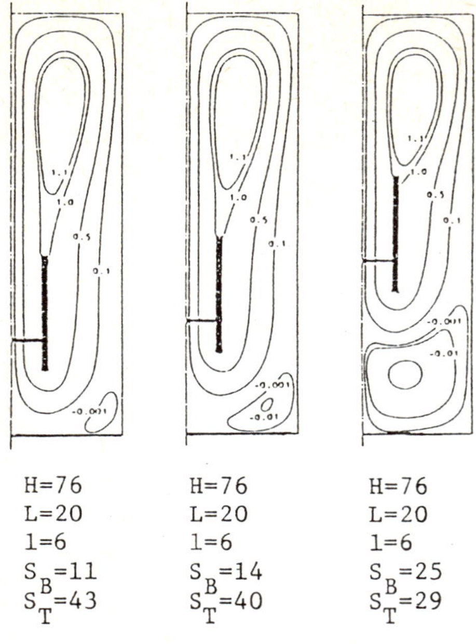

Fig. 13. Examples of calculated patterns.

by taking the inlet nozzle width, L_N, as the characteristic length and the mean velocity in it, U_0, as the characteristic velocity. Some examples of calculation results are shown in Fig. 13. It is clearly seen from the figure that the effect of the baffle plate exists. It is found that the most important parameters are the relative distance between the bottom and the lower end of the baffle, s_B/l, and the relative distance between the upper end of the liquid level, s_T/l. Fig. 14 shows the effects of these parameters on the fractions of area occupied by the circulating flow zones, A_T/A, A_B/A and $(A_T + A_B)/A$ (cf. Fig. 12,b). The effects of these two parameters can be mutually independent. The optimal range of parameters exists to minimize the total area of the circulating flow zones, and gives the best mixing in the liquid space. It is given by

$$s_T/l = 2 - 5, \text{ and } \quad s_B/l \simeq 1. \qquad (3)$$

Though the analysis is conducted for the two-dimensional case, the criteria (3) are considered as approximately applicable to the three-dimensional axisymmetrical cases, which are encountered in practical applications.

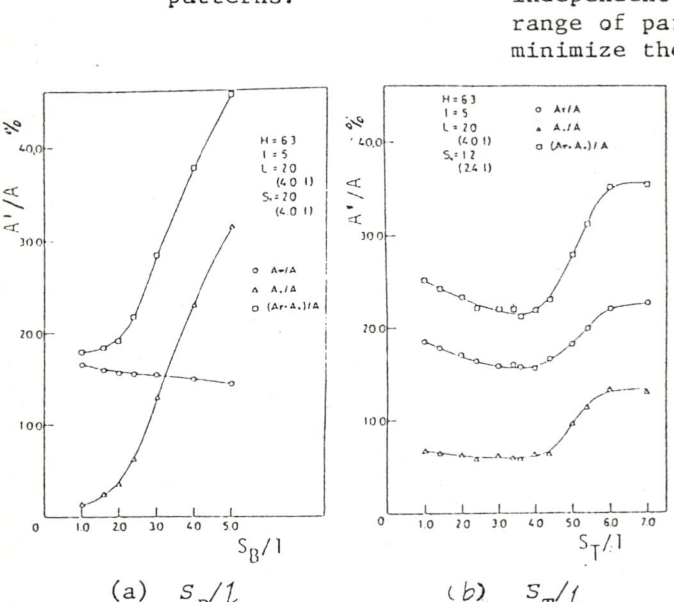

Fig. 14. Effects of the parameters.

5. DROPLET ENTRAINMENT IN STEAM SPACE.

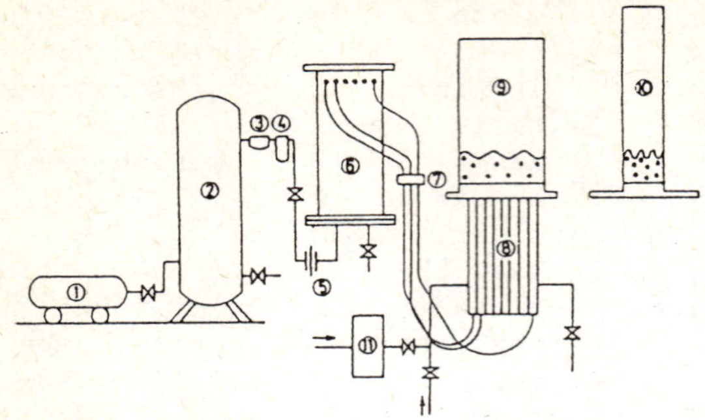

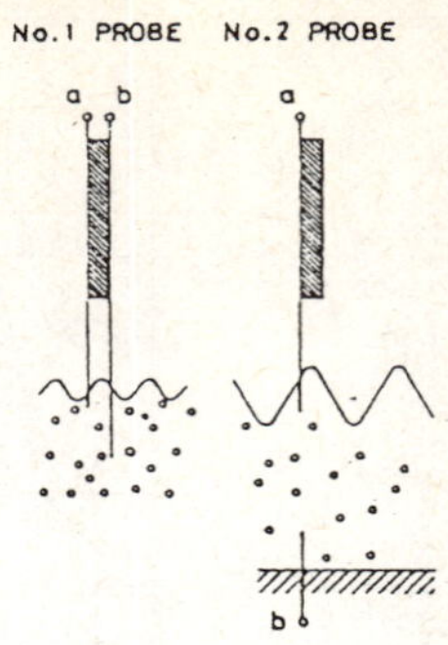

Fig. 15 General arrangement.
1: air compressor 2: air reservoir
6: air ditributor 7: throttle pieces
8: riser tubes 9: No. 1 drum
10: No. 2 drum 11: water purifier

Fig. 16 Water level probes.
No.1 for $U \leqq 1.0$m/s
No.2 for $U \geqq 1.0$m/s

Air is admitted uniformly to the bottom of the water-containig transparent cylinder, referred to as the No. 1 drum or the No. 2 drum, 300 mm ID, 1500 mm height or 150 mm ID, 2000 mm height. Because the surface wave is rough, the water level is defined as the position at which the tip of the probe as shown in Fig. 16 is in the water during 50 % of the time. The loading of air on the water surface is expressed by the velocity U m/s which is the amount of air dividedby the inner cross-section of the drum in the experiment. No. 2 drum is used for very high air loading of U above 1 m/s. The amount of entrained water and the droplet size distribution are measured in every test run at various height H above the predetermined water level h measured from the bottom of the drum.

As shown in Fig.17, the entrained droplet size is well represented by the following Rosin-Rammler ditribution, i.e.,

$$R = 100 \exp\left(-\frac{D_p}{D_e}\right)^n$$

where R: mass percentage of all droplets whose diameter exceeds D_p;
D_e : representative diameter corresponding to R = 36.8% ;
n : index for the uniformity of the distribution.

Evidently the space directly above the water level has larger droplets of less uniformity in size.

The representative diameter D_e of the droplet group diminishes first drastically and then suddenly slowly with the height above the water level as revealed by Fig.18 . The critical height H_c above which the droplet size ditribution is stable can thus be determined.

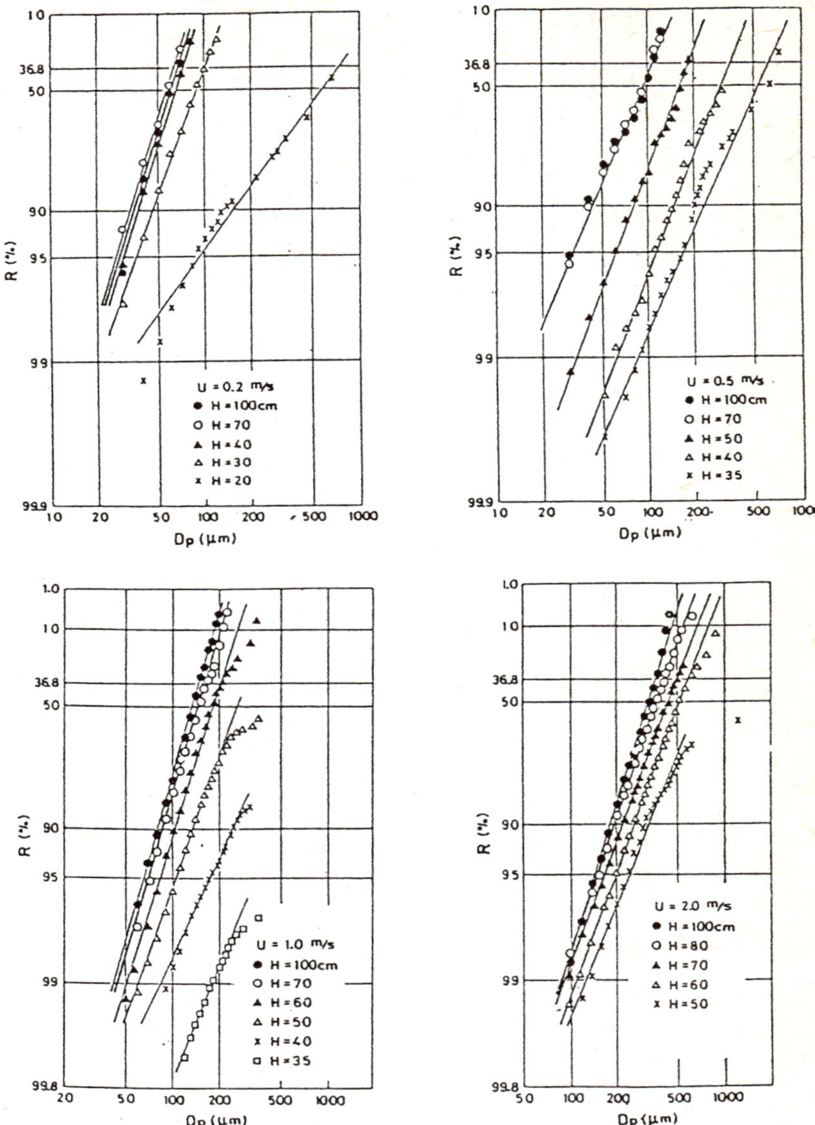

Fig. 17 Droplet size distribution in the space above the water level. U: air loading of the water surface, H; point of measurement above the water level.

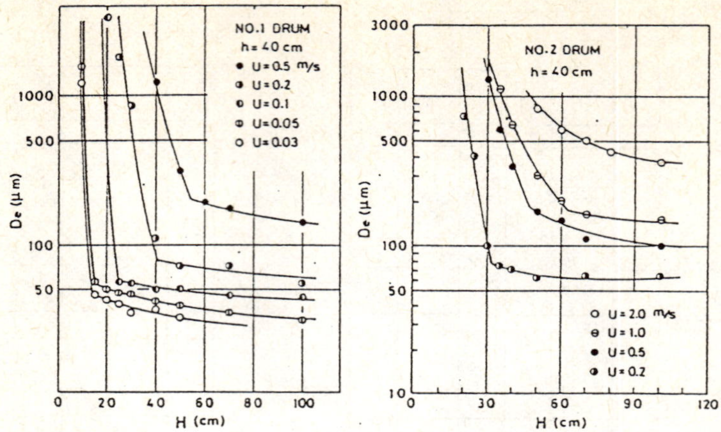

Fig. 18 Dependence of the representative diameter D_e on the height above the water level. The critical height H_c is discernible exept for the cases with exeptionally high air loadings.

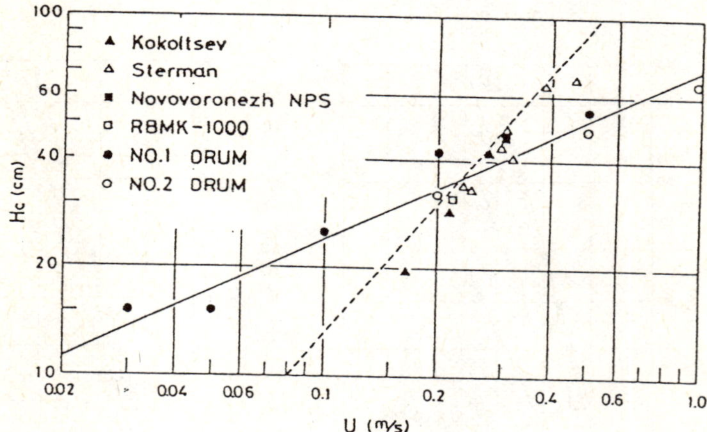

Fig. 19 Critical height H_c. Data and the dotted line by other authors are all for drums of steam boilers under pressure.

Obtained results are represented by the full line in Fig.19 and are compared with data by other authors obtained for steam boiler drums under pressure. The dotted line is by Kozlov [7]. The good agreement of the authors' line with the other data for steam and saturation water indicates the effectiveness of the presented method on the study of entrainment in actual steam accumulators.

It is an established fact that bubbles disappearing on the water

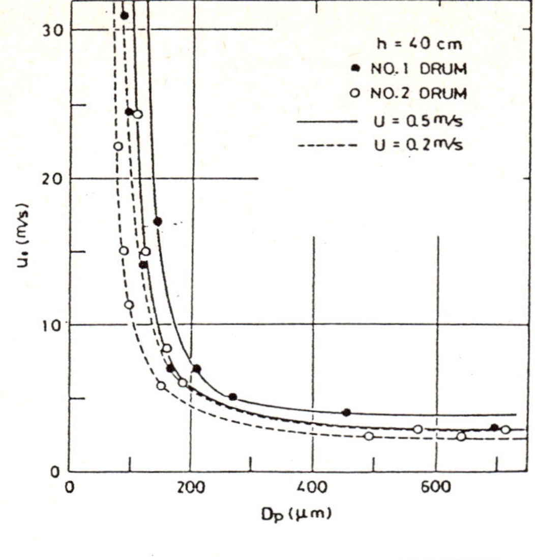

surface ejects droplets at speeds that depend on the size of the droplet. Under high air loading of the surface, the wave on the surface is so rough as to eject droplets of considerable sizes. Measurements on the upward ejected speed of droplets were made and is presented in Fig.20.

Fig. 20 Speed of ejected droplets.

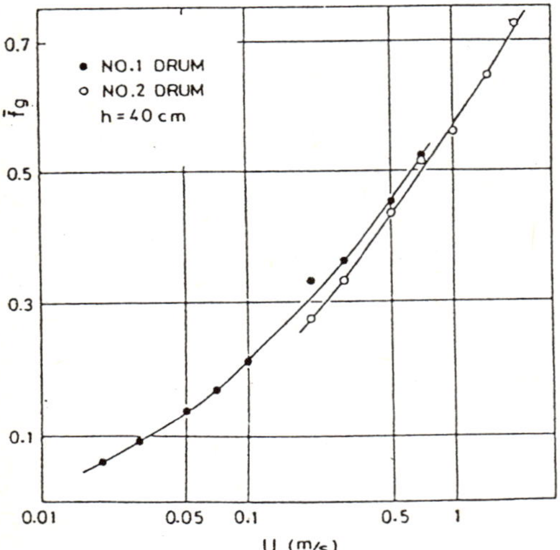

Fig. 21 Void fraction of the water space with the surface free.

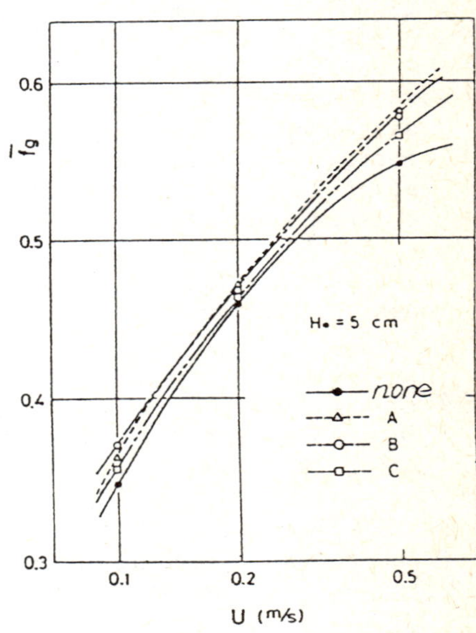

Fig. 22 Void fraction of the waterspace with a perforated plate at 5 cm below the free surface. The air speed through holes is from 0.58 to 2.93 m/s.

6. VOID FRACTION IN WATER SPACE.

Void fraction of the water space increases with the air loading through the water and raises the water level. Study was also made on the effect of air loading on water space void fraction. Only a limited amount of the result are given in Figs.21 and 22.

7. CONCLUSION.

The steam accumulator will surely find more application in the future electricity generating system because of its attractive features such as the easy site finding, comparable efficiency with the pumped storage hydraulic plant, and many others. It is suitable for the peak shaving of shorter duration, say one or two hours. Further study on rapid charging and discharging is expected to help its situation. Such study should naturally be supplemented by thermodynamic studies and automation.

REFERENCES

1. S.Nakanishi, M. Ozawa, S. Ishigai, S. Fujita, T. Matsumoto: "Transient behavior of pressurized hot liquid during depressurization". 16th Annual Symposium Proc., Heat Transfer Society of Japan, 1979, 364-366.

2. S, Ishigai, S, Nakanishi: "Investigations of the energy storage system using steam accumulators". Research on effective use of energy, Vol. 1, Reports of the special research project under grant in aid of scientific research of the Ministry of Eduation, Science and Culture, Japan. Jan. 1982, 191-198.

3. H. Sugita: "Research on separation of entrained water (in Japanese)". Doctorate thesis, Faculty of Engineering, Osaka University. 1980.

4. Md. Alagmir, J. H. Lienhard; "Correlation of pressure undershoot during hot water depressurization". Trans. ASME, J. Heat Transfer, 103C, 1981, 52-55.

5. Md. Alagmir, C. Y. Kan, J. H. Lienhard: "An experimental study of the rapid depressurization of hot water". Trans. ASME, J. Heat Transfer, 102C, 1980, 433-438.

6. P. J. Roache: "Coputational Fluid Dynamics", Hermosa Publishers, 1972.

7. Yu. V. Kozlov, V. F. Titov, V. I. Grishakov, V, B. Karasev, A. I. Dmitriev: "Use of precipitating separation of moisture in steam generators of nuclear power stations, and methods of calculatig it". Teploenergetica, 25-9, 1978, 61.

WORKSHOP PANELS:

UNRESOLVED PROBLEMS OF TWO-PHASE FLOW AND HEAT TRANSFER

MODELING TWO-PHASE SYSTEM AND INTERFACIAL TRANSFER

M. Ishii

Reactor Analysis and Safety Division
Argonne National Laboratory
Argonne, Illinois 60439

In general, two-phase flow problems are formulated in terms of the conservation equations. These conservation equations should be supplemented by various constitutive relations which specify transfer mechanisms within each fluid and at phase interfaces. The development of the two-phase flow constitutive relations depends heavily on experimental data and phenomenological modeling of these transfer mechanisms.

Initially Dobran posed questions on the sufficiency of the existing two-phase flow conservation equations and constitutive relations and methods to improve them if necessary.

The most important outstanding problems are related to the development of various constitutive relations and their verifications. Bankoff stressed the need of solving various limiting cases to find out the importance and shortcomings of a new formulation. The users of various models and constitutive relations should also be aware of a considerable spread usually existing between data and model predictions. Sometimes completely different mechanistic models can predict the same physical phenomena, thus the verification of a model cannot be easily accomplished on a small data base. The correct modeling of basic transfer mechanisms in two-phase flow depends heavily on the turbulent characteristics in the bulk fluids and at interfaces. The efforts to measure and understand the two-phase turbulence are necessary.

Both Kestin and Bankoff mentioned the roll of the entropy inequality for the development of the constitutive relations. Kestin emphasized its use for the modeling of dissipative phenomena and interfacial mass transfer terms. Certainly it should be a starting point for the interfacial transfer modeling. Lehay drew attention to the fact that some of the drag or virtual mass force models may not satisfy the second law of thermodynamics. The just setting or frame indifference requirement for constitutive relation was also cited. An assessment of various two-phase flow correlations and comparison with data were emphasized. For this purpose, Lehay proposed basic experiments which can isolate various effects and phenomena in two-phase systems. For example, the experiments on drag forces, turbulence and void behavior, adiabatic critical flow for transient force analyses, and wave propagation characteristics were recommended.

Collier stressed the importance of the two-phase flow regime criteria and non-equilibrium effects. The mechanical and thermal non-equilibrium become progressively important in transient phenomena which are often encountered in nuclear safety analysis. It is thought that the flow regimes under transient or entrance conditions can be quite different from the standard fully developed regimes. Under these conditions, the rate processes for changing flow regimes such as the coalescences of bubbles or entrainment of droplets should be modeled correctly. Persen showed an interesting movie on flow regime transitions. He drew attention to the relation between the transport processes and the structures of the flow and concluded that correct predictions of two-phase flow regimes were essential for analyses of practical problems.

FUTURE RESEARCH SUGGESTIONS ON BOILING AND CONDENSATION

J. G. Collier

UKAEA Safety and Reliability Directorate
Wigshaw Lane, Culcheth
Warrington WA 3 4 NE
United Kingdom

In introducing this session, Collier emphasized that boiling and condensing flows were characterized by a lack of hydrodynamic equilibrium. In many instances this was coupled with a lack of thermodynamic equilibrium. Boiling and condensing occurred within heat exchangers in a variety of situations:

(a) inside tubes, within tube bundles with both parallel and cross flows

(b) up-flow, down-flow and horizontal flows

(c) both steady and transient states

(d) with plain and enhanced heat transfer surfaces

In order to allow some structure to the discussion, he had listed a series of topics and had asked experts in each of these fields to give their views on where we should be concentrating our research efforts. This was an important aspect of the Workshop.

Sub-Cooled Boiling

F. Mayinger introduced this topic. He indicated that for engineering design it was necessary to be able to establish:

(i) the onset of subcooled boiling (point A)

(ii) the variation of void fraction (α) with length (z)

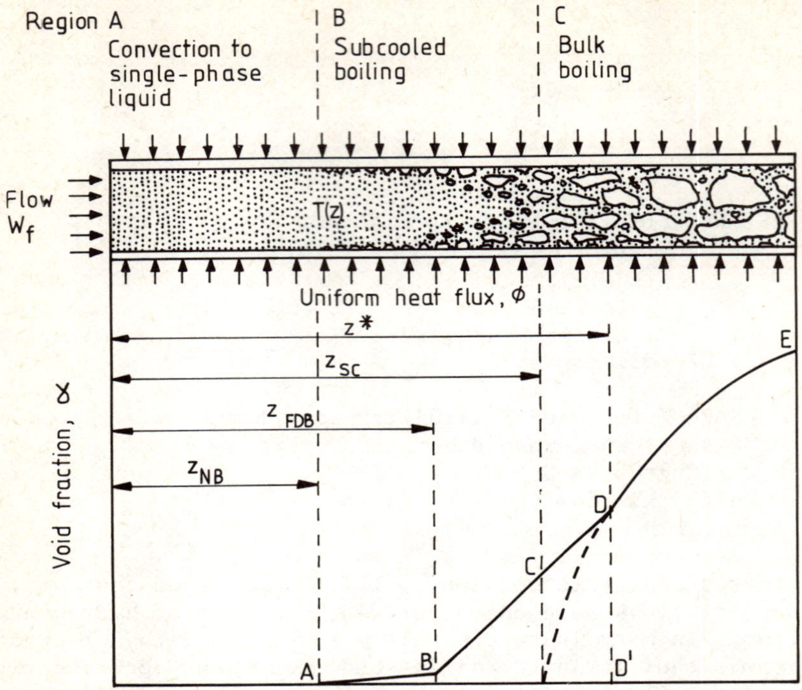

(iii) the variation of the pressure gradient (frictional and momentum) with length (z)

(iv) the heat transfer from the wall to the fluid

He ventured to suggest that the empirical methods generally available were probably adequate for engineering design. However, there were still considerable gaps in our understanding of the physical processes of subcooled boiling.

As examples he quoted:

(a) whether single phase turbulence theory could be considered adequate to predict the superheat at the wall and in the boundary layer near the wall?

(b) the problems of establishing improved criteria for the nucleation of vapor for a range of surface and flow conditions.

(c) the need for a better understanding of bubble growth in relation to the temperature field in the boundary layer and around the bubble and whether bubble departure is governed by inertial and/or buoyancy forces (low or high pressure).

(d) the delineation of the individual components of the local heat transport processes

$$\dot{q}_{FC} + \dot{q}_{MC} + \dot{q}_{EV} = \dot{q}_{WALL-FLUID} \qquad (1)$$

where $\dot{q}_{FC}$ is the convective heat flux to the single phase liquid flow

$\dot{q}_{MC}$ is the microconvective heat flux component to the liquid resulting from the bubble growth and departure and transported in the form of superheated liquid from the wall

$\dot{q}_{EV}$ is the heat flux which corresponds to the appearance of vapor at the wall and transported in the form of latent heat of evaporization from the wall

Mayinger also observed that the balance between the evaporation and bubble condensation leads directly to the void fraction profile.

(e) the motion and history, including condensation, of the detached bubbles which is needed to provide the variation of the local void fraction with distance from the heating surface.

Mayinger concluded by saying that there was a need for further detailed experimental studies to elucidate some of these points. In the ensuing discussion both Collier and Bankoff drew attention to the excellent early experimental studies of Gunther and Krieth [1]. Bankoff felt that the fact that the bubbles remain attached to the heat transfer surface whilst sliding along the surface at 80% of the free stream velocity may be important in ensuring that the liquid microlayer beneath the bubble is adequately renewed. Lahey cautioned that whereas present empirical methods for subcooled boiling may be adequate for steady state applications they may not be adequate for transient situations.

Users' Needs in Relation to Phenomenological Models

W. Kirchner presented his views as a user of the research information produced in the field of boiling and two-phase flow. He stressed the importance of understanding the transitions which occur in two-phase flow patterns. He made a plea for the application of more mechanistic models rather than the use of purely empirical correlations relating parameters. Whilst such correlations might prove adequate for engineering estimates they will not be satisfactory for transients. As an example of the type of mechanistic model which should be developed Kirchner cited the closed-form annular flow model developed by Hewitt, Hutchinson and Whalley [2]. He stressed the need for research workers to measure the important dependent properties of the flow and the importance of adequate instrumentation. Finally, he observed that not all two-phase flows were one or two dimensional and that, in larger apparatus in particular, multi-dimensional effects were very important, e.g. the flooding phenomenon in the downcomer region of a nuclear reactor.

Bankoff, in response to Kirchner's statements, felt that further work was needed on dynamic aspects which occurred in steady state flows as well as transients. He cited the study of bubble formation and growth in depressurizing liquids, the processes of "level swell," nucleation and critical flow which all needed further study.

Critical Heat Flux

D. C. Groeneveld gave his views on what had been achieved and still needed to be done in the area of critical heat flux evaluation. He cited the following achievements:

(i) a data base of 25,000 data points (mainly for water)

(ii) 400 CHF correlations

(iii) a mechanistic model for dryout in annular flow

(iv) the preparation of standard tables of CHF values (again for water)

(v) the concept of fluid-to-fluid scaling of CHF (Freon-to-Water)

(vi) graphical methods of predicting CHF (fluids other than water, e.g. Katto, Shah)

In discussing what still needs to be done Groeneveld proposed the following topics.

(i) Transient critical heat flux. Slow transients seemed to have little effect on the value of the critical heat flux but it was known that, for fast transients, the critical heat flux could increase. Relationships of the type

$$\frac{\left(\dot{q}_{CRIT}\right)_{TRANS}}{\left(\dot{q}_{CRIT}\right)_{S.S}} = fn\left(\frac{dp}{dt}\,,\ \frac{1}{\alpha}\frac{\partial\alpha}{\partial t}\,,\ ?\right)$$

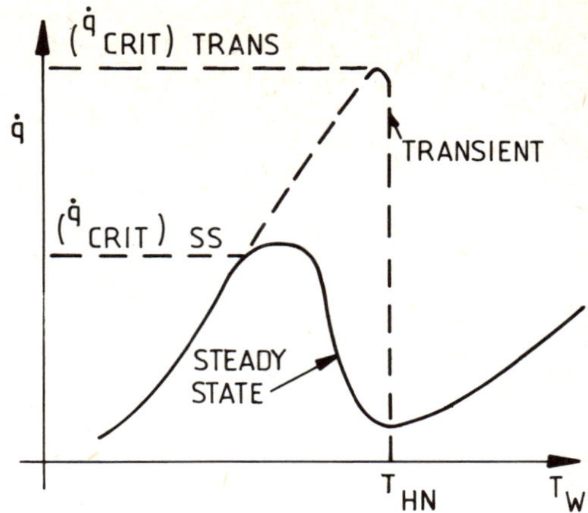

needed to be evaluated. The maximum transient critical heat flux will probably correspond to the point where the wall temperature reaches the homogeneous nucleation temperature T_{HN}

$$\text{ie } \dot{q}_{CRIT} \leq h_{NB} (T_{HN} - T_{SAT})$$

(ii) Downflow critical heat flux. There is a need for a method to predict the reduction in critical heat flux in down flow compared with up flow.

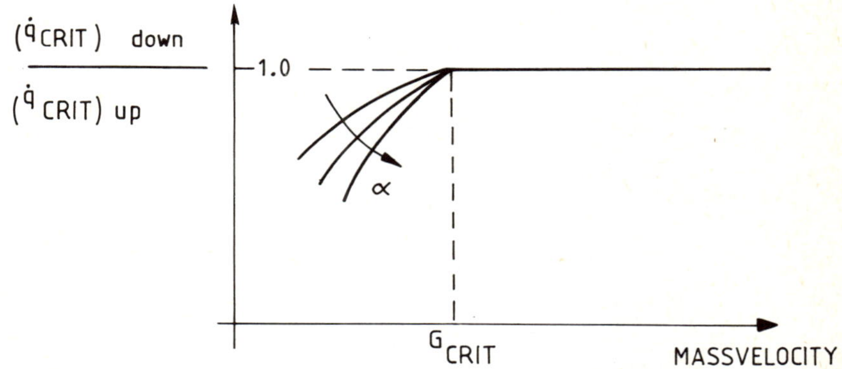

(iii) Critical heat flux under low flow conditions. Griffith had correlated the critical heat flux under low flow conditions as the ratio with the Zuber pool boiling CHF $(\dot{q}_{CRIT})_z$ plotted against the void fraction (α).

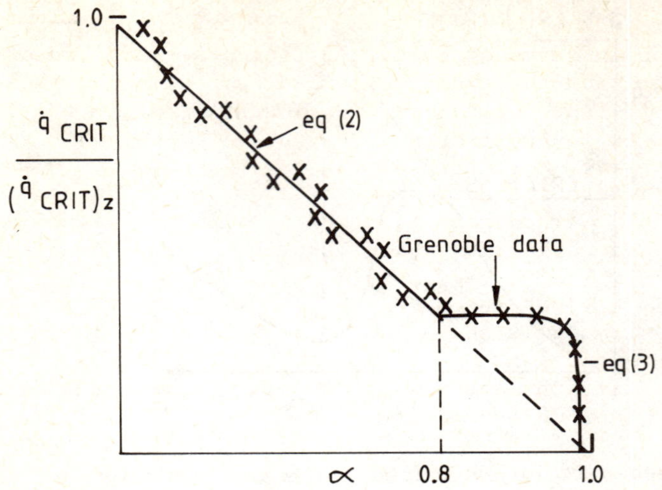

$$\dot{q}_{CRIT} = (\dot{q}_{CRIT})_z \; (1 - \alpha)$$
$$(\alpha \leq 0.8)$$

More recent data taken at Grenoble in France suggests that

$$\dot{q}_{CRIT} = (\dot{q}_{CRIT})_{\alpha=0.8} \left(\frac{1 - X}{1 - X_{\alpha=0.8}} \right)$$
$$(\alpha > 0.8)$$

(iv) Critical heat flux during unstable flow. Better information was needed of the effect of flow pattern transitions and inlet and outlet geometry effects on the critical heat flux.

Lahey in commenting upon Groeneveld's statement, felt that previous research on critical heat flux had been concentrated too much on simple geometries. Some empirical correlations were available for reactor fuel element rod bundles but these could not be extrapolated. There was a need for a much better understanding of "local" effects within rod bundle subchannels such as the role of turbulence on the process of "void drift."

Transition Boiling/Film Boiling

J. C. Chen indicated that during his earlier presentation he had been rather pessimistic about what had been achieved in the area of post-dryout heat transfer. For him the importance of mechanistic models was not only that such models are capable of predicting a number of dependent variables, but also that the model permits extrapolation in terms of geometry, fluid-to-fluid and steady state to transient. For the examples he had quoted in his paper, Chen observed that it had been possible to fairly well predict the wall temperatures in the post-dryout region whilst being considerably in error with respect to the amount of superheating in the vapor. He felt the post-dryout area was a good example of the need for mechanistic models. With regard to research topics, Chen made the following suggestions in the area of convective boiling heat transfer:

- Void fractions and slip ratios in dispersed-flow under post-CHF conditions
- Void fractions and slip ratios in inverted-annular-flow under post-CHF conditions
- Phase distribution and flow regime immediately downstream of CHF point
- Mechanism for droplet formation at CHF/Dispersed-Flow transition
- Droplet size spectrum and mechanisms that govern the size spectrum in dispersed flow; including effect of flow obstructions (spacers, probes, etc.)

W. Rohsenow identified the following topics in this area:

1. Effect of droplets in the vapor on wall-to-vapor heat transfer.
2. Measurement of and prediction methods for droplet size and distribution at the dryout location for:

   (a) annular flow

   (b) inverted annular flow

   (c) slow transient reflood in rod bundles

   (d) after rod spacers in rod bundles

Liquid Metal Boiling

H. Kattowski briefly reviewed the present state of knowledge with respect to pool boiling and forced convection boiling of liquid metals. Collier observed that, hydrodynamically, liquid metals behave like non-metallic fluids and that a good mechanistic model should be able to bridge the gap in heat transfer behavior between metallic and non-metallic fluids. For example, there were some similarities between water boiling at very low sub-atmospheric pressures and liquid metal pool boiling.

Boiling of Binary and Multi-Component Mixtures

John Thome gave the meeting his views on where research into boiling heat transfer of mixtures should concentrate. He identified the following topics:

(i) An examination of how the reduction in heat transfer coefficient for a mixture (as compared with the pure components) is affected by the level of heat flux. It might be expected that the percentage reduction will increase with heat flux level up to the critical heat flux.

(ii) An examination of the critical heat flux for mixtures including a resolution of the different observations regarding wires and flat plates.

(iii) The performance of enhanced surfaces, e.g. porous surfaces and finned surfaces in mixtures including their behavior at very low T's, e.g. 1-2°C.

(iv) A further study of incipient boiling superheats for mixtures.

(v) An extension of the present research for binary mixtures to include ternary mixtures. Dr. Thome observed that very little was published in the multi-component fluid area and what there was was contradictory.

(vi) The working up of a generalized method for pool boiling. The problem here was to get satisfactory expressions for the pool boiling of the pure components.

(vii) The extension of research to forced convection. In particular to see if the evaluation of the reduction in heat transfer coefficient based on ΔT_{bp} remains valid. Thome indicated that for one case he had examined it seemed to work for forced convection.

Condensation

J. Taborek, in his remarks on condensation, highlighted the

differences between what the classical Nusselt theories for condensation predict compared with what is actually observed to happen. He mentioned two examples:

(a) Condensation inside a tube. For condensation in down-flow inside a tube, theory would suggest that there exists a laminar flow region followed by a turbulent liquid film. In practice there is no laminar film region and the liquid film is highly disturbed right from initiation. For horizontal tubes Taborek stressed the importance of the flow pattern and the balance between shear and gravitational forces in determining what happened.

(b) Condensation within a tube bank. For condensation within a tube bank, Nusselt theory would indicate that the lower tubes within an array were inundated by condensate and that their effectiveness fell off as the number of tubes (n) varied to the negative one quarter power. In practice, for heavy liquid loadings near the beginning of the condenser, vapor shear was very important. As a result the condensate did not cascade from the higher to the lower tubes in the array but rather appeared in the form of a heavy rainstorm carried along by the vapor velocity. This vapor velocity also induced turbulence in the condensate enhancing the coefficient. Only near the outlet of the condenser where the vapor velocity was low would the theoretical flow pattern for condensate be valid.

Taborek suggested that for shell side condensation (average $\bar{h}$ between entry and exit) the simplest correlation has the form

$$\bar{h} = h_{Nu} \times \left({}^{1}/_{N} \right)^{n} \times \text{fn}\,(Re_v)$$

and produces reasonable results, even for large scale condensers

where h_{Nu} is the Nusselt theory for a single horizontal tube

N is the number of tubes in the array

n is a variable index, $[\text{fn}(Re_v)]$ ranging from 0 to Nusselt's value (0.25)

Re_v is the vapor Reynolds number.

However, a zonal approach is necessary for greater accuracy.

Conclusion

In closing the workshop panel on Boiling and Condensation, Collier thanked all the participants. He observed that there was no diminuation in the number of important research topics which had been identified and that research workers in this field would have plenty to keep them occupied over the forthcoming decades.

References

1. Gunther, F. C., and Krieth, F., Jet Prop Laboratory Progress Report No. 4-120, Pasadena (1950).

2. Whalley, P. B., Hutchinson, P., and Hewitt, G. F., The Calculation of critical heat flux in forced convection boiling, Paper B6.11, 5th IHCT, Tokyo, Japan (1974).

RECOMMENDED AREAS FOR FUTURE RESEARCH IN TWO-PHASE FLOW AND HEAT TRANSFER APPLICATIONS

R. K. Shah

Harrison Radiator Division
General Motors Co.
Lockport, New York 14094, USA

Research papers have been presented in this workshop covering almost all areas of two-phase flow and heat transfer. The results represent advances in some of the frontiers. The usefulness of this workshop will be enhanced if an attempt is made to identify clearly and concisely future research problems. This will benefit researchers and funding agencies by: (1) Providing a reference to researchers in planning of their research projects, (2) furnishing clearcut directions to the researchers for specific research problems leading to applications, and (3) providing research funding agencies a complete overview of possible areas of research related to heat transfer equipment and other engineering applications. Hopefully, the end product will be better designs of new equipment, improved designs of current equipment, and prediction of anticipated problems before the design is finalized. Major unresolved problems of two-phase fundamentals related to heat transfer equipment are summarized next based on the workshop discussion. The following individuals actively participated in the workshop session and/or provided written inputs: A. E. Bergles, J. G. Collier, F. Dobran, G. Farello, D. C. Groeneveld, M. Ishii, S. Kakaç, L. N. Persen, J. Reimann, W. M. Rohsenow, R. K. Shah, S. Sidemann, J. Taborek, V. Teschendorff, G. L. Wedekind, and J. W. Westwater.

Shell-and-Tube Exchangers

Two-phase flow and heat transfer problems have been investigated intensively during the past 30 years for many different applications of shell-and-tube exchangers. This has resulted in better understanding of two-phase phenomena in these exchangers. A number of important problems have been identified in the other panels of this workshop. A few additional problems are:

Flow regime maps. Many different flow regime maps exist for two-phase flow in a circular tube. Most of these maps are in dimensional form, are influenced by the test rig geometry, and are not general. Emphasis must be given to obtain generalized dimensionless flow regime maps for the important geometries found in practice. Existing flow regime maps are untested for large tubes and need to be verified experimentally. Flow regime maps are also needed for flow across circular tube bundles with horizontal, 45°, and vertical baffle cuts.

Two-phase pressure drop correlations. These correlations are needed for flow across circular tube bundles with horizontal, 45°, and vertical baffle cuts. While some data do exist, more work is needed, especially with hydrocarbons.

Condensation. The effect of number of tube rows in tube bundles is still poorly understood, especially if vapor flow is horizontal (vertical baffle cuts) which is usually the case in baffled shell side flow. Also needed to be investigated are the effect of vapor velocity across plain and finned tubes, and the fin retention effect as a function of the fin geometry and surface tension.

Boiling. It is now generally recognized that nucleate boiling will exist only for relatively small parts of industrial reboilers. Forced convective vaporization is usually dominant for both intube and crossflow boiling. More work is needed to clarify tube diameter and convective effects in crossflow boiling.

Flow distribution and header design. Vapor-liquid separation in headers of shell-and-tube exchangers and air coolers is poorly understood, but is very significant for design. Header size (depth) as a function of vapor velocity should be investigated.

Compact Heat Exchangers

Two-phase flow and heat transfer in compact heat exchangers is relatively an unexplored field. As a result, many problems need to be investigated to come up to the understanding and design capability of two-phase shell-and-tube exchangers. Here is a list of major unresolved problems.

Flow regime patterns and charts. Flow pattern maps are unavailable for compact noncircular continuous or interrupted flow passages. It is essential that a flow pattern study be conducted at least for some basic compact surfaces, such as plain triangular passages, rectangular passages, offset strip fins and louvered fins. For this purpose, scaled-up models of these fin geometries could be tested in transparent test sections to visualize the flow patterns and subsequently arrive at flow pattern charts.

Other important questions which need to be answered are: What is the flow pattern in a compact heat exchanger during flow boiling or flow condensation? For what fin configuration does a homogeneous flow result? Does any fin configuration result in plug flow, annular flow, or other types of flow known to occur in large straight round tubes? How do the various fin configurations affect the flow patterns?

Two-phase pressure drop. Homogeneous or separated flow models are used in computing the two-phase pressure drop in a circular tube. For compact noncircular passages, the surface tension effects are important in the corner regions at low flows. Also the friction contribution at the wall is not only due to the skin friction (surface shear), but also due to the form (or profile) drag at flow interruptions along the fins. Both the form drag and surface tension effects need to be incorporated in any two-phase pressure drop correlations for compact surfaces. At present, no guidelines are available to modify the available circular tube correlations for compact surfaces.

Fin efficiency. An accurate determination of the fin efficiency is important during two-phase flow in finned compact heat exchangers. This is because high heat transfer coefficients are generally associated with boiling or condensation, and the fin efficiency is reduced as the heat transfer coefficient h increases. Evaluation of the fin efficiency using the single-phase formulas may be substantially in error due to highly variable local coefficients, nonuniform temperature of the fluids flowing over the fins, and partially wetted fins. Basic understanding of local flow phenomena and wettability is essential to determine the local fin efficiency for stepwise design calculations.

Boiling correlations. Forced convective vaporization is a principal mode of heat transfer at low heat fluxes in plate-fin exchangers having offset strip fins. An analysis is needed to develop general correlations and computational methods for annular film flow with liquid entrainment induced by offset fins. For other fin geometries, flow-regime-based correlations for forced convective vaporization are needed.

Nucleate boiling is important at high heat fluxes. Two modes of flow boiling, forced convection and natural convection boiling, are of significant industrial importance. In both modes, nucleate boiling prevails due to very large wall superheat. No data exist for forced convection boiling and some data for natural convection boiling in compact surfaces. A systematic parametric study is needed for at least some basic fin geometries which includes measurements of the local pressure gradient, void fraction, and wall temperature. These data will permit local stepwise calculations of heat transfer and pressure drop in industrially important situations.

Condensation correlations. Very little information has been reported in the literature on condensation (gravity and/or vapor-shear controlled) in compact surfaces. The effect of surface tension in flow passage corners may not be negligible on heat transfer coefficients and condensate drainage characteristics. This needs to be investigated systematically for important fin geometries. Until condensation correlations are developed for compact surfaces, some basic experiments should be run to find out (1) for how small tube diameter (hydraulic diameter) the circular tube condensation correlations are valid, and (2) how the existing circular tube condensation correlations can be modified to take into account the noncircularity of flow passages.

All of the published data for flow boiling and flow condensation should be collected and analyzed to arrive at specific correlations for fin geometries investigated. These correlations then may be compared with those for the circular tube to show the limitations and range for use of each correlation.

Design of vaporizers and condensers. Current practice for design of vaporizers and condensers using compact surfaces is to use the same procedure as the tube side of shell-and-tube exchangers with the circular tube diameter replaced by the hydraulic diameter. Also, the single-phase Colburn factor j and Fanning friction factor f for the circular tube are replaced by those for the specific fin geometry in question. However, this is a very approximate method. The first question which arises is: Can we use the available theory for circular tube vaporizers/condensers for compact exchangers using the concept of hydraulic diameter? Industrial experience indicates the single-phase j and f data for plate-fin surfaces must be adjusted with some correction factors in order to match the predicted performance (using computer programs for vaporizers/condensers having circular tubes) with measured experimental performance. In the interim period (until further advances are made), correction factors for the single-phase j and f factors for compact interrupted surfaces should be developed.

Header design and flow maldistribution. The performance of a vaporizer/condenser is very sensitive to flow maldistribution with a two-phase mixture at the inlet. Current industrial practice is to feed the two phases (liquid and gas/vapor) separately and uniformly at the inlet so that a serious degradation in performance is minimized. The design of inlet and outlet distributors, and the location of inlet and outlet nozzle are very important for uniform flow distribution through the core region. At present no design guidelines or theoretical analyses are available for good header design. For uniform two-phase flow distribution, experimental investigation should be conducted to determine the flow resistance characteristics of the various forms of turns, combined with contractions and expansions as found in actual flow distributors of compact exchangers.

Other areas of investigation. Many forms of two-phase flow instabilities occur in cryogenic plate-fin exchangers. Some examples of these instabilities are density-wave, Ledinegg, geysering, and thermal acoustic oscillations. Practically no information exists on any form of these instabilities. Also, fouling and corrosion have received little attention. There is simply nothing published on multi-component two-phase flows and heat transfer in compact surfaces.

Heat Transfer Enhancement

Although the crowded schedule of this workshop has precluded discussion of heat transfer enhancement, it is important to include this topic among the recommendations for future work. The basic motivation is either to encourage a high heat flux by promoting nucleate boiling at low temperature difference or to accommodate a high heat flux while keeping the temperature difference low. The first objective applies to two-fluid heat exchangers in the power and process industries while the second objective applies to nuclear reactors and a variety of specialized applications in aerospace, electronics, and physics research.

Further understanding of the available "structured" surfaces for promoting nucleate boiling at low superheat ($\sim$1 K) is required. In particular, the influence of the controllable geometrical parameters must be resolved so that surfaces can be fabricated to give optimum performance for a particular application. New, reliable and cost-effective surfaces must be developed.

Renewed interest in high heat fluxes stems from the smallest boiling systems, development of high power microelectronic chips, and much larger systems, such as divertor targets and beam dumps for proposed thermonuclear fusion reactors. For the former, fluxes of 10^6 W/m^2 with refrigerant-like dielectric coolants may be required, while 10^8 W/m^2 with water are required for the latter. Both situations require forced convection subcooled boiling, but the fluxes are higher than those which can be easily achieved, given the realities of coolant distribution and available pressure drop. Enhancement techniques must be developed to accommodate these high fluxes.

Reactor Safety

Flow regime map for reactor accident analysis. Concern:

Adiabatic tube flow regime maps do not apply to fuel bundles during accident conditions.

Most heat transfer and fluid flow correlations depend strongly on the two-phase flow regime. Equations for flow regime boundaries have been suggested for vertical and horizontal flow; they are usually based on adiabatic tube data. Their application to spacer equipped bundles during transients at high heat flux is questionable. It is suggested that research into this be divided into (a) the high flow regime, where gravity effects are insignificant, and (b) the low flow regime (stratified flow, up and down vertical flow).

<u>Thermodynamic and heat transport properties at high temperatures</u>. Concern: Incorrect prediction of steam properties because of unwarranted extrapolation.

A limitation of some of the post-CHF correlations is their dependence on steam property equations. During a LOCA, sheath temperatures well in excess of 800°C are frequently postulated to occur and several heat transfer correlations require evaluation of steam properties at the sheath temperature. Current ASME steam property equations are valid only up to 800°C and should not be extrapolated. At higher temperatures, dissociation of steam (which depends on time and temperature) will occur, and will affect the steam properties.

SELECTED TOPICS OF INTEREST FOR FUTURE RESEARCH IN TWO-PHASE FLOW ANALYSES

M. Ishii

Reactor Analysis and Safety Division
Argonne National Laboratory
Argonne, Illinois 60439, USA

Considerable progress in understanding and predicting various two-phase flow phenomena has been made by a combination of rigorous model development, a number of small and large scale experiments and advanced computational techniques. However, there are many important outstanding problems which require further research in this area. Selected topics among them are given below.

Flow Regime and Interfacial Area

It is well-known that the transfer mechanisms between phases and between a two-phase mixture and structures depends on two-phase flow regimes. The concept of two-phase flow regimes implies the macroscopic characterization of interfacial structures. Because of these, the flow regime transition criteria and regime dependent two-phase flow correlations or constitutive relations have been developed in parallel. A number of flow regime transition criteria have been proposed in the past, however, much more effort on points such as listed below are necessary.

(1) Objective methods should be used for the determination of flow regimes and transitions. For this purpose, local two-phase flow probes and statistical analyses of local conditions become necessary.

(2) Often a flow regime transition is gradual and no definite transition point exists. A method should be established to deal with this situation. An introduction of an interfacial area correlation is recommended.

(3) A definition of a flow regime can depend on the size of the reference volume. A small scale flow regime and system global size flow regime can be quite different. Since most of the constitutive relations depend on the flow regimes, these effects should be carefully evaluated.

(4) Flow regime transition criteria under entrance or transient conditions should be developed.

The geometrical effects of interfacial structures may be modeled through the interfacial area concentration and void fraction. This is particularly true for a two-fluid formulation. The correlation for the interfacial area is under development and the following items are particularly important.

(1) Bubble number density and nucleation site density in bulk fluid (flashing flow).

(2) Bubble size, coalescence and disintegration correlations.

(3) Droplet size and size distributions in pool entrainment and inverted annular flow (post CHF condition).

(4) Development of interfacial transport equation.

Scaling of Two-Phase Systems

Similarity laws for scaling of two-phase systems are not well established. This introduces considerable difficulties in designing simulation experiments. The origin of the problem is the existence of a number of possible two-phase flow regimes and associated constitutive relations which can be quite different from each other. Furthermore, these constitutive relations often have inherent length scales based on the structures of the flow. For example, the bubble or droplet size, film thickness and wave amplitude and the inverse of the interfacial area concentration are the typical two-phase length scales. These different length scales introduce complications in terms of similarity laws for balance equations, flow regimes and interfacial transfer correlations.

Non-Equilibrium Effects

There are two particular conditions where non-equilibrium effects can be quite important and require further research. These are flashing flow leading to a choked condition and post CHF flow. In flashing flow, the nucleation characteristics in the bulk liquid as well as at the wall, and interfacial heat transfer mechanisms should be understood. Furthermore, the mechanical non-equilibrium between phases can also be important. Due to a large pressure gradient and considerable phase changes, the convective acceleration can be large. Therefore, the transient forces such as the virtual mass force may be important. There are unsolved questions related to the phase distribution, drag, and relative velocity between phases.

Counter-Current Flow

There is considerable uncertainty in predicting counter-current two-phase flow. There are basic questions on the mechanism of flooding, entrainment and entrance effects.

Data Base and Measurement

Recently, basic two-phase instrumentation techniques for research purposes have advanced considerably. However, practical local instrumentations for industrial applications at high pressure and temperature should be developed.

In terms of the basic experiments, measurements of two-phase turbulence, flow regimes, droplet and bubble sizes, local average velocity of each phase, interfacial area and interfacial transfers are needed. These will form data bases for the development of two-phase flow models and constitutive relations.

FURTHER RESEARCH SUGGESTIONS ON VARIOUS AREAS OF TWO-PHASE FLOW AND HEAT TRANSFER

S. Kakaç

Department of Mechanical Engineering
University of Miami
Coral Gables, Florida 33124, USA

The following individuals actively participated in the workshop session and/or provided written input: K. Akagawa (Japan), S. G. Bankoff (USA), M. Cumo (Italy), E. N. Ganić (Yugoslavia), M. Giot (Belgium), H. Hamaguchi (Japan), S. Kakaç (Turkey), J. Kestin (USA), R. T. Lahey, Jr. (USA), S. Nakanishi (Japan), M. Ozawa (Japan), L. N. Persen (Norway), W. A. Ragland (USA), T. Sakaguchi (Japan), and G. L. Wedekind (USA).

Their views on where we should be concentrating our research efforts on various areas of two-phase flow and heat transfer with their own research interests are presented below.

Two-Phase Flow Instabilities

The following individuals actively participated in the discussion and/or provided written inputs on two-phase flow instabilities: K. Akagawa, S. Kakaç, R. T. Lahey, Jr., S. Nakanishi, M. Ozawa.

K. Akagawa presented his views on flow instabilities and identified the following two problems for future study on "flow instabilities in gas-liquid two-phase flow system with compressible capacities."

Many studies have been conducted on two-phase flow instabilities and the classifications of the flow instabilities have also been presented, for instance, by Bouré et al., Kakaç, and others. In the two-phase flow instabilities in steam generating systems, the pressure-drop type oscillation and the flow pattern oscillation are important problems as well as the density-wave type oscillation. These oscillations are induced by steam generation in the channels. On the other hand, there are two-phase flow instabilities which are not due to the phase changes (evaporation or condensation) as described in the next section.

We wish to raise two problems for the future studies to be done on the two-phase flow instabilities under the circumstances described below.

(1) There is little information on two-phase flow instabilities in non-heated channels, which are encountered in many chemical plants. For instance, pressure oscillations and flow oscillations have been observed in non-heated channels which contain a mixing section of a gas and a liquid, and compressible volumes. Therefore, it is necessary to investigate these instabilities.

In our experiments on two-phase flow instabilities in non-heated single channel and parallel channels, flow instabilities which are combined pressure-drop type oscillation and flow pattern oscillations were observed. These oscillations differ from those in steam generating systems, but the characteristics are similar to the latter.

(2) Most of the studies on the two-phase flow instabilities have been concentrated to estimating the initiation of the flow instabilities or making flow instability maps. There is little information on the behavior or the profiles of the oscillations. In order to understand the physical mechanisms of the oscillations, an analysis using three dimensional coordinates (gas flow rate, liquid flow rate and pressure drop) is useful. The example will be shown in the next section. The curved surface which shows the static characteristics of the pressure drop has the nature of catastrophy as shown in Fig. 1. Based on the characteristic curved surface, the profiles of the pressure oscillations and the flow oscillations which have very complicated forms, can be well explained.

Akagawa also described the existence of oscillations in non-heated channels for air-water two-phase flow system:

Flow Oscillations in Air-Water Two-Phase Flow System: Main parts of the experimental setup are an overflow tank, a compressor, compressible volumes in an air and a water supply lines, a mixing section and a horizontal test section. This test section is a capillary glass tube of 3.18 mm I.D. and 3.1 m length. Two-phase pressure drop characteristics in this test tube is the curved surface represented by dotted lines in Fig. 1. Under the constant water volume flux, W_{lo}, the pressure drop increases, decreases and again increases with the increase in the gas volume flux, W_{go}. In the first increasing part of the pressure drop (region I), the flow

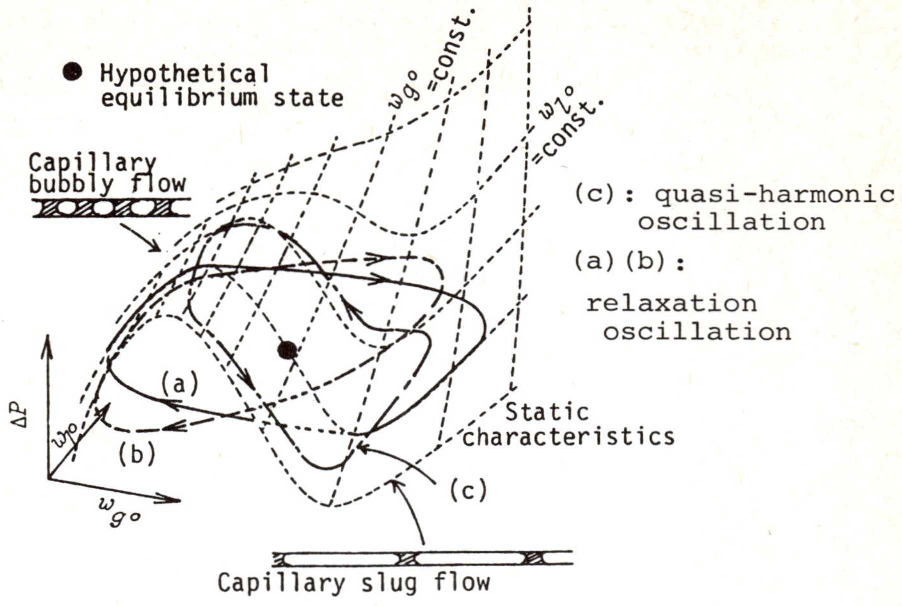

Fig. 1 Loci of limit cycle oscillations.

pattern is a capillary bubbly flow, and in the second increasing part (region III), the flow pattern is a capillary slug flow. The decreasing part of the pressure drop (negative slope region II) corresponds to the transition region between these two flow patterns.

When the hypothetical equilibrium state is in the negative slope region, two types of oscillation were observed depending on the amounts of two compressible capacities. One is the relaxation oscillation and the other is the quasi-harmonic oscillation. In general, the relaxation oscillations have large amplitudes of flow and pressure drop and long period. This is observed in the case of small amount of liquid side compressible volume and relatively large amount of gas side compressible volume. And the phase shift between the gas and the liquid flow oscillations is very small. By simple analysis, this oscillation is found to be characterized by the Van der Pol equation [1].

On the other hand, the quasi-harmonic oscillation has relatively small amplitudes of gas flow and pressure drop and their period is short. The phase shift between the gas and the liquid flow oscillations are relatively large.

These oscillation traces can be reproduced in the three dimensional coordinate system as shown in Fig. 1. The locus of the relaxation oscillation moves closely on the static characteristic surface in the regions I and III, and the locus leaves the characteristic surface in the region II, i.e. flow excursions occur. On the other hand, the locus of the quasi-harmonic oscillation moves closely on the characteristic surface and flow excursion does not occur. In the parallel channel system, we can observe similar behavior of the oscillation locus with the above mentioned results of the single channel experiment.

Thus, the existence of oscillations caused by the pressure drop characteristics and the flow pattern transition in non-heated channels are shown and also loci of the limit cycle oscillations in a three dimensional space are indicated in order to understand easily the mechanisms of the two-phase flow instabilities.

Akagawa also identified the following topics:

Shock phenomena in a two-phase flow. The pressure transients and the propagation velocity in a one-component two-phase flow induced by a rapid valve closure need to be investigated.

Flow instabilities in evaporators of cryogenic fluids. The characteristics of flow instabilities in evaporators of LN_2, LO_2, LH_2, and LNG should be investigated.

S. Kakaç gave his views on what needs to be done on two-phase flow instabilities, and he identified the following areas for experimental and theoretical studies:

Experimental Investigation

The two-phase flow applications in power or process engineering practice usually are parallel-channel systems operating between two common reservoirs. The most important difference between single and parallel-channel systems, as far as the stability is concerned, is the existence of different oscillation modes encountered in parallel-channel systems. When there is more than one channel present, the phase relationships between the oscillating flows in individual channels become an important concern. Previous studies by the authors on parallel-channel two-phase flow systems indicate that, depending on the particular type of instability and other conditions, the oscillations in parallel channels may be in phase or may be out of phase. When they are in phase, both the total flow rate and the individual channel flow rates oscillate together. In other instances, the individual channel flow rates oscillate with a certain phase lag between them, which may be, for example, 180° for two-channel systems, and 120° and 240° for three-channel systems. In such circumstances, the total flow rate stays constant while the individual channels oscillate. There can be similar phase relationships in pressure and temperature oscillations.

It is needed to investigate the effect of the following parameters on the two-phase flow instability characteristics of parallel-channel systems with different heat transfer surfaces:

a) Tube Heat Transfer Surface: The parallel-channel studies in the past were conducted with plain identical tubes. In recent years, there has been increased emphasis on techniques to enhance two-phase flow heat transfer in heat exchange equipment. Therefore, it is needed to examine the systems where the tubes have different heat transfer characteristics. For more effective investigation of the influence of the heater wall conditions on two-phase flow instabilities, the longer test tubes should be used. With longer heater tubes, the system would no longer be dominated by the exit conditions and, in addition to the effect of heater wall configuration on stability, it would be possible to investigate various other interesting phenomena such as the steady-state and the transient pressure profiles across the boiling region, the variation of the heat transfer coefficient along the tube, with proper instrumentation.

b) Tube Cross-Sectional Area: Two-phase instability experiments with different tube diameters should be carried out in order to investigate the effect of tube size combinations.

c) Heat Input: Experiments should be conducted with equal and unequal heat inputs into the channels in order to study the effect of the heat input combinations.

d) Inlet Throttling: The determination of the additional pressure drops necessary to stabilize the parallel channel system under different heater surface, tube size, heat input and flow conditions is needed for stability comparisons.

In addition, high speed photographic techniques may be used to investigate the two-phase flow regimes and their effect on two-phase flow instabilities in parallel channel systems with different heater surface conditions. For the improvement and modification of the existing two-phase flow models, it is of prime importance to observe the types of the flow regimes, such as bubbly, slug, churn, annular and mist flows during instabilities. Continuous films may be taken, using the existing high speed photographic facilities, of the flows through the parallel channels under steady-state conditions, at and near the instability thresholds, and under unstable operating conditions. As a result, it is expected that the two-phase flow behavior and the flow regimes could be observed clearly, and the effects of the various flow regimes on the instability will be better understood.

There is very limited study concerning the variation of flow patterns during the oscillations and their effect on the system instability. Due to the lack of useful correlations, according to our knowledge, there is not even one theoretical study which effectively deals with the effect of different flow patterns on two-phase flow instabilities. It is necessary to conduct photographic studies of two-phase flows both in steady-state and during oscillations, with the purpose of testing the existing correlations, or, if required, developing new ones. The drift-velocity model of Zuber and Findlay can be a powerful tool in studying two-phase flow instabilities if reliable expressions for the distribution parameter, C_o, and the vapor drift-velocity, U_{gj}, are available. Different investigators proposed different correlations but none of them has yet been generally accepted as being sufficiently accurate and universal. By using the available high speed photographic setup, it is possible to shed some light onto this controversy and more comprehensive ways of incorporating the flow pattern effects into the theoretical models can be developed as the research progresses.

Theoretical Investigations

The presently developed models, i.e., the homogeneous model and the drift-flux model, can be extended to parallel channels with augmented surfaces. The conservation equations for one-dimensional two-phase flows in parallel channel systems can be solved. The effects of the radial distribution of velocities, enthalpies and void fractions could be taken into account through the use of correlation coefficients. New alternative forms of the conservation equations that might be more useful for parallel channel flow systems could be formulated and solved.

In the theoretical investigation, the following should be taken into account: Variation of heat flux during oscillations, transients and with the flow regime should be accounted for by defining the heat transfer coefficient as a function of flow rate, flow regime, flow mode (stable or unstable), void fraction and heater characteristics. Also the changes in the coefficient of friction during oscillations, transients and with flow regime could be accounted for by defining the coefficient of friction as a function of flow rate, two-phase flow mode (stable or unstable), flow regime, void fraction and heater characteristics.

During some part (low flow rate) of the oscillatory cycles, superheat conditions might exist in the heater. Consequently, superheat conditions have to be included in the theory.

The point that separates the liquid region from the two-phase region undergoes a movement during transients and oscillations. This motion affects the pressure drop due to gravity and friction, and should accordingly be included in the study.

Visualization and photographic study of flow regimes around the stability-instability boundary will help the refinement of the modeling of the two-phase flow instabilities.

In the differential equations expressing the transient conservation laws of mass, momentum and energy, the flow parameters and fluid properties such as the velocity, density, pressure, temperature, enthalpy, heat transfer rate and coefficient of friction can be taken as variables. These differential equations have to be solved in finite difference forms to determine fluid properties and flow conditions at any point along the test channels in response to an arbitrary time variation in inlet plenum to exit plenum pressure-drop and/or inlet velocity.

The efforts should be concentrated on developing alternative analytical techniques for solving nonlinear problems. However, at the moment, there are no such methods at hand, or at least, they are too abstract to be conveniently applied to engineering problems. Because of this and also because of their importance in practical cases, there is a need to improve the existing numerical models so as to make them more accurate and more efficient.

Regarding the pressure-drop oscillations, our knowledge of the flow excursions which play an essential part in these oscillations is incomplete. The actual cycles for the pressure-drop oscillations usually differ from the ideal cycles predicted by the theory. This is due to an insufficient representation of the flow excursions in the theory. Further research is needed to remedy this situation.

So far, always two different numerical schemes have been constructed for the pressure-drop and the density-wave oscillations. However, the two mechanisms can be combined in a unified approach. It is possible to perform a similar unified numerical nonlinear analysis to obtain the superimposed oscillation cycles. A foreseeable difficulty is the treatment for the flow excursions.

R. T. Lahey discussed the application of bifurcation theory to the operation of boiling-water nuclear reactors. This theory deals with the branching, or bifurcation, of a stable steady solution of the governing equations to yield two or more solutions at a critical value of some parameter. If the bifurcation is subcritical, it is possible for the plant to be stable to small operating changes or outside disturbances, but unstable to large disturbances. This is a new and relatively unexplored field of great practical importance.

S. Nakanishi discussed the flow pattern transition instability and he made the following remarks: One decade has passed since the well known classification of two-phase flow instabilities was proposed by Bouré, Bergles and Tong [2]. Since then a great number of research works have been accumulated in this field and considerable progress has been achieved. So, the necessity for revision and expansion of this classification becomes very high. The flow pattern transition instability is one of the types of instabilities which is to be re-examined.

The relaxation instabilities including flow pattern transition instability seem vaguely understood as the identification and mathematical modelling of their mechanisms have not been satisfactorily established as yet except, probably, bumping type instabilities. Regarding the flow pattern transition instability, it is certain that the validity of its mechanism and nomenclature given by Bouré et al. is disputable. Flow pattern transition instability was observed in boiling two-phase flow in a forced circulation loop [3] and in a natural circulation loop [4]. We also reconfirmed its existence under both circulation conditions in a heat transfer test rig with Freon-113, and found two distinct properties; i) it requires high inlet subcooling, and ii) its oscillation period is independent of the transit time in the heating section [5]. According to Jeglic and Grace [3] and Bouré et al. [2], the mechanism of this instability is cyclic alternation of two flow patterns induced by discontinuous jump of void fraction at the transition from bubbly to annular flow regimes. In addition, it was shown for air-water adiabatic system that the frictional pressure loss curve abruptly decreases at the same transition [6]. If the above mechanism holds, the flow pattern transition instability should be observed in air-water system. However, we could not create this instability in our air-water rig in spite of all our efforts, while the density wave oscillation and the pressure drop oscillation were easily obtained. It is concluded that the anomaly of hydraulic resistance mentioned above cannot be the cause of the flow pattern transition instability without combination with some strong factor(s), and that the cyclic alternation of flow patterns is only the result of the instability. The true mechanism of this instability should be related to the dynamic response of the two-phase flow with subcooled boiling, though it was classified as a static instability in [2]. Nucleation origin might not hold for this case, as the oscillation does not have stochastic character. It is also questionable to adopt the instability as vapor core formation type instability as an alternative definition, since the observation by Chexal and Bergles [4] suggests that this is not the case. Identification of the detailed mechanism of the flow pattern transition instability will require more information about the phenomenon itself and better understanding of the dynamic characteristics of the subcooled boiling flow. Success in the modelling of the "flow pattern transition instability" will largely enrich not only the theory of two-phase flow instabilities but also that of subcooled boiling.

M. Ozawa and K. Akagawa drew attention to the importance of

limit cycle oscillations. There are many computer codes which predict the threshold condition of instabilities. On the other hand, investigations on the limit cycle oscillations, i.e., characteristics of the non-linear limit cycle oscillation, amplitude, period, etc., are few. In general, it is very difficult to analyze the nonlinear limit cycle oscillation without numerical methods. But by overcoming this difficulty one can determine the fundamental principles of oscillation. A preliminary approach can be found in [7, 8, 9].

S. G. Bankoff recommended the following areas for research in two-phase flow and heat transfer:

1. Upper limits of critical heat flux in nucleate boiling. Bankoff and Mason [10] showed that heat transfer coefficients of order 200 w/cm^2 °C could be obtained in condensing steam at the surface of a small steam bubble growing in a cold stream of water. This followed observations by Gunther [11] and Gunther and Kreith [12] of heat fluxes of the order of 3-5 kw/cm^2 in forced-convection subcooled nucleate boiling related to cooling of liquid-fuel rocket engines. Some work has been done on this high-flux mode of heat transfer, both in the West and in the Soviet Union, but much remains to be learned about the mechanism of such rapid heat transfer, as well as the upper limits of critical heat flux. Such devices as swirl flows, binary mixtures, roughened surfaces, and combinations thereof, all at very high velocities and subcoolings, need to be systematically explored. Further, there have been questions raised in recent years [13] as to the relative importance of latent heat transport through the bubbles, as compared to turbulent heat transfer from the wall to the liquid between the bubbles. Simplified analyses have been presented, but a more detailed experimental examination is needed.

2. Strong shock wave propagation through gas-liquid mixture. There have been many studies, both experimental and theoretical, of the propagation of weak shocks (either acoustic or finite-amplitude) through bubbly liquids, but strong shocks ($\Delta P/P \gg 1$) are much less well understood, particularly when the gas is a condensible vapor.

3. Flooding experiments. As shown in the review by Lee and Bankoff in this meeting, there is a large scatter of flooding data in the literature, for which a principal reason is the large variety at inlet and outlet geometries employed in various experiments. There is need for a standardized smooth entrance and exit system (porous medium entrance, with exit liquid being sucked off without disturbing the gas flow), and a standardized sharp-entrance and/or exit geometry. In this way the increased flooding tendency due to inlet/outlet conditions can be assessed, and a more realistic choice between various models in the literature can be made. The differences between steam-water and air-water flooding need further

examination, since this important question is not always appreciated. Flooding in complex geometries, such as multichannel flooding, flooding in tube bundles, and perforated-plate flooding (such as the upper tie-plate of a PWR) needs further work.

4. Bubble dynamics and critical flow. The critical flow of flashing mixtures, despite much previous work, is still not well understood. In particular, the growth of individual bubbles and the number of such bubbles as the mixture flows towards the throat region cannot at present be predicted accurately. Hence the void fraction at the throat, and the corresponding sonic velocity, cannot be predicted from first principles.

M. Cumo discussed the two-phase critical flow and jet forces: Considering the critical flows through channels of different geometry (D; L/D), starting from different inlet stagnation conditions (P_o, T_o, ΔT_{sub}), there are some characteristics which deserve further research. For flows characterized by small inlet subcooling, for instance, the phase distribution and slip within the discharge channel is of primary importance. The vapor is first generated close to the channel walls and a vapor annular layer is formed. It is not clear when this vapor migrates in the core of the channel and a reverse two-phase flow configuration is formed. Thus an "ad hoc" investigation is needed.

Also the phase distribution in two-phase jets requires a thorough investigation. Flashing liquid streams and droplets alter deeply the flow configuration and, consequently, the transversal pressure profiles and the forces against external targets. An insight of these two-phase jets, by means of different kinds of instrumentation, seems to be the first requirement to gather sufficient data for theoretical models.

M. Giot identified the following topics for further research in two-phase flow and heat transfer:

1. Spray cooling. Predictions of heat transfer between a hot plate and a liquid spray require a good knowledge of the hydrodynamics of liquid droplets when they hit the plate. Several studies conclude that some direct liquid-solid contacts even above Leidenfrost temperature are responsible for the high heat transfer rates. More basic knowledge of this phenomenon of direct contact would help in evaluating the average fraction of the surface area of the plate covered by the liquid phase.

2. Droplet vaporization. Underground coal gasification may require the cooling of a hot gas (e.g. 1000°C) at high pressure (e.g. 40 bar) by means of water sprays. As the gas is acid, vaporization must take place over a short distance from the sprayer, where the casing can be made of special materials. The existing

laws of spray droplet vaporization need to be extended and validated for these specific conditions.

3. <u>Critical flow of initially subcooled water</u>. Nucleation and bubble growth processes have a major influence on critical flow of an initially subcooled liquid. This influence can be clearly shown by comparing results obtained by means of a given test facility, where nucleation is enhanced or not. The mechanisms of heterogeneous nucleation of adiabatic superheated liquid flows should be further explored.

4. <u>Three-phase flow</u>. Only a small number of published studies deal with three-phase flow of air-water and coarse solid particles. Such flow is encountered for example in air-lift pumping of manganese nodules from the ocean. Much fundamental work is needed on drift velocities of the particles and of the gas bubbles, on friction laws, etc.

J. <u>Kestin</u> made the following remarks on multiphase flow research:

1. I am of the opinion that the most important outstanding problem in multiphase flow is that of creating a secure scientific basis for the fundamental governing equations. Let me explain my thoughts by making a comparison with classical fluid mechanics.

   The basic equations of Newtonian-fluid mechanics, covering such diverse phenomena as the effects of compressibility and chemical reactions, meet with general acceptance. They have been derived by the application of well-understood principles of mechanics and thermodynamics and have been found consistent with statistical mechanics (Boltzmann equation) in their overlapping domain of validity. These equations are valid for laminar flow, and we are sure that any discrepancies between calculation and experiment that may show up must be due to secondary causes: wrong property values, inaccuracies in instrumental indications, improper adjustment of boundary or initial conditions, etc.

   We believe that the effects of turbulence are also implied in the Navier-Stokes equations and learned from Reynolds how to treat them analytically. To be sure, a completely general treatment has still eluded us, but many more-or-less justified steps have been developed, and we can identify with assurance those that are almost certain to lead to results which agree with experiment.

   In the field of multiphase flow the situation is different. The preceding schema is not followed. The overwhelming proportion of "models" is restricted to a single spatial dimension, and the derivations are in practically all cases heuristic and intuitive. The effect of fluctuations is introduced in ways which I would characterize as idiosyncratic. As a result, we have a choice of a rather large number of "models" but no criteria which would allow us to prefer one "model" over another in a given set of circumstances.

The largely intuitive mode of derivation and the reliance on what one can only describe as capricious closure equations ("correlations" and "maps") makes it impossible to judge whether certain dissipative mechanisms which are sure to exist are important at all, or alternatively, under what conditions their effects must be included, or how to express them analytically. Here I would enumerate:

(a) The entropy production due to the transfer of mass between the phases.

(b) The effects of departures from unconstrained thermodynamic equilibrium (metastability) on the part of either phase present.

(c) The importance of the precipitation of particles of solid phases in one- and multi-component systems.

(d) The existence or absence of a justification for introducing "averaged" expressions for the transport properties of two- and multi-phase systems.

(e) The order of magnitude of the relaxation times which govern the above irreversible processes and their relation to the residence time of the fluid elements in a channel.

2. The complex of problems under the heading of choking.

(a) Can one analyze choking separately from the flow "model."

(b) How is choking related to the velocity of propagation of plane waves and how do relaxation processes affect it (frequency dependence of the speed of sound).

3. How far is it possible to transpose experimental results and empirical correlations obtained in air-water flows to systems in which liquid-vapor transitions induce very large compressibility effects?

4. Is it possible to establish similarity laws for the translation of results between different chemical species. How does the shape of the vapor-pressure curve (retrograde condensation) affect considerations of similitude?

Leif N. Persen drew attention to three areas within two-phase flow which need closer attention in our future research efforts:

1. Influence of system configurations on two-phase flow
2. A possible reorganization of flow regime maps
3. Instrumentation

1. Influence of system configurations on two-phase flow. Since my interest in two-phase flow has greatly been enhanced by the situation often found in offshore operations where gas and liquid may be transported in almost horizontal pipelines on the ocean bed until it finally goes into a riser which takes both gas and liquid up to the platform, it has become clear that the flow situation in the vertical riser is greatly dependent upon the geometry of the whole pipeline system. This includes the length of horizontal portion of the pipe, its deviation from the horizontal position, the geometry of the bend connecting the horizontal pipe with the riser, etc. The flow regime found in the vertical riser will change as these quantities are being changed. It will however also change as one goes further downstream in spite of the fact that the overall conditions for the flow do not change. This then shows that overall conditions are not alone defining the flow regime in the vertical riser, but that the geometry of the system that one operates will also give determining influences.

2. A possible reorganization of flow regime riser. The flow regime maps have so far in most cases been drawn with the superficial gas and liquid velocities as major coordinate axis. A trend away from this, whereby non-dimensional quantities are being used as coordinate axes for the flow regime maps, has been introduced by Duckler and Taitel in their presentations. Borders between the different regimes appear according to stability criteria which are being set up. It is however from an experimental point of view rather doubtful whether there are such sharp criteria. Usually the transition of one regime to another is rather slow and should rather be drawn in the flow regime map as a band instead of a line. In view of this, and in view of the remarks under 1., further studies of flow regime maps ought to be carried out bearing in mind that general behavior of the two-phase flow should be determined in terms of non-dimensional quantities.

3. Instrumentation. One will find that instrumentation for the experimental investigation of two-phase flow is rather difficult. Detailed studies of plug flow and churn flow in vertical risers are dependent on measuring techniques which will give adequate information about quantities which can be expressed analytically. Such instrumentation is at the moment being developed at the Institute of Applied Mechanics at the University of Trondheim and the results are encouraging for plug flow as well as for churn flow. The results of these efforts have been offered for presentation at IUATM Symposium on "Measuring Techniques in Gas-Liquid Two-Phase Flows," to take place in Nancy, France, July 5-8, 1983. Further development of instrumentation techniques in two-phase flow in general ought to

be encouraged.

W. A. Ragland and E. N. Ganič proposed the following suggestions for further research on flooding: Dukler et al. [14] and Maron and Dukler [15] discuss the importance of the location of the liquid feed entrance relative to the top of the test section. Dukler's analysis assumes flooding to be a film phenomenon versus an entrainment phenomenon with "split flow" as one of the solutions for film thickness at the flooding point. The location of the liquid inlet must be different than the exit for internal recycle to take place in the film. By assuming that flooding is related but not identical to entrainment, it should be possible to reconcile the end effects due to a length of test section above the liquid feed. In this case, the entrained flow is responsible for the internal recycle. For a given distribution of droplets at a given axial location, it should be possible to apply the results of research on droplet deposition to develop a distribution of droplets along the entire downstream section of the tube. The droplet distribution would be a function of gas velocity, droplet size, radial position (in polar geometries), and axial position. In addition, phase change effects may be added. The distribution of any internal recycle due to entrainment then becomes a convolution of a droplet deposition function and an entrainment source function. The use of such an approach may require a relaxing of the criteria of no liquid out the top of the tube before flooding since droplet deposition implies only an asymptotic limit once a finite amount of liquid has been entrained.

Levy and Healzer [16] have proposed applying mixing length theory to a wavy, turbulent liquid-gas interface for cocurrent, annular flow. The analysis divides the flow into three regions: gas core, wall film, and wavy gas-liquid transition region. It may be useful to combine such an approach with Dukler's analysis [14, 15]. In such an analysis, Dukler's analysis could provide the equations of liquid motion in the film. Work on droplet deposition, discussed above, may aid in the analysis of the gas core. The transition region would present the remaining and perhaps most difficult element of the analysis. In this region, entrainment could be based on a model such as that of Ishii and Grolmes modified for counter-current flow. Some statistical data on the shape of the film would then provide the final part of the solution.

In closing, they stated that, although flooding may never yield to a complete analytical solution, there are several areas of research which should reduce the uncertainty in the present level of knowledge such that accurate predictions may be made. The most useful areas of experimental research appear to be in the development of diagnostic techniques to determine the interface shape and liquid velocity profile within the film. Analytically, this is a good time for reflection to determine if solutions to other fluid problems might be applied to flooding.

T. Sakaguchi, H. Hamaguchi and M. Ozawa presented the following topics as their research interest:

1. Measurement of flow rate in two-phase flow. After changing a two-phase flow into a uniform flow (bubbly flow or mist flow), the flow rate will be measured using a standard orifice. It may be possible to measure the flow rate by the characteristics of the pressure drop when the uniform flow is generated. We think this experiment has not been reported by other investigators.

2. Flow pattern conversion or flow pattern control in two-phase flow. In connection with problem No. 1, we are developing a converter to change slug flow into bubbly flow. Then we will also develop converters for other flow patterns. This will be applicable to industrial two-phase flow systems, for example, in the measurement of the flow rates of gas and liquid phases.

3. Hydrodynamic analysis of disturbance waves. There is no known method to classify a wave on liquid film in annular flow into a ripple and a disturbance wave. We are going to make definitions of both waves, and we will try to explain the phenomena in the disturbance wave using the concept of a solitary wave.

4. Hydrodynamic analysis of slug flow. We are measuring a velocity distribution in a liquid slug and the momentum of a liquid in slug flow to analyze the flow mechanism of slug flow, and we are studying a large bubble as a basic study of slug flow.

5. Dynamic behavior in two-phase flow. When gas or liquid flow rate is stepwisely changed in two-phase flow, the flow shows extraordinary phenomena. These transient phenomena such as a transient slug and propagation of a liquid level have been analyzed.

G. L. Wedekind gave the meeting his views on the future direction for two-phase flow research: Although the complexity of the numerous physical mechanisms which are simultaneously involved in two-phase flow systems sometimes appear to be overwhelming, considerable progress in our understanding has been made over the years by a substantial research effort which has been aimed at a variety of problems and applications. The fact that a generalized theoretical approach for two-phase flows has not been developed should not be too surprising when we consider that single phase turbulent flow is still in a similar state of incompleteness.

In attempting to gain some insight into the future direction for two-phase flow research, it would seem helpful to draw upon experience in the development of single phase flow theory. The present state of the art for single phase flow is the result of both experimental and theoretical research, with the greatest advances appearing to have occurred when both types of research were closely related. Another observation is that, although the general equations of motion for a single phase flow have been developed, the value and need for condition specific or special case models and constitutive relationships still exist. The type of models and relationships referred to here include potential flow theory, Bernoulli's model, Prandtl's viscous boundary layer theory, one-dimensional gas dynamics, perfect gas model, Stokes' viscosity model, etc. These models can be considered to be special cases of a more generalized model, yet a conceptual understanding of these condition specific models, along with the physical insight they provide, are necessary ingredients not only to an understanding of the generalized models, but to the formulation of appropriate models for application specific processes and systems.

It seems evident that a proper development of two-phase flow theory will require a similar structure. There is a need for the development of both condition specific models and for more generalized models, with a continuous effort being made to relate the two. It is important, however, that much care be given to distinguish the difference between condition specific and hardware specific research. Condition specific theoretical models provide physical insight into the phenomena under consideration which extends beyond the particular experimental apparatus that were used for their development or verification.

The potential value of condition specific theoretical and experimental two-phase flow research can best be illustrated by considering the impact which Prandtl and Von Karman's concept of a boundary layer have had on the whole area of viscous flow theory.

It is recognized that one school of thought might argue that with the advent of the high speed digital computer, the general equations of motion along with the continuity and energy equations and a complete set of constitutive relationships for each fluid component would be all that is necessary for completing our understanding of two-phase flow theory. From a pedagogical perspective, however, this would be insufficient. Good design engineers and future researchers need physical insight into the principal physical mechanisms involved in specific applications of two-phase flow phenomena. Although limited to specific conditions, simplified models often provide this needed insight with greater clarity than the more general models. The various models, from the more general to the simplistic, actually complement one another, where each has its particular part to play in an overall understanding of the field.

It is also recognized that there will probably always be a need for empirical correlations associated with such aspects of two-phase flow as pressure drop, heat transfer, void fraction, etc. Considerable care must be exercised to keep these results from being system specific.

This does not mean that system specific research should be discouraged in general. In fact, there are certain two-phase flow phenomena, particularly transient or dynamic phenomena, which require specific coupling effects of other system components before they manifest themselves. Certain two-phase flow instabilities fall into this category. However, considerable effort must be made to extract from such experimental research the principal physical mechanisms which contribute to the observed phenomena, enabling the formulation of an appropriate theoretical model, or more basic experimental research to be initiated which would lead to such a model.

From my particular perspective, research related to both condition specific models and their generalization should be considered fundamental research. Both types of research should be encouraged, with an emphasis placed upon relating the various models and establishing a better understanding of their individual limitations. To some extent, this has been taking place; however, more of it should be encouraged. Although this approach is important for steady flow phenomena as well, it is particularly important in transient flow systems where dynamic and instability phenomena occur.

It is one thing to pinpoint specific areas of two-phase flow where more research is needed, it is another thing to be able to place a level of priority on each area. There are two specific areas of two-phase flow which have received relatively little attention. One has to do with two-phase condensing flow transients, the other is transient pressure drops in either evaporating or condensing flows. There still seems to be the need for continued clarification, physical insight, and theoretical models directed at predicting the stability criteria for both evaporating and condensing flow instabilities. Since most transient models utilize constitutive relationships like heat transfer, pressure drop, void fraction, etc., which have been developed from steady state experimental and theoretical analyses, research associated with the improvement of these relationships should be supported. This would include the development of more reliable and accurate instrumentation and measurement techniques.

In conclusion, although admittedly the challenge is great, I believe some form of direction can be given to the overall progress of two-phase flow research by applying some of the criteria and the perspective outlined above. It should be recognized, however, that this particular perspective is limited at best, and needs to be integrated into the perspective of many other researchers if a more complete view is to be achieved.

References

1. Ozawa, M. et al., Oscillatory flow instabilities in gas-liquid two-phase flow systems, in Nuclear Reactor Safety Heat Transfer, Ed. S. G. Bankoff, Hemisphere Publishing Corp., N.Y. (1982).

2. Bouré, J. A., Bergles, A. E., and Tong, L. S., Review of Two-Phase Flow Instability, ASME Paper 71-HT-42 (1971).

3. Jeglic, F. A., and Grace, T. M. Onset of Flow Oscillations in Forced Flow Subcooled Boiling, NASA-TN-D 2821 (1965).

4. Chexal, V. K., and Bergles, A. E., Two-Phase Instabilities in a Low-Pressure Natural Circulation Loop, AIChE Symp. Ser., No. 131, Vol. 69 (1973), 37-45.

5. Ozawa, M., Nakanishi, S., and Ishigai, S., Flow Instabilities in Multi-Channel Boiling Systems, ASME Paper 79-WA/HT-55 (1979).

6. Inoue, A., and Aoki, S., Fundamental Studies for Pressure Drop in Two-Phase Flow in a Vertical Pipe, Trans. JSME, Vol. 32, No. 238 (1966), 940-947.

7. Ozawa, M. et al., Bulletin of the JSM, 22-175 (1979), 1763.

8. Ozawa, M. et al., Bulletin of the JSM, 22-170 (1979), 1113.

9. Ozawa, M. et al., Proc. 1980 ICHMT, Int. Seminar, Dubrovnik, Nuclear Reactor Safety and Heat Transfer, Hemisphere Pub. Corp.

10. Bankoff, S. G. and Mason, J. P., AIChE Journal, 8, 30-33 (1962).

11. Gunther, F. C., Report 4-75, Jet Propulsion Laboratory, Cal. Tech. (1950).

12. Gunther, F. C. and Kreith, F., Report 4-120, Jet Propulsion Laboratory, Cal. Tech. (1950).

13. Plesset, M. S. and Prosperetti, A., Int. J. Heat Mass Trans., 21, 725-734 (1978).

14. Dukler, A. E. and Smith, L., Report No. NUREG/CR-0617 (1979).

15. Maron, D. Moalem and Dukler, A. E., Letters in Heat and Mass Transfer, 8, pp. 453-463 (1973).

16. Levy, S. and Healzer, J. M., ASME Journal of Heat Transfer, 103, pp. 492-500 (1981).

INDEX